Metal Oxide Nanocomposite Thin Films for Optoelectronic Device Applications

Scrivener Publishing
100 Cummings Center, Suite 541J
Beverly, MA 01915-6106

Publishers at Scrivener
Martin Scrivener (martin@scrivenerpublishing.com)
Phillip Carmical (pcarmical@scrivenerpublishing.com)

Metal Oxide Nanocomposite Thin Films for Optoelectronic Device Applications

Edited by

Rayees Ahmad Zargar

*Department of Physics, Baba Ghulam Shah Badshah University,
Rajouri (J&K), India*

Scrivener Publishing

WILEY

This edition first published 2023 by John Wiley & Sons, Inc., 111 River Street, Hoboken, NJ 07030, USA and Scrivener Publishing LLC, 100 Cummings Center, Suite 541J, Beverly, MA 01915, USA
© 2023 Scrivener Publishing LLC
For more information about Scrivener publications please visit www.scrivenerpublishing.com.

All rights reserved. No part of this publication may be reproduced, stored in a retrieval system, or transmitted, in any form or by any means, electronic, mechanical, photocopying, recording, or otherwise, except as permitted by law. Advice on how to obtain permission to reuse material from this title is available at http://www.wiley.com/go/permissions.

Wiley Global Headquarters
111 River Street, Hoboken, NJ 07030, USA

For details of our global editorial offices, customer services, and more information about Wiley products visit us at www.wiley.com.

Limit of Liability/Disclaimer of Warranty
While the publisher and authors have used their best efforts in preparing this work, they make no representations or warranties with respect to the accuracy or completeness of the contents of this work and specifically disclaim all warranties, including without limitation any implied warranties of merchantability or fitness for a particular purpose. No warranty may be created or extended by sales representatives, written sales materials, or promotional statements for this work. The fact that an organization, website, or product is referred to in this work as a citation and/or potential source of further information does not mean that the publisher and authors endorse the information or services the organization, website, or product may provide or recommendations it may make. This work is sold with the understanding that the publisher is not engaged in rendering professional services. The advice and strategies contained herein may not be suitable for your situation. You should consult with a specialist where appropriate. Neither the publisher nor authors shall be liable for any loss of profit or any other commercial damages, including but not limited to special, incidental, consequential, or other damages. Further, readers should be aware that websites listed in this work may have changed or disappeared between when this work was written and when it is read.

Library of Congress Cataloging-in-Publication Data

ISBN 978-1-119-86508-7

Cover image: Pixabay.Com
Cover design by Russell Richardson

Set in size of 11pt and Minion Pro by Manila Typesetting Company, Makati, Philippines

Printed in the USA

10 9 8 7 6 5 4 3 2 1

Contents

Preface xvii

Part I: Nanotechnology 1

1 Synthesis and Characterization of Metal Oxide Nanoparticles/
  Nanocrystalline Thin Films for Photovoltaic Application 3
  *Santosh Singh Golia, Chandni Puri, Rayees Ahmad Zargar
  and Manju Arora*
  1.1 Present Status of Power Generation Capacity
      and Target in India 4
  1.2 Importance of Solar Energy 4
  1.3 Evolution in Photovoltaic Cells and their Generations 6
      1.3.1 First Generation Photovoltaic Cell 7
      1.3.2 Second-Generation Photovoltaic Cell 8
      1.3.3 Third Generation Photovoltaic Cell 9
      1.3.4 Fourth Generation Photovoltaic Cell 11
  1.4 Role of Nanostructured Metal Oxides
      in Production, Conversion, and Storage
      in Harvesting Renewable Energy 12
  1.5 Synthesis of Nanostructured Metal Oxides
      for Photovoltaic Cell Application 13
      1.5.1 Chemical Vapor Deposition Method 14
      1.5.2 Metal Organic Chemical Vapor Deposition Method 15
      1.5.3 Plasma-Enhanced CVD (PECVD) Method 15
      1.5.4 Spray Pyrolysis Method 16
      1.5.5 Atomic Layer Deposition or Atomic Layer
            Epitaxy Method 17
      1.5.6 Chemical Co-Precipitation Method 17
      1.5.7 Sol-Gel Method 18
      1.5.8 Solvothermal/Hydrothermal Method 19
      1.5.9 Microemulsions Method 20

v

		1.5.10 Microwave-Assisted Method	20
		1.5.11 Ultrasonic/Sonochemical Method	21
		1.5.12 Green Chemistry Method	22
		1.5.13 Spin Coating Method	22
		1.5.14 Dip Coating Method	23
		1.5.15 Physical Vapor Deposition (PVD) Methods	24
		1.5.16 Pulsed Laser Deposition Method	25
		1.5.17 Sputtering Method	26
		1.5.17.1 Radio Frequency (RF) Sputtering Method	26
		1.5.17.2 DC Sputtering Method	27
		1.5.18 Chemical Bath Deposition Method	27
		1.5.19 Electron Beam Evaporation	28
		1.5.20 Thermal Evaporation Technique	29
		1.5.21 Electrodeposition Method	30
		1.5.22 Anodic Oxidation Method	30
		1.5.23 Screen Printing Method	31
	1.6	Characterization of Metal Oxide Nanoparticles/Thin Films	31
	1.7	Conclusion and Future Aspects	32
		References	32
2	**Experimental Realization of Zinc Oxide: A Comparison Between Nano and Micro-Film**		**45**
	Rayees Ahmad Zargar, Shabir Ahmad Bhat, Muzaffar Iqbal Khan, Majid Ul Islam, Imran Ahmed and Mohd Shkir		
	2.1	Introduction	45
	2.2	Approaches to Nanotechnology	46
	2.3	Wide Band Semiconductors	48
	2.4	Zinc Oxide (ZnO)	48
		2.4.1 Crystal Structure of ZnO	49
	2.5	Properties of Zinc Oxide	50
		2.5.1 Mechanical Properties	50
		2.5.2 Electronic Properties	50
		2.5.3 Luminescence Characteristics	50
		2.5.4 Optical Band Gap	51
	2.6	Thin Film Deposition Techniques	52
		2.6.1 Thin and Thick Film	52
		2.6.2 Solution-Cum Syringe Spray Method	52
	2.7	Procedure of Experimental Work	53
	2.8	Calculation of Thickness of Thin ZnO Films	54
	2.9	Structural Analysis	54

		2.9.1 XRD (X-Ray Diffraction)	56
		2.9.2 SEM (Scanning Electron Microscope)	56
	2.10	Optical Characterization	56
		2.10.1 UV Spectroscopy	58
		2.10.2 Photoluminescence (PL) Spectroscopy	59
	2.11	Electrical Characterization	60
		2.11.1 Resistivity by Two-Probe Method	60
	2.12	Applications of Zinc Oxide	62
	2.13	Conclusions and Future Work	63
		References	63
3	**Luminescent Nanocrystalline Metal Oxides: Synthesis, Applications, and Future Challenges**		**65**
	Chandni Puri, Balwinder Kaur, Santosh Singh Golia, Rayees Ahmad Zargar and Manju Arora		
	3.1	Introduction	66
	3.2	Different Types of Luminescence	67
		3.2.1 Photoluminescence	68
		3.2.2 Thermoluminescence	69
		3.2.3 Chemiluminescence	70
		3.2.4 Sonoluminescence	70
		3.2.5 Bioluminescence	71
		3.2.6 Triboluminescence	71
		3.2.7 Cathodoluminescence	71
		3.2.8 Electroluminescence	72
		3.2.9 Radioluminescence	72
	3.3	Luminescence Mechanism in Nanomaterials	73
	3.4	Luminescent Nanomaterials Characteristic Properties	74
	3.5	Synthesis and Shape Control Methods for Luminescent Metal Oxide Nanomaterials	75
		3.5.1 Chemical Vapor Synthesis Method	75
		3.5.2 Thermal Decomposition Method	76
		3.5.3 Pulsed Electron Beam Evaporation Method	77
		3.5.4 Microwave-Assisted Combustion Method	78
		3.5.5 Hydrothermal/Solvothermal Method	79
		3.5.6 Sol-Gel Method	80
		3.5.7 Chemical Co-Precipitation Method	81
		3.5.8 Sonochemical Method	81
		3.5.9 Continuous Flow Method	82
		3.5.10 Aerosol Pyrolysis Method	83
		3.5.11 Polyol-Mediated Methods	83

		3.5.12 Two-Phase Method	84
		3.5.13 Microemulsion Method	85
		3.5.14 Green Synthesis Method	85
	3.6	Characterization of Nanocrystalline Luminescent Metal Oxides	86
	3.7	Applications of Nanocrystalline Luminescent Metal Oxides	87
	3.8	Conclusion and Future Aspects of Nanocrystalline Luminescent Metal Oxides	88
		References	89
4	**Status, Challenges and Bright Future of Nanocomposite Metal Oxide for Optoelectronic Device Applications**		**101**
	Ajay Singh, Sunil Sambyal, Vishal Singh, Balwinder Kaur and Archana Sharma		
		Abbreviations	102
	4.1	Introduction	103
	4.2	Synthesis of Nanocomposite Metal Oxide by Physical and Chemical Routes	105
		4.2.1 Synthesis of Metal Oxides Nanoparticles by Chemical Technique	105
		4.2.2 Synthesis of Metal Oxides Nanoparticles by Physical Technique	107
		4.2.3 Synthesis of Metal Oxides by Mechanical Technique	109
	4.3	Characterization Techniques Used for Metal Oxide Optoelectronics	109
		4.3.1 X-Ray Diffraction (XRD)	109
		4.3.2 Scanning Electron Microscopy (SEM)	109
		4.3.3 Transmission Electron Microscopy (TEM)	110
		4.3.4 Rutherford Backscattering Spectrometry (RBS)	110
		4.3.5 Fourier-Transform Infra-Red (FTIR)	110
		4.3.6 Raman Spectroscopy	111
		4.3.7 Luminescence Technique	111
	4.4	Optoelectronic Devices Based on MOs Nanocomposites	111
		4.4.1 Light-Emitting Device	111
		4.4.2 Photodetector	112
		4.4.3 Solar Cell	113
		4.4.4 Charge-Transporting Layers Using Metal Oxide NPs	116
		4.4.5 MO NPs as a Medium for Light Conversion	116
		4.4.6 Transparent Conducting Oxides (TCO)	116
	4.5	Advantages of Pure/Doped Metal Oxides Used in Optoelectronic Device Fabrication	117

4.6	Parameters Required for Optoelectronic Devices Applications	118
4.7	Conclusion and Future Perspective of Metal Oxides-Based Optoelectronic Devices	119
	Acknowledgement	119
	References	120

Part II: Thin Film Technology 129

5 Semiconductor Metal Oxide Thin Films: An Overview — 131
Krishna Kumari Swain, Pravat Manjari Mishra and Bijay Kumar Behera

5.1	Introduction	132
5.1.1	An Introduction to Semiconducting Metal Oxide	133
5.1.2	Properties of Semiconducting Metal Oxide	134
5.1.3	Semiconducting Metal Oxide Thin Films	135
5.1.4	Thin Films Deposition Method	135
5.1.4.1	Physical Vapor Deposition (PVD) Method	136
5.1.4.2	Evaporation Methodology	137
5.1.4.3	Thermal Evaporation	138
5.1.4.4	Molecular Beam Epitaxy	139
5.1.4.5	Electron Beam Evaporation	139
5.1.4.6	Advantages and Disadvantages of PVD Method	140
5.1.4.7	Sputtering Technique	140
5.1.4.8	Advantages and Disadvantages of Sputtering Technique	141
5.1.4.9	Chemical Vapor Deposition (CVD)	142
5.1.4.10	Photo-Enhanced Chemical Vapor Deposition (PHCVD)	143
5.1.4.11	Laser-Induced Chemical Vapor Deposition (LICVD)	144
5.1.4.12	Atmospheric Pressure Chemical Vapor Deposition (APCVD)	144
5.1.4.13	Plasma Enhanced Chemical Vapor Deposition (PECVD)	144
5.1.4.14	Atomic Layer Deposition (ALD)	145
5.1.4.15	Electrolytic Anodization	145
5.1.4.16	Electroplating	145
5.1.4.17	Chemical Reduction Plating	146
5.1.4.18	Electroless Plating	146

		5.1.4.19	Electrophoretic Deposition	146
		5.1.4.20	Immersion Plating	146
		5.1.4.21	Advantages and Disadvantages of CVD Process	147
		5.1.4.22	Sol-Gel Method	147
	5.1.5	Application of Semiconducting Metal Oxide Thin Films		148
		5.1.5.1	Photovoltaic Cells	148
		5.1.5.2	Thin-Film Transistors	149
		5.1.5.3	Computer Hardware	149
		5.1.5.4	LED and Optical Displays	149
	5.1.6	Limitations of Semiconductor Thin Films		149
5.2	Conclusion and Outlook			150
	Acknowledgement			150
	References			150

6 Thin Film Fabrication Techniques — 155
Lankipalli Krishna Sai, Krishna Kumari Swain and Sunil Kumar Pradhan

6.1	Introduction	156
6.2	Thin Film – Types and Their Application	157
6.3	Classification of Thin-Film Fabrication Techniques	157
6.4	Methodology	159
	6.4.1 Thermal Evaporation	160
	6.4.2 Molecular Beam Epitaxy	161
	6.4.3 Electron Beam Evaporation	161
	6.4.4 Sputtering Technique	163
	6.4.5 Chemical Vapor Deposition (CVD)	167
	6.4.6 Atomic Layer Deposition (ALD)	171
	6.4.7 Liquid Phase Chemical Formation Technique	171
	6.4.8 Electrolytic Anodization	172
	6.4.9 Electroplating	172
6.5	Advantages of CVD Process	173
6.6	Comparison Between PVD and CVD	174
6.7	Conclusion	174
	References	175

7 Printable Photovoltaic Solar Cells — 179
Tuiba Mearaj, Faisal Bashir, Rayees Ahmad Zargar, Santosh Chacrabarti and Aurangzeb Khurrem Hafiz

7.1	Introduction	179
7.2	Working Principle of Printable Solar Cells	180

7.3		Wide Band Gap Semiconductors	181
	7.3.1	Cadmium Telluride Solar Cells (CIGS)	181
	7.3.2	Perovskite Solar Cells	181
	7.3.3	Solar Cells Based on Additive Free Materials	182
	7.3.4	Charge-Carrier Selective Layers That Can Be Printed	184
7.4		Metal Oxide-Based Printable Solar Cell	184
7.5		What is Thick Film, Its Technology with Advantages	186
	7.5.1	Thick Film Materials Substrates	187
	7.5.2	Thick Film Inks	187
7.6		To Select Suitable Technology for Film Deposition by Considering the Economy, Flexibility, Reliability, and Performance Aspects	188
	7.6.1	Experimental Procedure for Preparation of Thick Films by Screen Printing Process	189
	7.6.2	Quality of Printing	190
	7.6.3	The Following Factors Contribute to Incomplete Filling	191
7.7		Procedures for Firing	191
	7.7.1	Thick Film Technology has Four Distinct Advantages	192
7.8		Deposition of Thin Film Layers via Solution-Based Process	193
	7.8.1	Approaches for Coating	194
	7.8.2	Casting	194
	7.8.3	Spin Coating	194
	7.8.4	Blade Coating	195
		Conclusion	198
		References	198

8 Response of Metal Oxide Thin Films Under Laser Irradiation   203
Rayees Ahmad Zargar

8.1		Introduction	203
8.2		Interaction of Laser with Material	205
8.3		Radiation Causes Modification	206
8.4		Application Laser Irradiated Films	207
8.5		Wavelength Range of Radiation	208
8.6		Laser Irradiation Mechanism	209
8.7		Experimental Procedure	211
	8.7.1	Thin Film Technologies	211
	8.7.2	What is Thick Film, Its Technology with Advantages	212
	8.7.3	Experimental Detail of Screen Printing and Preparation of $Zn_{0.80}Cd_{0.20}O$ Paste for Coated Film	212

xii Contents

        8.7.4   Variation of Optical Properties   213
        8.7.5   Electrical Conduction Mechanism   216
        8.7.6   Conclusion and Prospects   217
        References   217

Part III: Photovoltaic and Storage Devices   221

9 Basic Physics and Design of Photovoltaic Devices   223
Rayees Ahmad Zargar, Muzaffar Iqbal Khan, Yasar Arfat, Vipin Kumar and Joginder Singh

  9.1   Introduction: Solar Cell   224
  9.2   Semiconductor Physics   225
  9.3   Carrier Concentrations in Equilibrium   227
  9.4   p-n Junction Formation   229
  9.5   Process of Carrier Production and Recombination   229
  9.6   Equations for Poisson's and Continuity Equation   230
  9.7   Photovoltaic (Solar Power) Systems   231
  9.8   Types of Photovoltaic Installations and Technology   232
  9.9   Electrical Characteristics Parameters   233
        9.9.1   Open-Circuit Voltage   233
        9.9.2   Density of Short-Circuit Current (Jsc)   234
        9.9.3   Fill Factor Percentage (FF%)   234
        9.9.4   Power Conversion Efficiency (η)   235
        9.9.5   Dark Current   235
        9.9.6   Standard Test Conditions   235
  9.10 Basic p-n Junction Diode Parameters   236
  9.11 Conclusion   237
        References   237

10 Measurement and Characterization of Photovoltaic Devices   239
Saleem Khan, Vaishali Misra, Ayesha Bhandri and Suresh Kumar

  10.1   Introduction   240
  10.2   Electrical and Optical Measurements   242
  10.3   Current–Voltage (I-V) Characterization   242
  10.4   Quantum Efficiency   246
  10.5   Hall Effect Measurements   248
  10.6   Photoluminescence Spectroscopy and Imaging   252
  10.7   Electroluminescence Spectroscopy and Imaging   254
  10.8   Light Beam Induced Current Technique (LBIC)   255
  10.9   Electron Impedance Spectroscopy (EIS)   255
  10.10 Characterization by Ellipsometry Spectroscopy   257

10.11	Conclusion	258
	References	258

11 Theoretical and Experimental Results of Nanomaterial Thin Films for Solar Cell Applications 263
Muzaffar Iqbal Khan, Rayees Ahmad Zargar, Showkat Ahmad Dar and Trilok Chandra Upadhyay

11.1	Introduction	263
11.2	Literature Survey	266
	11.2.1 Zinc Oxide (Z_nO)	266
	11.2.2 Tin Oxide (S_nO_2)	267
	11.2.3 Cadmium Oxide (CdO)	268
	11.2.4 Nickel Oxide (NiO)	271
	11.2.5 Magnesium Oxide (MgO)	272
	11.2.6 Aluminium Oxide (Al_2O_3)	274
	11.2.7 Cobalt Oxide (CoO)	275
	11.2.8 Tungsten Oxide (WO_3)	276
11.3	Theoretical and Experimental Results	277
	11.3.1 Theoretical Model	277
	11.3.2 Sellmeier Model	277
	11.3.3 Optical Properties Derive from the Above Equations	280
11.4	Experimental Results of Optical Properties	282
11.5	Conclusions	285
	Acknowledgement	286
	References	286

12 Metal Oxide-Based Light-Emitting Diodes 295
Shabir Ahmad Bhat, Sneha Wankar, Jyoti Rawat and Rayees Ahmad Zargar

12.1	Introduction	296
12.2	Structure of LEDs	297
12.3	Working Principle of LEDs	298
12.4	Selection of Material for Construction of LEDs	299
12.5	Basic Terminology Involved in Fabrication of LEDs	300
	12.5.1 Color Rendering Index (CRI)	300
	12.5.2 CIE Color Coordinates	301
	12.5.3 Forster Resonance Energy Transfer (FRET)	302
12.6	LEDs Based on ZnO (Zinc Oxide)	302
12.7	Transition Metal Oxide-Based LEDs	307
	12.7.1 Electronic Structure of TMO Films	308
12.8	Lanthanide-Based OLEDs	310

12.9	Conclusion	312
	References	312

13 Metal Oxide Nanocomposite Thin Films: Optical and Electrical Characterization — 317
Santosh Chackrabarti, Rayees Ahmad Zargar, Tuiba Mearaj and Aurangzeb Khurram Hafiz

13.1	Introduction	318
	13.1.1 Classification of the Nanocomposites	319
13.2	Nanocomposite Thin Films (NCTFs)	320
13.3	Materials Used for Preparation of NCTFs	320
	13.3.1 Gold	320
	13.3.2 Platinum	322
	13.3.3 Silver	323
	13.3.4 Palladium	323
	13.3.5 Boron	324
	13.3.6 Nickel	324
	13.3.7 Titanium	324
	13.3.8 ZnO	324
	13.3.9 Iron Oxide	325
	13.3.10 Silicon (Si)	325
	13.3.11 Cobalt	326
	13.3.12 Molybdenum	326
	13.3.13 SnO_2 (Tin Oxide)	326
	13.3.14 CuO	326
	13.3.15 Tungsten	327
	13.3.16 CdS	327
	13.3.17 Graphene Oxide	327
	13.3.18 Carbon Nanotubes	328
13.4	Methods of Preparation of NCTFs	328
	13.4.1 Cold Spray Approach	329
	13.4.2 Sol–Gel Approach	329
	13.4.3 Dip Coating	329
	13.4.4 Spray Coating and Spin Coating	329
	13.4.5 Electroless Deposition	329
	13.4.6 *In Situ* Polymerization Approach	330
	13.4.7 Chemical Vapor Deposition Approach	330
	13.4.8 Physical Vapor Deposition Approach	330
	13.4.9 Thermal Spray Approach	330
	13.4.10 Solution Dispersion	330

13.5	Applications		331
	13.5.1	Gas Sensors	331
	13.5.2	Batteries	331
	13.5.3	Solar Cells	331
	13.5.4	Antennas	331
	13.5.5	Optoelectronics	331
13.6	Examples		332
	13.6.1	Graphene-Based Metal Oxide Nanocomposites	332
	13.6.2	Carbon-Based Metal Oxide Nanocomposites	332
	13.6.3	Silicon-Based Metal Oxide Nanocomposites	333
13.7	Laser Irradiation Sources		333
	13.7.1	Some of the Factors Affecting Laser–Material Interactions	334
		13.7.1.1 Laser Wavelength	335
		13.7.1.2 Laser Intensity	335
		13.7.1.3 Laser Interaction Time	335
		13.7.1.4 Surface Roughness of the Material	336
13.8	Functional Characterization Techniques		336
	13.8.1	Electrical Characterization	336
	13.8.2	Optical Characterization	338
13.9	Conclusion		341
	References		342

14 Manganese Dioxide as a Supercapacitor Material — 361
Mudasir Hussain Rather, Feroz A. Mir, Peerzada Ajaz Ahmad, Rayaz Ahmad and Kaneez Zainab

14.1	Introduction		362
	14.1.1	Classifications of Supercapacitors	364
	14.1.2	Electrochemical Double Layer Capacitor	364
	14.1.3	Pseudocapacitors	365
	14.1.4	Hybrid Capacitors	365
14.2	Supercapacitor Components		366
	14.2.1	Electrode	366
	14.2.2	Electrolytes	367
	14.2.3	Separators	368
	14.2.4	Current Collector	368
14.3	Methods for MnO_2 Nanoparticles		369
	14.3.1	Hydrothermal Route	369
	14.3.2	Sol-Gel Technique	370
14.4	Doped-MnO_2 Materials		372

	14.4.1	Titanium-Doped with MnO_2	372
	14.4.2	Vanadium-Doped with MnO_2	373
	14.4.3	Tin-Doped with α-MnO_2	373
	14.4.4	Al, Cu, Mg-Doped with MnO_2	374
	14.4.5	Co- and Ni-Doped with MnO_2	375
	14.4.6	Ag-Doped with MnO_2	377
	14.4.7	Bismuth-Doping with Additives	377
14.5	MnO_2 with Polymer Composites		378
	14.5.1	Polypyrrole-Coated with MnO_2	378
	14.5.2	Polyaniline-MnO_2	379
	14.5.3	Polybithiophene-Coated MnO_2	380
14.6	Nanocomposites		381
	14.6.1	Graphene-MnO_2-Polyaniline	382
	14.6.2	MnO_2-Carbon Nanocomposite	382
14.7	Conclusion		384
	References		386
Index			**399**

Preface

Metal Oxide Nanocomposite Thin Films for Optoelectronic Device Applications provides insight into the fundamental aspects, latest research, synthesis route development, preparation, and future applications of metal oxide nanocomposite thin films.

The fabrication of thin film-based materials will be important to the future production of safe, efficient, and affordable energy. Such devices are used to convert sunlight into electricity. Thin film devices allow excellent interface engineering for high-performance printable solar cells. Their structures are highly reliable, stand-alone systems that can provide megawatts. They have been used as power sources in solar home systems, remote buildings, water pumping, megawatt-scale power plants, satellites, communications, and space vehicles. With these mass applications, the demand for photovoltaic devices skyrockets every year.

Alternately, nanocomposite film coating has also revolutionized the fields of materials science and applied physics, thus progressing as a unified discipline for scientific industries. For the first time, this comprehensive handbook presents the emerging scenario of deposited techniques for the synthesis of metal oxide nanocomposites. The handbook is divided into three sections (Nanotechnology; Thin Film Technology; Photovoltaic and Storage Devices). It covers important topics like different metal oxide properties, scale-up processes, low-cost synthesis methods, various characterizations, and their respective physics. It will be an important volume for academic researchers and those in industries, exploring the applications of nanoparticles in semiconductors, power electronics, and more.

This book covers the basics of advanced nanometal oxide-based materials, their synthesis, characterization, and applications, and all the updated information on optoelectronics. Topics discussed include the implications of metal oxide thin films, which are critical for device fabrications. It provides updated information on the economic aspect and toxicity, with great focus paid to display applications, and covers some core areas of nanotechnology, which are particularly concerned with optoelectronics and the available technologies. The book concludes with insights into the role of nanotechnology and the physics behind photovoltaics.

My thanks go to Wiley and Scrivener Publishing for their continuous support and guidance.

Rayees Ahmad Zargar

Part I
NANOTECHNOLOGY

1

Synthesis and Characterization of Metal Oxide Nanoparticles/Nanocrystalline Thin Films for Photovoltaic Application

Santosh Singh Golia[1], Chandni Puri[1,2], Rayees Ahmad Zargar[3] and Manju Arora[1]*

[1]*CSIR-National Physical Laboratory, Dr. K.S. Krishnan Marg, New Delhi, India*
[2]*Academy of Scientific and Innovative Research (AcSIR), Ghaziabad, India*
[3]*Department of Physics BGSBU, Rajouri (J&K), India*

Abstract

The rapid depletion of natural resources especially fossil fuels due to excessive consumption has raised the severity of problem which encouraged researchers to explore existing and new advanced nanomaterials for application in non-renewable energy devices. The promising photovoltaic devices are suitable alternative for meeting energy demand as compared to other water, wind, and tidal energy resources. The extensive work is going on the economic, lightweight, and flexible organic dye sensitized and inorganic-organic hybrid type solar cell materials like nanosized pure and doped transition metal oxides or their nanocomposites because of their excellent stability, electronic, high resistance, and thermal properties. They can be easily prepared by different chemical and physical methods and plays role of light absorbers, transparent electrodes, and hole and electron transport layers in solar cells. The large intrinsic mobility of charge carriers in pure metal oxides has been utilized as charge collectors which improve the charge selectivity of electrodes due to their particular band gap energy. Metal oxides in nanocrystalline thin film form helps in optimizing light distribution within devices for efficient light absorption. In this chapter briefly introduced the present status of power generation capacity and target in India, importance of solar energy, evolution of photovoltaic cell, nanostructured metal oxides synthesis, characterization and application in photovoltaic cells is discussed.

Keywords: Nanoparticles, synthesis, characterizations, solar cell applications

---

*Corresponding author: manjuarorain@gmail.com

Rayees Ahmad Zargar (ed.) Metal Oxide Nanocomposite Thin Films for Optoelectronic Device Applications, (3–44) © 2023 Scrivener Publishing LLC

1.1 Present Status of Power Generation Capacity and Target in India

India has emerged as the world's third-largest energy consuming country owing to enormous rise in the usage of domestic consumer products as well as at workplace electronic/electrical gadgets to make life easier. The energy consumption has been doubled since 2000 as reported in India Energy Outlook 2021 [1]. This demand increases with sharply with time due to enormous rise in population, industrialization, and development of new smart cities with time. These are the key factors which encourage addition/demand of more consumer products/appliances and at offices in day-to-day life. Recently, Government of India, Ministry of Power has published the status of power at a glance received from Central Electricity Authority (CEA) which states that at present the total installed power generation capacity 3,99,497 (MW) generation is fulfilled 2,36,109 MW (59.1%) by Fossil Fuel and 1,63,388 MW (40.9 %) by Non-Fossil Fuel sources [2]. Coal, lignite, gas and diesel come under fossil fuel resources, while new and renewable energy (hydro (big and small), wind, solar, biomass, waste to energy) and nuclear energy resources are categorized under non-fossil fuel. The new and renewable energy resources contribute in the production of 1,56,608 MW power i.e. 39.2% and nuclear power plants produce 6780 MW (1.7%) of the total installed power generation setups. These details of the installed power generation capacity (MW) from each fossil and non-fossil fuel are presented in the following Figure 1.1.

The electricity generation target fixed for 2021–22 financial year was 1356 Billion Unit (BU) in fossil fuels through thermal energy produced 1155.200 BU, hydro plants: 149.544 BU; Nuclear plants: 43.020 BU; and 8.236 BU was import from Bhutan to overcome the demand from different Sectors of Society.

1.2 Importance of Solar Energy

India receives ~5,000 trillion kWh per annum energy from Sun over its land with many parts have an exposure of 4–7 $kWh.m^{-2}$ per day [3]. Solar energy is effectively captured for large/small scale power generation all over India in villages, remote areas, urban and upcoming smart cities as an alternative to fuel sources owing to its plenty availability. Government of India has categorized under National mission to cope up with the rising electricity demand by people with time. National Institute of Solar Energy

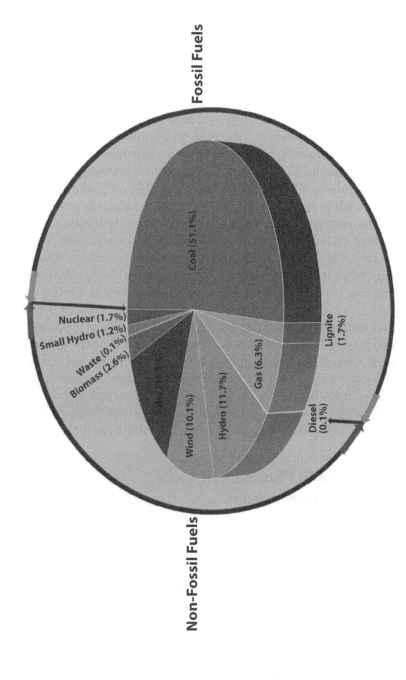

Figure 1.1 3D-π chart for installed power capacity (MW) of fuels for 2021–22 in India.

has reported that ~ 748 GW solar power can be produced by utilizing 3% of the unused/barren land for the installation Solar PV modules along with grid connected power generation capacity within stipulated time schedule over the years. This clean, and eco-friendly renewable energy resource needs initial investment to start i.e. installation of solar panels and storage system setup depending upon the power requirement of individual/society/ offices/village/city etc. with merits of off-grid/on-grid distribution at low/ normal temperature. The first off-grid practical application of photovoltaic solar panels/cells is to provide power for orbiting satellites, spacecraft in space and pocket calculators etc. But now, the photovoltaic modules connected to grid for large scale power generation are in demand which require inverter to convert the direct current (DC) electricity obtained from photovoltaic cells modules(PCMs) to AC electricity. The off-grid power generation by PCMs is suitable for lighting remote areas, lamp posts along roads/streets, emergency telephones, remote sensing applications. They require minor maintenance of cleaning panels from time to time for efficient absorption of sunlight and battery/storage devices contacts cleaning. At present, the awareness and need of people has started using solar energy driven lighting on roads/homes/offices/hospitals/industries, heating and cooling appliances in some regions/states of India with no harmful emissions. The availability of electricity in turn simplifies the life of people especially residing in villages/remote areas. The people have started their own skill/ancestral work/business/initiate existing or new small and medium scale industries at their own places as a source of income which in turn has improved their standard of living and thinking. So, just by providing electricity to villages/towns/remote areas, one can restrict the migration of people to cities in search of job and forced to live away from their families.

1.3 Evolution in Photovoltaic Cells and their Generations

Photovoltaic Cells (PCs) are the basic constituent of solar power technology which convert sunlight to direct current (DC) electricity and utilized directly or through batteries to power devices. The solar cell is like a p-n junction diode, having two layers i.e. p-type and n-type of doped semiconductors and available in different sizes as well as shapes to utilize the maximum effective surface area and minimize contact resistance [4–7]. n-type semiconductor layer coated with antireflection coating for sufficient absorption of sunlight and embedded with metal which acts as front negative electrode and p-type semiconductor with conductive oxide layer play role of back positive electrode. When, these p-type and n-type semiconductor layers are placed one over another to establish contact between them for inducing electric potential. Electrons from n-type semiconductor layer moves across the junction and diffuses to the p-type semiconductor layer that create a static positive charge in n-type layer. Similarly at the same time, the holes from p-type layer diffuse across the junction and develop a static negative charge. Both these escaped free electrons and holes pair up and disappear. At a particular stage, the depletion zone is formed at the p-n junction which restricts the migration of free charge carriers. These separated static positive and negative charges create an electric field across the depletion zone. The built-in electric field supplies voltage required for driving the unidirectional current through an external circuit. When the semiconductor layer absorbs photon energy equivalent or more from sunlight than its band gap, valence electrons from valence band (VB) are excited to conduction band (CB) and leave behind positively charged vacancies i.e. holes in VB. Hence, these transitions lead to the flow of electrons toward the negative end and holes toward the positive direction. The atoms collide and initiate free electrons migration for generating the current of electricity. The major issue for PCs is the very high cost of materials and installation units. This problem encouraged researchers for the development of new alternative economic materials which can markedly enhance the efficiency, durability, technical viability of solar cells/modules/panels. With the evolution in new materials for solar cell fabrication are classified into four generations.

1.3.1 First Generation Photovoltaic Cell

The conventional first generation photovoltaic cells belong to wafer technology in which silicon (Si) or gallium arsenide (GaAs) semiconductors doped with different element are used for photovoltaic cell development. Si has an indirect energy band gap value about 1.12 eV, while GaAs has direct energy band around 1.43 eV at ambient temperature with outstanding optical properties. Thicker single crystalline (Float Zone (FZ)) and polycrystalline (Czochralski (CZ)) Silicon wafers doped with group III (B, Al) for p-type and group V (P, As) for n-type periodic table elements wafers have been used in photovoltaic cells for efficient absorption of sunlight, while highly efficient GaAs photovoltaic cell requires lesser material. The developed photovoltaic cells have maximum efficiency around 28%, 21%, and 16% respectively. GaAs-based photovoltaic cells have maximum efficiency ~44%. The extensive R&D work has been reported on the synthesis of these materials and the commercially available technologies for photovoltaic cell fabrication. Both these substrates are very costly owing to their processing cost. The huge abundance and cost effectiveness of Si as compared to rare elements Ga and As has encourages wide range usage of polycrystalline silicon photovoltaic cell panels for large scale production to meet about 90% [8] of demand in spite of the low photovoltaic cell efficiency.

1.3.2 Second-Generation Photovoltaic Cell

These photovoltaic cells are based on thin film technology in which the two heterojunction layers are deposited in between two contact layers which require material of little micrometer thickness deposited on very economic transparent conductive oxide coated glass/ceramic substrate. Extensive work was carried out during 1990 to 2000 period on these type photovoltaic cells and cheaper than first generation photovoltaic cells. The copper-indium-gallium-diselenide (CIGS) and CdTe chalcopyrite structured thin films have been used for the photovoltaic cells fabrication and reported the highest efficiency of photovoltaic cell 22.6% and 22.1% [9–11]. Copper-zinc-tin-sulfide/selenide (CZTSSe) photovoltaic cells have achieved higher efficiency of 12.6%. However, for reducing the cost and utilization of abundant, nontoxic material with rare-metal free absorber layer, a substitute for CIGS is necessary [9–13]. When compared with silicon photovoltaic cells, CZTSSe absorbers are highly efficient due to reduced energy losses and can harvest more photons from the sun. However, the actual efficiency levels in comparison with silicon are still lagging. These generations of photovoltaic cells, made up of the above-mentioned thin films exhibit a high photovoltaic absorption coefficient, requiring a layer thickness of ~2.5 µm, compared to ~170 to ~250 µm for silicon photovoltaic cells [14]. CIGS-based photovoltaic cell consists of a multilayered thin film deposited onto a substrate made of glass, metal, or polymer foil. The p- and n-type semiconductor materials used in this PV cell are CIGS and cadmium sulfide, respectively. They are sandwiched between the molybdenum positive electrical back contact and the transparent conductive oxide which acts as negative electrical contact. Alternatively, Cu_2ZnSnS_4 (CZTS) is another prominent thin-film photovoltaic cell material which is an inexpensive, abundant, nontoxic material, and also a rare-metal free low-cost absorber layer suitable to be used as a substitute for CIGS absorber layer in the thin-film photovoltaic cells [15]. The economic amorphous and microcrystalline Si thin films are also considered under second generation photovoltaic cells.

The front illumination side of photovoltaic cell consists of an intrinsic a-Si:H passivation layer and a p-doped amorphous silicon layer as emitter. These layers are deposited by plasma enhanced chemical vapor deposition (PECVD) technique and an antireflective transparent conductive oxide (TCO) top layer by physical vapor deposition (PVD) method. The charge collection layer is screen-printed through metallic contacting grid. While, on the back side electron collecting stack has been used, which comprises of PECVD derived an intrinsic a-Si:H passivation layer, n-type amorphous silicon along with PVD deposited TCO layer and metallic contacting layer. The minimal material requirement, low temperature processing, light weight, flexibility of films for fitting on curved surface/panels, light weight and flexible textile or polymeric sheets and lower cost/Watt are some of the advantages of the second generation photovoltaic cells. While, the low efficiency of these photovoltaic cell in comparison to first generation photovoltaic cell is their disadvantage.

1.3.3 Third Generation Photovoltaic Cell

This generation photovoltaic cells are different from the earlier two first and second generations because they use different nanomaterials and the high surface to volume ratio of nanomaterials as well as their tendency to acquire different morphological shapes e.g. quantum dots, nanotubes, nanowires, nanopillars, nanocrystals, etc. This category is comprised of the organic dye sensitized, photoelectrochemical, conjugated conducting polymer, and polymer PVCs which are used in various space and territorial applications. Their design is different from the normal semiconductor p-n junction device for the separation of photogenerated charge carriers. These PVCs have multiple energy levels which means multiple carrier pair generation and these carriers can be captured before thermalization process [16, 17]. That's why third generation PVCs are also known as hot carrier solar cells with the advantages of very high limiting efficiencies. The fundamental operational principle of these PVCs, carrier thermalization processes minimize or restrict the energy loss and use the excessive thermalization energy as increased voltage in the external circuit and sustain the high absorption as well as the consequent external current of a small bandgap material. So, by tuning bandgap energy, on can capture light from different parts of the solar energy. These PVCs are developed on flexible substrates to achieve better interactions among molecule-to-molecule [18].

Dye sensitized solar cells (DSSCs) are considered as alternative for conventional silicon solar cells owing to their high throughput, ease of fabrication and cost effectiveness. In the photochemical cell, the dye molecules semiconductor layer is formed in between photosensitized anode and electrolyte. The extensive work has been carried on these type solar cells by using nanocrystalline metal oxide and their composites. The non-hazardous and stable Anatase phase TiO_2 nanoparticles porous layer has been used as anode layer and dye molecules absorbs the incident sunlight to promote electron transport from a lower level to an excited one. The excited electron is injected by the dye into the semiconducting TiO_2 anode layer and the chemical diffusion of electrons from the TiO_2 layer to the conductive indium tin oxide (ITO)/tin oxide (SnO_2) layer forms an outer circuit. The electrons come back to the cell to complete the circuit and for returning the dye back to its "normal" state via electrolyte solution and maintains the movement of electrons through the cell. This cell is sandwiched in between the two conducting overlapped glass plates. Platinum electrode act as the cathode part and a mixture of iodide/tri-iodide redox couple plays the role of electrolyte. The usage of nanomaterials in DSSC helps in harvesting the excessive energy in the absorbed photons for multiple exciton generation which in turn markedly enhances the energy conversion efficiency of a solar cell. Hence the broad band semiconducting oxide nanomaterials play a vital role in energy conversion efficiency and control the energy sustainability of solar PV. The major disadvantage of DSSC is the synthesis and chemical stability of the dye and liquid organic electrolytes. A lot work has been reported on the DSSCs which involves the use of different nanocrystalline pure, mixed and functionalized metal oxides [19–35].

To overcome the limitations of DSSCs, the economic organic-inorganic halide i.e. perovskites and QDs used in-place of dyes. First time, these perovskite-based PV solar cells were fabricated by Miyasaka *et al.* in 2009 [36] with maximum power conversion efficiency of ~ 3.8%. Now the efficiency reported for such nano solar cells is more than 20% which is greater than that of organic thin film solar cell. The highest efficiencies reported for perovskite solar cells have been mainly obtained with methyl ammonium lead halide materials. For example, lead halide perovskite materials, methyl ammonium lead iodide with band gap of about 1.5-1.6 eV with a light absorption spectrum up to a wavelength of 800 nm has been measured and widely used as a light capturer in solar cells. The perovskite-based 3rd generation solar cells are categorized into two types (i) comprised of compact TiO_2 and mesoporous TiO_2 layers and (ii) vapor-deposited perovskite solar cells based on small molecule organic solar cell and having only compact TiO_2 layer. Perovskite solar cells contain-

40 nm compact TiO_2 layer on transparent conducting oxide. When the perovskite ($CH_3CH_2NH_3PbX_3$) in the porous layers absorb light, electron hole pairs are produced with electrons on the conduction band (3.93 eV) and holes on the valence band (-5.43 eV). The electron hole pairs are separated swiftly by introducing electrons into TiO_2 (-4.0 eV) and transporting holes to the platinum counter electrode (-5.0 eV) forming a photocurrent in the device. The third-generation devices can be easily fabricated using simple industrial technologies which are capable of fabricating polymer solar cells. Though stability and performance of this generation of solar cell devices are limited and still at lab-scale, they have great potential and can be easily commercialized in the near future. The concentrated solar cell is also one of the promising technologies used in 3rd generation PVCs fabrication. These cells work on the principle of concentrating large amount of solar radiation on its small area. Though the semiconducting material cost is very high, this way one can drastically reduce the cost. This setup needs very good quality integrated perfect optical system with condensing levels from about 10–1000 suns. This technology will definitely emerge as very attractive acceptable technique in future.

1.3.4 Fourth Generation Photovoltaic Cell

These photovoltaic cells involve composite technology. They use mixture of polymers with nanoparticles to form single thin multispectral layer which are further stacked to produce economic with enhanced efficiency multi-spectrum photovoltaic cells [37]. Such, tandem PV cell consists of upper and lower solar cells with buffer layer in between them and they try absorbing the maximum or entire wavelength region of solar spectrum. The charge carriers produced by the absorption of sun light upper cell are collected by the electrodes and photocurrent moves through the buffer layer present in between these two PV cells [38]. The tandem cell contains at least two p-n junctions with cells composed of materials that absorb different photon energies. The upper cell absorbs the higher energies and the lower cell would absorb the lower energies which are not absorbed by the upper cell. The tandem cells have higher efficiency because they absorb more photons of the photovoltaic spectrum for energy conversion. This photovoltaic technology has been used for space application. Tandem photovoltaic cells are composed of the periodic table groups III and V elements compounds e.g. GaAs, InP, GaSb, GaInP, and GaInAs etc. The highest reported efficiency of these PV cells containing two and three cells are around 43% and 48% respectively.

1.4 Role of Nanostructured Metal Oxides in Production, Conversion, and Storage in Harvesting Renewable Energy

Nanomaterials have drawn considerable interest in the field of renewable energy production, conversion, and storage devices e.g. PV cells, fuel cells, batteries, thermal plants, hydrogen generation and new hi-tech storage devices etc. due to their high catalytic, mechanical, electrical, and optical performance as compared to bulk materials [39]. The particle size, morphological shape, large surface-to-volume ratio, and grain boundaries of nanomaterials have markedly enhanced efficiency of energy harvesting devices [40]. Nowadays, by the incorporation of nanotechnology and nanomaterials high power, stable energy generation, conversion, storage setups for large- and small- scale production as well as power distribution are in use all over the world. The nanostructured wide band metal oxides in thin/thick film form are used as active (transport) layer or passive (transparent conducting oxide electrode) layer in PV cells [41, 42]. Such nano dimensional metal oxides remarkably increase the visible region energy absorption via multiple reflections from the normal sunlight for efficient conversion to electricity, reduces electron-hole recombination process and ease of tuning band gap for their dual usage as absorber or window layer in thin films/heterojunction PV cells [43]. They minimize discharging rate of storage devices. The wide band gap energy semiconducting metal oxides exhibit optoelectronic properties and used as charge transport layers (CTLs) in PV devices. The n-type wide band gap ZnO, TiO_2, SnO_2 metal oxides with low work-function allows electron transport in the device and blocks hole transport. They are used as electron transport layers (ETLs). While, p-type wide-bandgap metal oxides like NiO, MoO_3, V_2O_5 and WO_3 etc. with high work function permit hole transport and blocks electron. They can be utilized as hole transport layer (HTL) in solar cell. But, some of these oxides suffer with a problem of not blocking electron transport efficiently. So, one needs very stable metal oxides for CTLs application which work efficiently throughout PV cells life. It's active layer/ETL, active layer/HTL, cathode/ETL, and anode/HTL interface contacts should be very stable and they should remain intact for long-time operations. The last one is that during PV cell working, the active layers are in the excited states and they are mostly sensitive to surrounding atmospheric oxygen or moisture. Hence, the effect of environmental stability on device performance can be greatly enhanced if CTLs also play role as barriers for oxygen and moisture diffusion. It means the optoelectronic properties of nanostructured metal oxides as CTLs viz. work function, band structure, conductivity, intragap states and optical properties parameters should be optimized for exploiting their usage in PV cells application.

1.5 Synthesis of Nanostructured Metal Oxides for Photovoltaic Cell Application

The various chemical and physical methods are employed for the synthesis of nanostructured metal oxides with desired stoichiometry, structure, particles shape (e.g. nanospheres, nanorods, nanowires, nanotubes, nanoflakes, nanoribbons, nanoflowers, nanosheets etc.) and size, surface morphology in random and regular ordered arrangements of agglomerated ensembles in thin films etc. as per device requirement. The chemical methods are widely used as compared to physical methods owing to their low cost, ease of controlling stoichiometry by optimizing synthesis parameters, appropriate selection of reactants and synthesis process. In physical methods, the cost of substrates, instrumentation and their maintenance are very high. Chemical methods involved in the deposition of nanocrystalline metal oxides thin films are summarized under two categories (i) evaporation, sputtering and gas-phase-transport-based techniques and (ii) the solution deposition method. The chemical vapor deposition (CVD) [44–47], metal organic chemical vapor deposition (MOCVD) [48–52], plasma enhanced chemical vapor deposition (PECVD) [53–55], spray pyrolysis (SP) [56–61], atomic layer epitaxy (ALE) and atomic layer deposition (ALD) [62–69] comes under gas-phase methods. While, chemical co-precipitation [70–75], sol-gel [76–81], solvothermal [75, 82–87]/hydrothermal [88–90], microemulsion [91–95], microwave assisted [96–101], ultrasonic/sonochemical [102–104], green chemistry [106–111] methods are used for obtaining nanoparticles of desired size and morphology whose dispersions are used in spin [112–114] and dip-coating [77, 115, 116] techniques for depositing thin films. Physical vapor deposition [117–124] methods are pulsed laser deposition (PLD) [125–128], radio frequency (RF) sputtering [129–132] and dc magnetron sputtering [133–136]. The key difference between PVD and CVD methods is in the vapor used for film deposition. In PVD, the of vapor consists of physically discharged atoms and molecules which merely condenses on the substrate. While for CVD, the vapor undergoes a chemical reaction on the substrate to form a thin film. In addition to these methods, the other methods like chemical bath deposition (CBD) [137–142], electron beam evaporation (EBE)

[143–147], thermal or vacuum evaporation [148–156], electrochemical deposition [157–161] and anodic oxidation [162–167]. The screen printing [168–174] technique has been used for the deposition of metal oxide thick films on different substrates.

1.5.1 Chemical Vapor Deposition Method

Chemical Vapor Deposition (CVD) method undergoes series of chemical reaction viz. gas phase reaction chemistry, thermodynamics, kinetics and transport phenomena, film nucleation and growth process and reactor setup at specific process parameters like temperature, pressure and reaction rates for mass as well as energy transport for thin film deposition [44–47]. These factors control the chemical reaction between the reactant and substrate to obtain good quality of films of desired composition and physical properties. The tentative CVD reactions are pyrolysis, reduction, oxidation, compound formation, disproportionation, and reversible transfer etc. The basic steps followed in all types of CVD methods are listed below:

- Convection or diffusion of reactant in a gas phase to the reaction chamber
- Chemical and gas phase reactions produce reactive species and by-products
- Transport of the reactants via boundary layer to the substrate surface
- Chemical and physical adsorption of the reactants on the surface of substrate
- Heterogenous reactions at the surface results in the formation of a solid film
- The volatile by-products are desorbed by diffusion through the boundary layer to main gas stream
- The gaseous by-products are removed from the reactor via convection and diffusion process.

CVD methods are categorized on the basis of energy sources (thermal, laser, photon) and reactors used for initiating the chemical reaction. The device application decides the selection of appropriate process/reactor as per substrate material requirement and coating materials, surface morphology, film thickness, uniformity, availability of precursors, and their cost. Nanocrystalline TiO_2 and Sb doped SnO_2 thin films prepared by CVD method for application in DSSCs was reported by P.S. Shinde and C.H. Bhosle [46] and J.I. Scott et al. [47] respectively.

1.5.2 Metal Organic Chemical Vapor Deposition Method

Metal organic chemical vapor deposition (MOCVD) method is also known as metal organic vapor phase epitaxy (MOVPE) and generally used for synthesizing III-V and II-IV semiconducting compounds [48]. The III-IV compounds are widely in various optoelectronic and electronic devices for large scale production. Now, the epitaxial growth of thin films by MOCVD technique has been also used in the preparation of metals, dielectrics and multifunctional nanomaterials for many new state-of-art advanced technologies. In MOCVD setup, all sources are vaporized and moved into the reactor with carrier gas. MOCVD reactors have the following three different geometries (i) high-speed rotation vertical reactor, close-coupled showerhead (CCS) reactor and planetary rotation horizontal reactor in its system. The injected gases should be ultra-pure and have finely controlled entry into the system. The reaction steps involved for producing III-V semiconductors are given as:

- Vaporization of the group III and group V precursors into gas phase to transport them towards substrate for diffusion via the phase interface
- Adsorption of molecules/atoms at the surface of substrate
- Molecules/atoms are directed to crystallization zones for chemical reaction between group III and V salts
- Decomposition and/or desorption of the by-products by diffusion and convection out methods from the boundary layer
- Removal of the by-product out from the reactor by evacuation process.

The work on MOCVD processed textured ZnO, CdO and TiO_2 thin films for solar cells applications have been reported [49–52].

1.5.3 Plasma-Enhanced CVD (PECVD) Method

Plasma-enhanced CVD (PECVD) method deposit a thin film from gaseous state to a solid state on the substrate. The chemical reaction takes place after the creation of a plasma in the reactor chamber which subsequently deposit thin film on the substrate surface. PECVD uses an electrical source of energy to generate plasma to undergo reaction process. The electrical energy produces ionic species and radicals through homogeneous chemical reactions to actuate heterogeneous chemical reactions for thin films formation on the substrate. The main advantage of PECVD over thermal CVD method is the very low operating temperature i.e. near to the ambient temperature [53]. The use of plasma to activate the gas phase chemistry opened many new reaction routes for the film deposition at low temperatures. H. Huang et al. [54] and L. Mazzarella et al. [55] deposited SnO_2 and nanocrystalline silicon thin films by PECVD method to use then in solar cells.

1.5.4 Spray Pyrolysis Method

In spray pyrolysis method, the thin film deposition involves the spraying of precursor solution onto a heated surface of substrate for the reaction of chemical constituents to produce desired chemical compound [56, 57]. This process completes by following anyone of the three ways:

(i) The first way, the droplets of the solution sprayed by atomizer fall on the hot surface, the solvent evaporates and components then react in the dry state.
(ii) The second way, the solvent evaporates on the way when the drops are moving towards the hot surface and the dry solid gets attached to the hot surface by decomposition.
(iii) The third way, the processes involved are the vaporization of solvent as the droplets are transported towards the substrate and consequently undergoes heterogeneous reaction with the solution components. The substrate temperature, carrier gas flow rate, nozzle-to-substrate distance, and the solution amount along with its concentration are some of the parameters which controls the quality of deposited film. Out of these, the substrate temperature is found to be the key parameters for depositing thin film by spray pyrolysis method because it is involves the drying of droplets, decomposition, crystallization, and grain growth processes. The main part of the spray pyrolysis deposition setup is the atomizer to form aerosol from the precursor solution. The chemical reactants are chosen in such a manner that the products other than the desired compound should evaporate at the deposition temperature. In spray pyrolysis, the reaction in the vapor phase at moderate high temperature takes place in normal air atmosphere for depositing transparent conducting thin films of nanocrystalline metal oxides in aqueous and non-aqueous medium such as Sb-doped SnO_2 [58], CdO [59], porous TiO_2 [60] and Ce-doped CdO [61]

as compact layers on the substrate surface having conductive or non-conductive films for solar cells application. Spray pyrolysis method has been widely used by the researchers owing to its simplicity and cost effectiveness.

1.5.5 Atomic Layer Deposition or Atomic Layer Epitaxy Method

Atomic Layer Deposition (ALD) or Atomic Layer Epitaxy (ALE) Method is also known as pulsed CVD. This process is based on the principle of self-limiting surface chemical reactions between two gaseous precursors and permits the layer-by-layer deposition of ultrathin thin films even up to atomic levels [62–64]. Such deposited conformal films have high aspect ratio structures, uniform thickness over large surface areas, at comparatively low temperatures and at moderate pressure. This method can be used in fabrication of hybrid nanostructures for many applications like in magnetic recording heads, optoelectronic, complementary metal-oxide-semiconductor (CMOS) transistors, integrated circuits (ICs), micro-electro-mechanical systems (MEMS micro-electro-mechanical systems (MEMS) and passivation/antireflection coatings etc. ALD method has been successfully used in the preparation of high-k metal gate transistors which markedly reduces the size of ICs chips. By this method, the ultrathin layers of ZnO [65–69] and V_2O_5 [60] are deposited in solar cells. The metal oxide thin layers deposited on CdTe/ITO interface imparts the role of barrier layer in solar cells [66].

1.5.6 Chemical Co-Precipitation Method

In the chemical co-precipitation methods, the sparingly soluble products from aqueous solutions (or other solvents) separates out after mixing reactant solutions followed by thermal decomposition to obtain the desired metal oxide/oxides. This method undergoes the nucleation, growth, Ostwald ripening, and/or agglomeration processes simultaneously [70–74]. Under the high-supersaturating condition, at first very fine large number of small crystallites as sparingly soluble precipitates appear due to initiation of nucleation process which swiftly agglomerate to form bigger sized thermodynamically stable nanoparticles under growth process. During this process, the smaller particles have been used to form bigger sized nanoparticles via Ostwald ripening. In this step, the particle size, morphology, surface/volume ratio and properties of the final products are optimized in the absence of stabilizer. The synthesis of desired stoichiometric metal oxide nanoparticles (MONPs), the precipitated nanoparticles are subjected to drying, calcination/sintering and annealing processing. While in some cases, MONPs are directly obtained by precipitation reaction. The agglomeration of MONPs has been drastically reduced by the following two ways (i) steric repulsion by capping ligands and (ii) Electrostatic (van der Waals) repulsions via chemisorption process. In steric repulsion the MONPs are capped with bigger sized organic molecules/polymeric chains as surfactants which encapsulate each particle and restricts agglomeration process. On the other hand, in electrostatic repulsion process, the charged species such as H^+, OH^- or other ionic species present at the surface of MONPs are involved in stabilization and reducing agglomeration. But MONPs capped with insulating surfactant organic molecules hampers it's charge transport layer property owing to the variations in the energy barriers for charge carrier transport layers [71].

D. Ouyang et al. [72] studied the usage of cost-effective metal oxide nanocrystals synthesized by solution-method as Carrier Transport Layers in organic and perovskite solar cells for improving their efficiency and life. M. Bhogaita and D. Devaprakasam [73] investigated the hybrid photoanode based on TiO_2-ZnO prepared by co-precipitation route as better alternative to individual TiO_2 or ZnO photoanode for dye-sensitized solar cell using phyllanthus reticulatas pigment sensitizer for effectively enhancing device efficiency by reducing charge recombination process. M. Yin et al. [74] prepared low cost non-hazardous Cu_2O by co-precipitation method from cuprous acetate with an aim for photovoltaic applications because of its high optical absorption coefficient, low band gap i.e. ~ 2.2 eV. While, M.M. Rashad et al. reported nano-powdered SnO_2 synthesis by chemical co-precipitation and solvothermal methods and their use in Dye Sensitized Solar Cells (DSSCs) for increasing solar cell efficiency.

1.5.7 Sol-Gel Method

Sol-gel one of the extensively used method in which metal alkoxides or metal salts as reactants are used to undergo hydrolysis reaction to form the "sol" i.e. uniform dispersion of fine particles and then follows by the condensation process to get "gel" a cross-linked network of particles [76–78]. The gel is generally aged for several hours to days and dried to obtain porous Metal Oxide hydroxide/MONPs [78, 79]. Such crystalline porous MONPs are subjected to calcination at high temperature in the temperature controlled furnace to achieve the desired structure, morphological shape and size of MONPs. However, the calcination at high temperature increases the particle size and decreases the surface area of the nanoparticles because of the heat induced enlargement effect.

The conventional aqueous sol–gel approach is suffered from high reaction rates and unable to control the hydrolysis and aggregation processes. In addition to this, the aqueous sol–gel approach needs annealing at high temperature to form crystalline phase of sample. On the other hand, non-aqueous sol–gel approach was used to overcome these problems [78, 80, 81]. It follows various chemical reactions, like hydrolysis, alcoholysis, aminolysis, and halide elimination to prepare colloidal oxide nanocrystals. As per the reaction mechanism pathways, metal precursors and their concentration, activation reagents, and reaction parameters such as temperature (includes pre- and post- heat treatment of the materials used), pH of solution, duration and nature of solvents were cautiously chosen to control the reaction kinetics and crystallization processes.

M. Szindler et al. [76] reported the synthesis of ZnO nanopowder by sol-gel method and characterized by various analytical techniques with an aim to use this as photoanode in DSSCs. S. Thiagarajan et al. [77] discussed the synthesis of metal oxide nanoparticles by sol-gel route and described in detail the synthesis of ZnO, Al doped ZnO, Iron oxides, tin oxide, and copper oxide by this method for application in the energy conversion devices. B. Ludi et al. described the mechanism behind the formation of ZnO nanoparticles in benzyl alcohol by the non-aqueous sol-gel method [81].

1.5.8 Solvothermal/Hydrothermal Method

In the solvothermal method, the precursor solutions are treated in closed reaction vessel/autoclave at temperatures more than the boiling points of solvents used. The pressure developed in the vessel by the solvent vapors raises the boiling point of the solvent and form highly crystalline materials [82–84]. Solvothermal as well as hydrothermal methods are the most powerful and widely used methods to synthesize nanomaterials because of the high reproducibility, simple procedure, and the easy scale-up merits. The large production of the nanomaterials is possible by using continuous reactors. However, the solvothermal methods generally require long reaction time ranging from several hours to days. Therefore, simple and rapid synthetic method for the preparation of metal oxide nano-assemblies always attracts researchers. In this context, C.T. Dinh [84], X. Liang et al. [85], Z. Li [86] and R. Krishnapriya [87] prepared by solvothermal method shape controlled, nanocrystalline TiO_2 particles, colloidal metal oxide as charge transport layer in LEDs and solar cells, WO_3 as hole transport layer in perovskite solar cells and ZnO as photoanodes in DSSCs respectively.

While, J. Beusen et al. [88], I.T. Papadas et al. [89], and X.D. Ai [90] prepared TiO_2 nanoparticles amorphous layer, $CuGaO_2$ nanoparticles and ZnO whiskers shaped particles respectively by hydrothermal method for application in solar cells.

1.5.9 Microemulsions Method

The microemulsion method deals with thermodynamically stable transparent solution with low viscosity formed spontaneously from the surfactant, water, and oil mixed solution. A water-in-oil microemulsion is a solution of nano sized water droplets stabilized by surfactant molecules monolayer [91–93]. The microemulsion has been used as a nanoreactor for preparing uniform size nanoparticles via chemical reactions confined within the aqueous core. Hence, the nanoparticles formed by this method are stabilized by the surrounding surfactant layer acting and controls the aggregation of as prepared nanoparticles [91]. By the addition of solvent, such as acetone or ethanol, the precipitate are formed by adding acetone or ethanol which are collected either by filtration or centrifuging the mixture. The droplet size has diameter in about 2–20 nm range which is far more than the surfactant monolayer thickness. This reveals that the enlargement in water or oil droplet size takes place during the chemical reaction of reactants for nanoparticles formation. The particle size can be easily controlled by changing water concentration and different morphologies are obtained by varying surfactant concentration. It means, in microemulsions monodispersed droplets exist in a dynamic state which undergo continuous collisions with each other to form bigger sized droplets followed by their breakup. This spontaneous reaction results in the formation of micelles with precipitate in it. The monodispersed metal oxide nanoparticles derived by microemulsion method have very fine in nature [92]. Nanostructured ZnO for solar cell applications are also prepared by microemulsion method [94, 95].

1.5.10 Microwave-Assisted Method

Microwave (MW) assisted method utilizes MW irradiation to proceed for chemical reactions. In this method, the reactants are heated through microwave dielectric heating which depends upon material characteristic property to absorb MW energy and transforms it into heat. In the liquid phase synthesis of nanomaterials, MW irradiation initiates heating either by dipolar polarization or by ionic conduction. The dipolar polarization mechanism is governed by the rotation of polar solvent molecules or reactant salts constituting dipoles in the reaction mixture. While, charged particles like free ions or ionic species present in material follows ionic conduction. When irradiated at MW frequencies, the dipoles in the sample try to align themselves in the direction of the applied electric field. As the electric field oscillates, the molecular dipoles accept the alternating electric field streamlines and loses energy as heat by the molecular friction and dielectric loss [96, 97]. The amount of heat generated is directly proportional to the ability of dipoles alignment with respect to the applied MW frequency. Whereas, in ionic conduction, the ions/charged particles present in the sample vibrates on applying MW field. During this process, the vibrating molecules collide with their adjacent molecules and produce heat. On comparing the two mechanisms in terms of heat generation, it is found that the ionic conduction mechanism is better than dipolar rotation mechanism. Microwave-assisted method in conjunction with other chemical method has been used for the synthesis of pure and complex metal oxides. This method has several advantages also viz. efficient source of heating elevate reaction rates for faster synthesis; ease of size and shape control by optimizing instrument and reaction process parameters; selective heating as per materials requirement to MW; reproducibility of reaction products is better than by conventional heating. M. Fathy *et al.* [98] and H.-E. Wang *et al.* [99] reported the synthesis of nanocrystalline TiO_2 and reduced graphene oxide by MW-assisted synthesis for usage in economic DSSCs. L.C. Nehru and C. Sanjeeviraja [100] used MW assisted combustion for the preparation of nanocrystalline SnO_2. H. Sun *et al.* [101] prepared ZnO nanocrystals for DSSCs by MW-assisted hydrothermal method.

1.5.11 Ultrasonic/Sonochemical Method

This synthesis method involves high-intensity ultrasound sonication of metal salt and oxygen in solution wherein physical effects such as acceleration and collision of particles change the particle size, distribution, morphologies and compositions [101]. Reactive radicals and atoms react with metal ions in the solution to form MeOs and this is one of the best methods to deposit NPs on polymer matrix, polystyrene sphere and silica particles. The work has been reported on ultrasonically produced nanocrystalline metal oxides e.g. ZnO [102], TiO_2, WO_3, and V_2O_5 [103] in non-aqueous solvent, and NiO [104] for sensor, photovoltaic, electrochromic, catalysis, magnetic resonance imaging (MRI) and drug delivery applications.

22 Metal Oxide Nanocomposite Thin Films

1.5.12 Green Chemistry Method

Interdisciplinary eco-friendly Green chemistry method is based on the development of technologies for more efficient chemical reactions to reduce or prevent the environmental pollution. The competence of "green" chemistry is divided into two main areas. The first one deals with the processing and application of environmentally hazardous waste and by-products of the chemical industry. The second more promising area is concerned with the development of new industrial processes to eradicate or minimize the formation and usage of toxic products [105]. "Green" chemistry method permits to produce the desired material via safest possible route by choosing appropriate raw materials and chemical reaction methods with zero or minimum waste, which do not require the hazardous chemicals. The plant extracts, algae, fungi, bacteria, etc. are generally used in the "green" synthesis of metal and metal oxide NPs [105–107]. The "green" synthesis method is sensitive to pH, pressure, temperature, solvent used. The phytochemicals like ascorbic acids, phenols, carboxylic acids, terpenoids, amides, flavones, aldehydes, ketones etc. are found in the roots, leaves, stems, and fruits of the plant extracts which are involved in reducing the metal salts to metal NPs [107]. Different group from various laboratories have adopted "green" chemistry route for producing metal oxide nanoparticles [108–111].

1.5.13 Spin Coating Method

The spin coating method has been extensively used wet coating method for the synthesis of nanocrystalline thin films by the researchers and industries owing to its low cost and simplicity. This method has been used for preparing photo-resist coating, microfabrication and in the development of nanoparticle assemblies. The precursor, the target material is mixed with a volatile solvent, which evaporates during the spinning process. The thin films are deposited on flat surface by using few drops of precursor in solution or sol-gel solution form on substrate rotated at the rate of 1000–5000 rotations per minute to spread the fuel by centrifugal force. The thickness of deposited film varies with respect to the applied angular spinning speed and the concentration of the solution as well as solvent used during the film deposition process. One can achieve the thickness of the films less than 10 nm by this method. This process can be used to deposit very thin layer of nanocrystalline metal/metal hydroxide on the surface of the substrate. The optimum thick of deposited film as required for particular application is obtained by repeating the process number of times. The thickness of the deposited is measured by ellipsometry method. The pure and doped nanostructured zinc oxide thin films deposition by spin coating method was reported [112–114]. The size, shape and surface morphology of the deposited nanoparticles is optimized by the selection of appropriate precursors and their concentration, solvents and synthesis parameters like rotation rate, pH of solution and sintering temperature etc. These wide band gap II-VI films are used for different applications in solar cells.

1.5.14 Dip Coating Method

The economic dip coating method deals with the deposition of thin films from the solution of precursors in pure or sol-gel form. The substrate is immersed into a coating bath generally consists of a polymeric or particulate sol. Upon withdrawal from the bath, the substrate retains a boundary layer. At the same time, the solvent begins to evaporate. This process is comprised of three steps:

(i) Immersion and dwell time: The substrate is immersed into the precursor solution at a constant speed and keeps in it for a certain dwell time, so that the substrate will have sufficient time to interact with the precursor solution for complete wetting.
(ii) Deposition & Drainage: By pulling the substrate upward at a constant speed with the help of motor to form a thin layer of precursor solution and the drainage of extra liquid from the substrate surface.
(iii) Evaporation: In this step, the solvent evaporates by heating from the fluid during drying process of the coated thin film. Then, the coated film is heated at about 200°C or higher temperature to remove organic impurities and formation of the oxides film of desired chemical composition. The rate of evaporation is sensitive to the withdrawal velocity and drying conditions which affects the microstructure of dip-coated film. When the drying front reduces into the pores of the particulate network, then capillary pressures pertaining to pore size develops for the densification of structure. It means, the evaporation of the solvent i.e. drying of film is a very crucial step for this wet-coating technique because cracks are often developed in the film which in turn influences the film quality. Hence, this step demands the precise control over drying conditions.

As this is the coating process, such deposited films clearly indicate the link between the structure of the precursor solution or sol, respectively, and their microstructure. The variations in nanostructured metal or metal oxide deposited films characteristic physical and chemical properties by dip coating method depend upon the sol concentration, dwell time, capillary pressure and temperatures used for drying, decomposition and sintering of films. By the introduction of new advanced versions of dip coating setups like angular dependent dip coating and the evaporation-induced self-assembly (EISA-process) which promote the fast production of patterned porous or nanocomposite thin films. Nanocrystalline ZnO and cobalt-based metal oxide thin films deposition by sol-gel dip coating method and their characterization by various analytical techniques have been reported by R.A. Zargar, C.D. Dushkin [115] and A. Amri *et al.* [116].

1.5.15 Physical Vapor Deposition (PVD) Methods

PVD techniques are used for the deposition of thin films in vacuum in which the source material is vaporized into the gas phase to deposit on substrate. The source material is at first vaporized either by evaporation or sputtering sources in atoms/molecules form and then followed by condensation and nucleation process of these atoms/molecules takes place on substrate via vacuum or low pressure gaseous or plasma environment. Plasma or ions are basic the components of the vapor phase. The reactive gas is also introduced to the vapor in some cases during the film deposition process. The atoms or molecules are transported in the form of a vapor under a vacuum or low pressure gaseous or plasma environment to the surface of the substrate to initiate condensation process. PVD processes have been used for depositing thin films from a few nanometers to micro-meter thickness range. The evaporation based PVD techniques produces very pure films with better control over film thickness [117, 118]. The composition and microstructure of the thin film can be changed by ions/radiation bombardment of the vapors during film growth which causes sputtering, re-condensation, and nucleation of the film atoms with fast surface movement of the atoms around the surface of the film. Sputtering techniques are based on the usage of ionized species accelerated to the target which on hitting the target surface to eject it's material for depositing a desired thin film on the substrate [119]. These economic techniques are extensively used owing to good repeatability for uniformity and tolerance in film thickness over large area of substrate e.g. as antireflection coatings [120] in PV cells. The uniformity in film thickness depends upon the configuration and environment between the source and the substrate materials, along with the source emission properties [121, 122]. The following parameters such as substrate holder geometry, vacuum pressure, system temperatures (substrate and material flux temperatures), and angular configuration of the deposition material to the substrate are optimized to obtain good quality with uniform thickness of thin films. One can also use this method for multilayers, thick films, free-standing structure and in hybrid form in conjunction with other deposition techniques.

The composition of thin films deposition in a vacuum chamber is sensitive to the residual pressure (in 10^{-3} - 10^{-9} Pa range), partial vapors of water, oxygen, nitrogen and hydrocarbons in high vacuum range and carbon monoxide, hydrogen and helium in ultrahigh vacuum range. The characteristic properties of the vaporized species of the film material by evaporation or sputtering [98] are controlled by the flux J, kinetic energy E, sizes (atoms, clusters) and ionization state (neutral/ionic) which in turn pertain to material(s) and on the type as well as operating parameters of the source. The kinetic energy, E, for thermally evaporated species is in 100–200 MeV range and 5-10 eV for sputtered species at varying substrate potential. The movement of free species emitted by the sources towards the substrate facilitates the manipulation of their electrical charge and kinetic energy. In addition to these factors, one should also take into the consideration about the desorbed or residual atmospheric gases/vapors species and the tentative products formed by the possible interactions between the source and the material. The increase of the pressure inside the vacuum chamber presents direct evidence for the desorption of gases or vapors on operating sources for the film deposition on substrate.

Y. Yan *et al.* [123] and G. Jimenez-Cadena *et al.* [124] deposited different nanostructured in nanowires and nanosheet morphologies of ZnO on ITO coated glass substrate by PVD process for application in DSSCs and characterized them various analytical techniques to confirm their formation with desired morphology. The solar cell efficiency was increased from 0.1% to 0.5% on using dye N197 as sensitizer [124].

1.5.16 Pulsed Laser Deposition Method

In pulsed laser deposition (PLD) method uses a high-power pulsed laser for the vaporization of target material deposited into a plasma plume to deposits on the substrate. It's etching process is very complicated because as the laser power gets absorbed by the target and ejects electrons, atoms, ions, and clusters. The film deposition efficiency depends upon the type of target material, laser pulse energy, repetition rate, distance between target and substrate, substrate temperature, etc. PLD method is also used to ablate many materials and combinations by choosing the appropriate laser wavelength to match the absorption properties of the target. Nanostructured ZnO [125], TiO_2 [126], Sb-doped SnO_2 [127] and Cu_2O [128] synthesized by pulsed laser deposition technique which can be used for solar cell application.

1.5.17 Sputtering Method

In Sputtering process, the individual atom emits out from the source surface by atomic collision cascades on exposure via high energy ion bombardment with better adhesion to substrate and getting thicker film in a vacuum. At low pressure plasma of <0.67 Pa, the sputtered particles are in line of sight. While in higher plasma pressure of 0.67 to 4 Pa range, the energetic particles sputtered or reflected from the sputtering target are thermalized by gas phase collision before they reach the substrate surface. The different sputtering methods such as diode sputtering (cathode or radio frequency), reactive sputtering, bias sputtering, magnetron sputtering and ion-beam sputtering are used for the deposition nanocrystalline metal oxide films on different substrates as per application demand. In magnetron sputtering process, the magnetic field has been employed to control the secondary electrons motion escaping from the target surface due to ion bombardment. The configuration and strength of the magnetic field array decides the rate of ionization efficiency current to supply the target material for high deposition rate at the substrate. The improved ionization efficiency in magnetron sputtering method permits the sputtering process at a lower pressure of 100 Pa and lower voltage of –500 V as compared to the normal sputtering process at 10 Pa pressure and –2 kV to 3 kV voltage. The magnetron sputtering units are classified into two categories (i) conventional (balanced): the plasma is confined to the target region and (ii) unbalanced magnetron sputtering: all the magnetic field lines are not closed, some moves towards the substrate, while some adopts the closed field path. Conductive and insulating materials thin films are deposited by using radio frequency (RF) and direct current (DC) power supply respectively in magnetron sputtering setups. While, magnetic arrays in magnetron sputtering configuration can be changed *in situ* without hampering the electromagnet [122, 129–136].

1.5.17.1 Radio Frequency (RF) Sputtering Method

This sputtering method uses the electrical potential of the current in the vacuum environment at radio frequencies to restrict accumulation of charge on some sputtering target materials. In RF sputtering, the cathode (the target) which is to become the thin film coating and an anode is connected in series with a blocking capacitor in between. The capacitor is part of an impedance-matching network that provides the power transfer from the RF source to the plasma discharge. The cathode is bombarded by high voltage in a vacuum chamber leading to high energy ions sputtering off atoms as a thin film covering the substrate to be coated at a fixed frequency of 13.56 MHz even on insulating substrate and withstand plasma in low 0.13 to 2 Pa pressure range in the entire chamber. For solar cell applications, nanocrystalline NiO and TiO_2 thin films have been used as hole transport layer in perovskite solar fabrication [129–132].

1.5.17.2 DC Sputtering Method

In DC sputtering setup, a pair of planar electrodes (as the cold cathode and anode) has been used. The inert argon gas has been usually used as working gas inside the deposition chamber DC voltage is applied between the conducting cathode (target material) and anode (substrate) for the glow discharge to generate gaseous ions which are accelerated towards the target material, and sputtering takes place to deposit a thin film on the surface of the substrate material. Lot of work is reported on the deposition of nanocrystalline metal oxide thin films for usage in photovoltaic cells [133–136] which involves the production of TiO_2, $TiO_2/CuO/Cu_2O$ and CuO thin films. CdO belongs to the family transparent conducting oxide like SnO_2, In_2O_3, $In_2O_3{:}SnO_2$ (ITO) and ZnO. This is highly transparent in the visible range along with high electrical conductivity and carrier concentration derived from an inherent non-stoichiometry which make it suitable for different applications in solar cells. This film are also used as antireflection coating in solar cells. TiO_2 thin films are already widely in use as n-type semiconducting layer in solar cells. Copper oxide and titanium dioxide semiconducting thin films are used in photovoltaic cells due to their low cost, availability in abundance and non-toxic nature. While, copper oxides are used as absorber layer in heterogeneous solar cell owing to their optimal direct band gap in the range of 2.1–2.6 eV for Cu_2O and 1.3–2.3 eV for CuO and act as p-type and n-type semiconducting layers respectively in solar cells.

1.5.18 Chemical Bath Deposition Method

Chemical bath deposition (CBD) method is also called as controlled precipitation or solution growth method, or chemical deposition. This method has been used for the depositing of metal chalcogenide and metal oxide thin films. The CBD method needs only hot plate with magnetic stirrer, low-cost reactants and a large number of insoluble substrates can be coated in a single run with a proper jig design. Electrical conductivity of the substrate is not the prime condition for the film deposition but, it should have access for the solution. This method does not require high temperatures to grow textured, pin hole free and uniform thin film on the substrate. Hence, by using this method one can inhibit the oxidation and corrosion of metallic substrates and by optimizing film synthesis parameters one can obtain desired orientation and grain structure of the film. The film formation takes place only when the ionic product increases over the solubility product [137, 138]. Water or paraffin bath with constant magnetic stirring assembly hot plate has been used for heating the chemical bath to the required temperature. C.D. Lokhande $et\ al.$ [139] studied the work on microstructure dependent performance of nanocrystalline metal oxide thin films. R.S. Mane $et\ al.$ [140] investigated the nanocrystalline TiO_2/ZnO thin films for the development of DSSCs. While, S. Sagadevan $et\ al.$ [141] and S. Ebrahimiasl $et\ al.$ [142] explored the synthesis, optimization of synthesis parameters and characterization SnO_x (x = 1,2) thin films derived by the CBD method.

1.5.19 Electron Beam Evaporation

Electron beam evaporation method involves subsequent vacuum condensation of metals and nonmetals on the substrate [143–147]. This is a very simple technique in which the material is bombarded and heated by an electron beam. The electrons are accelerated by 2-6 KV to get high energy with large number of process parameters to achieve very high coating growth rates e.g. up to several micro-meters per minute. The evaporation takes place at ultra-high vacuum for deposition of the controlled composition of uniform thickness of the thin film from specific deposition rate. In this method, magnetic field is also applied to focus and bend the electron trajectory. The film deposition rate is of the order of 0.1–100 μm min^{-1} i.e. more than other PVD methods. The reactive evaporation method has been used for depositing metal oxides thin films. In this method, the evaporation of metal atoms from a target material in a chamber is carried out in the presence of reactive gas (e.g. Oxygen) at a partial pressure to undergo chemical reaction to form a compound/metal oxide thin film on the surface of the substrate. Reactive evaporation can form stoichiometric oxides, alloys. A.F. Khan $et\ al.$ [143], F. Ali $et\ al.$ [144] and M.I. Hossain $et\ al.$ [145] used e-beam evaporation method for the depositing thin films of SnO_2,

WO_{3-x} as electron transport layer, and (NiO, SnO_x, MoO_x) metal oxides as carrier transport layer for large scale perovskite solar cells respectively. D. Mergel *et al.* [147] deposited TiO_2 thin films by reactive evaporation at 300 °C and at various pressures and deposition rates. In some cases activated reactive evaporation method has been employed in which the evaporation of atom from the surface of the target material takes place in the presence of plasma which react with the reactive gas within the chamber to form compounds with fast deposition rate and better adhesion properties. The plasma enhances the rate of reaction between the target and the substrate, ionization of both coatings of metals and gas atom in the vapor phase. When the substrate is kept at a negative bias voltage then it is known as biased activated reactive evaporation.

1.5.20 Thermal Evaporation Technique

Thermal (or vacuum) evaporation is one of the oldest method for film deposition and it's growth on solid substrate and still widely used in the research laboratories and industries. This method completes in the following three steps basic (i) the vapor is created from the source material at very high temperature by either by sublimation or boiling, (ii) the vapor ejected from the source material is transferred to the substrate in vacuum with minimum interference and (iii) condensation of the vapor takes place on the surface of substrate to form a solid thin film. These steps are repeated number of time for the growth and nucleation of thin film growth. The process is generally performed under high vacuum pressure (HV), and the trajectory of the movement of the target material to the substrate follows straight path. Vapor flux is created by heating the surface of source material to a sufficiently high temperature in a vacuum. The flux then condenses at the surface of the substrate material to produce a thin film. The vacuum environment helps in developing safe zone for effectively reducing gaseous contaminants during the deposition process to around or below acceptable level and permits the evaporated atoms to follow collision less transfer from the source on the substrate surface. The gas pressure is mostly in 0.0013 Pa to 1.3×10^{-9} Pa range depending on the degree of the contamination in the deposition setup, with the mean free path not less than 5 mm. The thermal vaporization rate is very high compared to other PVD processes [148, 149]. Tungsten wire coils are normally used as the source of thermal heating or by using high energy electron beam for heating the target material to an elevated temperature. The major problems encountered in this method are the source of material to be vaporized and its purity. M. Jin *et al.* [150], A. Wisistoraat [151] and H. Suhail *et al.* [152] deposited thin films of (ZnO and ZnO:Al), ZnOnanopolypods,

CdO by thermal evaporation method respectively for optoelectronic and sensing applications. While, C. Kaito *et al.* [153], M. Asemi *et al.* [154], L.M. Franco Arias *et al.* [155] and M.C. Rao, O.M Hussain [156] prepared SnO_2, Al-Doped ZnO, TiO_2 and WO_3 thin films by vacuum evaporation method respectively for application in solar cells and electrochromic devices.

1.5.21 Electrodeposition Method

The conventional electrodeposition method works on the principle of electrolysis in which chemical reaction occurs at electrode–electrolyte interface by passing electric current through an electrolyte. When electrolysis is carried out in the electrolyte, metal is deposited on the cathode and at the same time anode is dissolved in the solution. The amount of dissolution and deposition is determined by the quantity of electricity. Electrodeposition involves the movements of metallic ions in solution towards a cathode in an applied electric field. The ions accept electrons and deposited on the cathode as atoms or molecules [157–159]. This simple and low-cost technique has been also used for the synthesis of noble metal and metal oxide nanomaterials via electrochemical reaction either by reduction or oxidation of precursors on the conductive substrates. In this context, nanostructured ZnO films [159–161] and TiO_2/dye hybrid films [162–164] were electrochemically deposited and explored their properties for various applications.

1.5.22 Anodic Oxidation Method

The anodic oxidation is also a part of electrochemical deposition method which helps in growing nanotube arrays of metal oxides by the oxidation of metal. This process is comprised of three steps. In the first step, metal is generally dissolved in F- ions containing acidic electrolyte to generate large number of metal ions. In the second step, the electric current starts increasing to form metal oxide film which is partially etched to form irregular pits/pin holes. In the final third step, the H- anions start clustering at the bottom of fine micropores, then metal oxide at the base dissolves and metal as substrate surface is exposed to the electrolyte and helps in getting metal oxide nanotube arrays. The anodization potential, rate of potential rise, oxidation time, pH of solution, F- ions concentration, and aqueous or organic solution medium are some key parameters to be optimized to get better quality and uniform structured nanocrystalline metal oxide films. J. Lin *et al.* [165] presented the overview on anodic nanostructures for solar cells application. D.-J Yang [166] and B. Wang *et al.* [167] reported the synthesis of TiO_2 nanotubes for DSSCs and CuO films respectively.

1.5.23 Screen Printing Method

The economic screen-printing is very facile and recommended as mass production technique for thin and thick films [168]. The precursor, organic binder and solvents are mixed to thixotropic paste which is applied on top of the fixed screen (a finely woven mesh of either stainless steel wires or polymer fibers mounted under tension on a metal frame) so that openings/pores of the screen are filled with the paste in the flood stroke. The screen should not come in contact with the substrate. In the screen printer, screen act as a carrier of the to-be printed pattern. In the print stroke, the screen comes in contact with substrate. Then, the squeegee induces pressure on screen and pushes the paste to pass through the screen mesh on the substrate. Squeegees are available in U- or V -shapes, and made up of different hardness and materials. Polyurethane squeegees having hardness in 60-80 shore range are normally used in the screen printers. The backside is covered with a photosensitive emulsion layer on the screen pattern is irradiated with UV light through a film having pattern. The pattern is thus transferred onto the substrate and the deposited film acquire thickness in 5-50 μm range. For electronic application mostly glass substrate is used. To obtain good quality of screen-printed films, one should optimize the web material, the emulsion, the printing procedure and the printing paste [168]. This involves the web tension and thickness, the mesh count, the emulsion thickness, the squeegee hardness, shape and pressure, the printing velocity, the accuracy of the positioning and the paste parameters. For the nanocrystalline metal oxide thin/thick film deposition by screen printing, the sol-gel or other chemical routes are used to produce metal oxide nanoparticles or alkoxides are used for preparing precursor paste with binder and suitable solvent [169, 170]. Finally, the films of desired chemical compositions are obtained by sintering them at a particular temperature for fixed duration. The synthesis and characterization of screen-printed titania films for application in DSSCs has been nicely described by the authors in the following references [171–174].

1.6 Characterization of Metal Oxide Nanoparticles/Thin Films

The as-prepared nanocrystalline metal oxides in nanoparticle or thin/thick film for are primarily characterized by X-ray diffraction (XRD), FTIR, Raman, X-ray photoelectron spectroscopic (XPS) techniques to reveal their structural details. While Scanning electron microscopy (SEM) and high-resolution transmission electron microscopy (HRTEM) techniques are used for imaging surface morphologies, size, and shape of constituting nanoparticles/arrays. The stoichiometric composition of the samples is derived from EDS data. UV-Vis, Diffused reflectance and photoluminescence (PL) spectroscopic investigations are used to get band structural details. The I-V characteristics and direct current (dc) conductivity studies provide information about the electrical characteristic properties of the prepared nanocrystalline metal oxides. These investigations help in confirming the formation of samples with desired properties for photovoltaic cells application.

1.7 Conclusion and Future Aspects

In this book the authors tried to give the overview of energy requirement, severity of problem, evolution in photovoltaic cell generations with time, role of nano metal oxides in photovoltaic cells, synthesis and characterization of nanocrystalline metal oxides in nanoparticles or in thin/thick film forms. The preparation of new nanocrystalline metal oxides or modifications in existing materials in particle or film form for solar cells application is an area of interest for researchers. This work will always remain in demand because the need of energy consumption with time never cease. The development of new cost effective synergic materials for energy production is really a very interesting and challenging problem for the present and future.

References

1. India energy outlook 2021 – analysis, IEA, 2021.
2. Power sector at a glance ALL INDIA, Government of India | Ministry of Power (powermin.gov.in), 2022.
3. Current status, Ministry of New and Renewable Energy, Government of India (mnre.gov.in), 2022.
4. PVPS I, A snapshot of global PV. Report IEA PVPS T1-29, 2016.
5. Tao, M., Inorganic photovoltaic photovoltaic cells: Silicon and beyond. *Electrochem. Soc. Interface*, 30, 5, 2008.
6. Ranabhat, K. *et al.*, An introduction to photovoltaic cell technology. *J. Appl. Eng. Sci.*, 14, 481, 2016.
7. Sze, S.M. and Irvin, J.C., Resistivity, mobility and impurity levels in GaAs, Ge, and Si at 300 K. *Solid-State Electron.*, 11, 599, 1968.

8. Clavero, C., Plasmon-induced hot-electron generation at nanoparticle/metal-oxide interfaces for photovoltaic and photocatalytic devices. *Nat. Photonics*, 8, 95, 2014.
9. Simya, O.K., Mahboobbatcha, A., Balachander, K., A comparative study on the performance of Kesterite based thin film photovoltaic cells using SCAPS simulation program. *Superlattices Microstruct.*, 82, 248, 2015.
10. Hossain, M.I., Prospects of CZTS photovoltaic cells from the perspective of material properties, fabrication methods and current research challenges. *Chalcogenide Lett.*, 9, 231, 2012.
11. Wang, W., Winkler, M.T., Gunawan, O., Gokmen, T., Todorov, T.K., Zhu, Y., Mitzi, D.B., Device characteristics of CZTSSe thin-film photovoltaic cells with 12.6% efficiency. *Adv. Energy Mater.*, 4, 1301465, 2014.
12. Siebentritt, S., Why are kesterte photovoltaic cells not 20% efficient—review. *Thin Solid Films*, 535, 1, 2013.
13. Bag, S., Gunawan, O., Gokmen, T., Zhu, Y., Todorov, T.K., Mitzi, D.B., Low band gap liquid-processed CZTSe photovoltaic cell with 10.1% efficiency. *Energy Environ. Sci.*, 5, 7060, 2012.
14. Shoori, K. and Kavei, G., Copper indium gallium diselenide-CIGS photovoltaic cell technology a review. *Int. Mater. Phys.*, 1, 15, 2013.
15. Simya, O.K., Mahboobbatcha, A., Balachander, K., Compositional grading of CZTSSe alloy using exponential and uniform grading laws in SCAPS-1D simulation. *Superlattices Microstruct.*, 92, 285, 2016.
16. Luque, A. and Hegedus, S. (Eds.), *Handbook of photovoltaic science and engineering*, Wiley, London, 2003.
17. Dubey, S., Sarvaiya, J.N., Seshadri, B., Temperature dependent photovoltaic (PV) efficiency and its effect on PV production in the world – a review. *Energy Proc.*, 33, 311, 2013.
18. Jordehi, A.R., Parameter estimation of solar photovoltaic (PV) cells: A review. *Renew. Sustain. Energy Rev.*, 61, 354, 2016.
19. Ossai, A.N., Alabi, A.B., Ezike, S.C., Aina, A.O., Zinc oxide-based dye-sensitized solar cells using natural and synthetic sensitizers. *Curr. Res. Green Sustain. Chem.*, 3, 100043, 2020.
20. Anta, J.A., Guillén, E., Tena-Zaera, R., ZnO-based dye-sensitized solar cells. *J. Phys. Chem. C*, 116, 11413, 2012.
21. Lanlan, L., Renjie, L.K., Fan, T.P., Effects of annealing conditions on the photo-electrochemical properties of dye-sensitized solar cells made with ZnO nanoparticles. *Solar Energy*, 84, 844, 2010.
22. Roy, P., Kim, D., Paramasivam, I., Schmuki, P., Improved efficiency of TiO_2 nanotubes in dye sensitized solar cells by decoration with TiO_2 nanoparticles. *Electrochem. Commun.*, 11, 1001, 2009.
23. Xin, X., Scheiner, M., Ye, M., Lin, Z., Surface-treated TiO_2 nanoparticles for dye-sensitized solar cells with remarkably enhanced performance. *Langmuir*, 27, 1459414598, 2011.

24. Nakade, S., Matsuda, M., Kambe, S., Saito, Y., Kitamura, T., Sakata, T., Wada, Y., Mori, H., Yanagida, S., Dependence of TiO_2 nanoparticle preparation methods and annealing temperature on the efficiency of dye-sensitized solar cells. *J. Phys. Chem. B*, 106, 10004, 2002.
25. Chen, W., Qiu, Y., Zhong, Y., Wong, K.S., Yang, S., High-efficiency dye-sensitized solar cells based on the composite photoanodes of SnO_2 nanoparticles/ZnONanotetrapods. *J. Phys. Chem. A*, 114, 3127, 2010.
26. Banik, A., Mohammad, M.S., Qureshi, M., Efficient energy harvesting in SnO_2-based dye-sensitized solar cells utilizing nano-amassed mesoporous zinc oxide hollow microspheres as synergy boosters. *ACS Omega*, 3, 14482, 2018.
27. Li, Z., Zhou, Y., Sun, R., Xiong, Y., Xie, H., Zhou, Z., Nanostructured SnO_2 photoanode-based dye-sensitized solar cells. *Chin. Sci. Bull.*, 59, 2122, 2014.
28. Ghosh, R., Brennaman, M.K., Uher, T.O., Myoung-Ryul, S., Edward, T., McNeil, L.E., Meyer, T.J., Lopez, R., Nanoforest Nb_2O_5 photoanodes for dye-sensitized solar cells by pulsed laser deposition. *ACS Appl. Mater. Interfaces*, 3, 3929, 2011.
29. Suresh, S., Unni, G.E., Satyanarayana, M., Sreekumaran Nair, A., Mahadevan Pillai, V.P., Silver nanoparticles-incorporated Nb_2O_5 surface passivation layer for efficiency enhancement in dye-sensitized solar cells. *J. Colloid Interface Sci.*, 524, 236, 2018.
30. Jin, X., Liu, C., Xu, J., Wang, Q., Chen, D., Size-controlled synthesis of mesoporous Nb_2O_5 microspheres for dye sensitized solar cells. *RSC Adv.*, 4, 35546, 2014.
31. Prabhu, N., Agilan, S., Muthukumarasamy, N., Senthil, T.S., Enhanced photovoltaic performance of WO_3 nanoparticles added dye sensitized solar cells. *J. Mater. Sci.: Mater. Electron.*, 25, 5288, 2014.
32. Biswas, R. and Chatterjee, S., Effect of surface modification via sol-gel spin coating of ZnO nanoparticles on the performance of WO_3 photoanode based dye sensitized solar cells. *Optik*, 212, 164142, 2019.
33. Yang, W., Xu, X., Li, Z., Yang, F., Zhang, L., Li, Y., Wang, A., Chen, S., Construction of efficient counter electrodes for dye-sensitized solar cells: Fe_2O_3 nanoparticles anchored onto graphene frameworks. *Carbon*, 96, 947, 2016.
34. Reda, S.M., Synthesis of ZnO and Fe_2O_3 nanoparticles by sol–gel method and their application in dye-sensitized solar cells. *Mater. Sci. Semicond. Process.*, 13, 417, 2010.
35. Kumar, D.K., Kříž, J., Bennett, N., Chen, B., Upadhayaya, H., Reddy, K.R., Sadhu, V., Functionalized metal oxide nanoparticles for efficient dye-sensitized solar cells (DSSCs): A review. *Mater. Sci. Energy Technol.*, 3, 472, 2020.
36. Kojima, A., Teshima, K., Shirai, Y., Miyasaka, T., Organometal halide perovskites as visible-light sensitizers for photovoltaic cells. *J. Am. Chem. Soc.*, 131, 6050, 2009.

37. Martí, A. and Luque, A., *Next generation photovoltaics: High efficiency through full spectrum utilization*, p. 136, CRC Press, Boca Raton, FL, USA, 2003.
38. Neuhaus, D.H. and Munzer, A., Industrial silicon wafer solar cells. *Adv. Optoelectron.*, 2007, 24521, 2007.
39. Rishabh Anand, E., Chapter 3 Essentials of nanotechnology, first ed, MedTec, Scientific International, India, 2015.
40. Gogotsi, Y., *Nanomaterials handbook*, CRC Press, Boca Raton, FL, 2006.
41. Gerling, L.G., Mahato, S., Morales-Vilches, A., Masmitja, G., Ortega, Voz, P.C., Alcubilla, R., Puigdollers, J., Transition metal oxides as hole-selective contacts in silicon heterojunctions solar cells. *Sol. Energy Mater. Sol. Cells*, 145, 109, 2016.
42. Fortunato, E., Ginley, D., Hosono, H., Paine, D., Transparent conducting oxides for photovoltaics. *MRS Bull.*, 32, 242, 2007.
43. Singh, R.S., Rangari, V.K., Sanagapalli, S., Jayaraman, V., Mahendra, S., Singh, V.P., Nano-structured CdTe, CdS and TiO_2 for thin film solar cell applications. *Sol. Energy Mater. Sol. Cells*, 82, 315, 2004.
44. Pedersen, H. and Elliott, S.D., Studying chemical vapor deposition processes with theoretical chemistry. *Theor. Chem. Acc.*, 133, 1476, 2014.
45. Hitchman, M.L. and Jensen, K.F. (Eds.), *Chemical vapor deposition: Principles and applications*, Elsevier. Academic Press, San Diego, 1993.
46. Shinde, P.S. and Bhosale, C.H., Properties of chemical vapour deposited nanocrystalline TiO_2 thin films and their use in dye-sensitized solar cells. *J. Anal. Appl. Pyrolysis*, 82, 83, 2008.
47. Scott, J.I., Martinez-Gazoni, R.F., Allen, M.W., Reeves, R.J., Optical and electronic properties of high quality Sb-doped SnO_2 thin films grown by mist chemical vapor deposition. *J. Appl. Phys.*, 126, 135702, 2019.
48. Wright, P.J., Crosbie, M.J., Lane, P.A., Williams, D.J., Jones, A.C., Leedham, T.J., Davies, H.O., Metal organic chemical vapor deposition (MOCVD) of oxides and ferroelectric materials. *J. Mater. Sci.: Mater. Electron.*, 13, 671, 2002.
49. Wenas, W.W., Yamada, A., Konagai, M., Takahashi, K., Textured ZnO thin films for solar cells grown by metalorganic chemical vapor deposition. *Jpn. J. Appl. Phys.*, 30, L441, 1991.
50. Gulino, A., Dapporto, P., Rossi, P., Fragala, I., A liquid MOCVD precursor for thin films of CdO. *Chem. Mater.*, 14, 1441, 2002.
51. Hongsingthong, A., Yunaz, I.A., Miyajima, S., Konagai, M., Preparation of ZnO thin films using MOCVD technique with D_2O/H_2O gas mixture for use as TCO in silicon-based thin film solar cells. *Sol. Energy Mater. Sol. Cells*, 95, 171, 2011.
52. Oussidhoum, S., Hocine, D., Chaumont, D., Crisbasan, A., Bensidhoum, M.O.T., Bourennane, E.-B., Moussi, A., Lesniewska, E.A., Geoffroy, N., Belkaid, M.S., Optimization of physicochemical and optical properties of nanocrystalline TiO_2 deposited on porous silicon by metal-organic chemical vapor deposition (MOCVD). *Mater. Res. Express*, 6, 125917, 2019.

53. Hellegouarc'h, F., Arefi-Khonsari, F., Planade, R., Amouroux, J., PECVD prepared SnO_2 thin films for ethanol sensors. *Sens. Actuators B*, 73, 27, 2001.
54. Huang, H., Tan, O.K., Lee, Y.C., Tse, M.S., Preparation and characterization of nanocrystalline SnO_2 thin films by PECVD. *J. Cryst. Growth*, 288, 70, 2006.
55. Mazzarella, L., Morales-Vilches, A.B., Korte, L., R. Schlatmann, R., Stannowski, B., Versatility of nanocrystalline silicon films: From thin-film to perovskite/c-Si tandem solar cell applications. *Coatings*, 10, 759, 2020.
56. Patil, P.S., Versatility of chemical spray pyrolysis technique. *Mater. Chem. Phys.*, 59, 185, 1999.
57. Mooney, J.B. and Radding, S.B., Spray pyrolysis processing. *Annu. Rev. Mater. Sci.*, 12, 81, 1982.
58. Fauzia, V., Yusnidar, M.N., Lalasari, L.H., Subhan, A., Umar, A.A., High figure of merit transparent conducting Sb-doped SnO_2 thin films prepared via ultrasonic spray pyrolysis. *J. Alloys Compd.*, 720, 79, 2017.
59. Lokhande, B.J., Patil, P.S., Uplane, M.D., Studies on cadmium oxide sprayed thin films deposited through non-aqueous medium. *Mater. Chem. Phys.*, 84, 238, 2004.
60. Okuya, M., Nakade, K., Kaneko, S., Porous TiO_2 thin films synthesized by a spray pyrolysis deposition (SPD) technique and their application to dye-sensitized solar cells. *Sol. Energy Mater. Sol. Cells*, 70, 425, 2002.
61. Mohammed, A.S., Kafi, D.K., Ramizy, A., Abdulhad, O.O., Hasan, S.F., Nanocrystalline Ce-doped CdO thin films synthesis by spray pyrolysis method for solar cells applications. *J. Ovonic Res.*, 15, 37, 2019.
62. George, S.M., Atomic layer deposition: An overview. *Chem. Rev.*, 110, 111, 2009.
63. Johnson, R.W., Hultqvist, A., Bent, S.F., A brief review of atomic layer deposition: From fundamentals to applications. *Mater. Today*, 17, 236, 2014.
64. Garcia-Alonso, D., Potts, S.E., van Helvoirt, C.A., Verheijen, M.A., Kessels, W.M., Atomic layer deposition of B-doped ZnO using triisopropyl borate as the boron precursor and comparison with Al-doped ZnO. *J. Mater. Chem.*, 3, 3095, 2015.
65. Lujala, V., Skarp, J., Tammenmaa, M., Suntola, T., Atomic layer epitaxy growth of doped zinc oxide thin films from organometals. *Appl. Surf. Sci.*, 82-83, 34, 1994.
66. Chambers, B.A., MacDonald, B.I., Ionescu, M., Deslandes, A., Quinton, J.S., Jasieniak, J.J., Andersson, G.G., Examining the role of ultra-thin atomic layer deposited metal oxide barrier layers on CdTe/ITO interface stability during the fabrication of solution processed nanocrystalline solar cells. *Sol. Energy Mater. Sol. Cells*, 125, 164, 2014.
67. Ukoba, O.K. and Jen, T.-C., Review of atomic layer deposition of nanostructured solar cells. *Int. Conf. Eng. Sustain. World, J. Phys.: Conf. Ser.*, 1378, 4, 042060, 2019.
68. Tynell, T. and Karppinen, M., Atomic layer deposition of ZnO: A review. *Semicond. Sci. Technol.*, 29, 043001, 2014.

69. Costals, E.R., Masmitja, G., Almache, E., Pusay, B., Tiwari, K., Saucedo, E., Raj, C.J., Kim, B.C., Puigdollers, J., Martin, I., Voza, C., Ortega, P., Atomic layer deposition of vanadium oxide films for crystalline silicon solar cells. *Mater. Adv.*, 3, 337, 2022.
70. Cushing, B.L., Kolesnichenko, V.L., O'Connor, C.J., Recent advances in the liquid-phase syntheses of inorganic nanoparticles. *Chem. Rev.*, 104, 3893, 2004.
71. Liang, X., Bai, S., Wang, X., Dai, X., Gao, F., Sun, B., Ning, Z., Ye, Z., Jin, Y., Colloidal metal oxide nanocrystals as charge transporting layers for solution-processed light-emitting diodes and solar cells. *Chem. Soc. Rev.*, 46, 1730, 2017.
72. Ouyang, D., Huang, Z., Choy, W.C.H., Solution-processed metal oxide nanocrystals as carrier transport layers in organic and perovskite solar cell. *Adv. Funct. Mater.*, 29, 1804660, 2018.
73. Bhogaita, M. and Devaprakasam, D., Hybrid photoanode of TiO_2-ZnO synthesized by co-precipitation route for dye-sensitized solar cell using phyllanthus reticulatas pigment sensitizer. *Sol. Energy*, 214, 517, 2021.
74. Yin, M., Wu, C.-K., Lou, Y., Burda, C., Koberstein, J.T., Zhu, Y., O'Brien, S., Copper oxide nanocrystals. *J. Am. Chem. Soc.*, 127, 9506, 2005.
75. Rashad, M.M., Ibrahim, I.A., Osama, I., Shalan, A.E., Distinction between SnO_2 nanoparticles synthesized using co-precipitation and solvothermal methods for the photovoltaic efficiency of dye-sensitized solar cells. *Bull. Mater. Sci.*, 37, 903, 2014.
76. Szindler, M., Szindler, M.M., Boryl, P., ZnO nanocrystalline powder prepared by sol-gel method for photoanode of dye sensitized solar cells application. *J. Achiev. Mater. Manuf. Eng.*, 88, 12, 2018.
77. Thiagarajan, S., Sanmugam, A., Vikraman, D., Facile methodology of sol-gel synthesis for metal oxide nanostructures, in: *Recent Applications in Sol-Gel Synthesis*, U. Chandra (Ed.), p. 68708, Intechopen Book, United Kingdom, 2017.
78. Niederberger, M., Nonaqueous sol–gel routes to metal oxide nanoparticles. *Acc. Chem. Res.*, 40, 793, 2007.
79. Tseng, T.K., Lin, Y.S., Chen, Y.J., Chu, H., A review of photocatalysts prepared by sol-gel method for VOCs removal. *Int. J. Mol. Sci.*, 11, 2336, 2010.
80. Lee, J.H., Zhang, S., Sun, S.H., High-temperature solution-phase syntheses of metal-oxide nanocrystals. *Chem. Mater.*, 25, 1293, 2013.
81. Ludi, B., Suess, M.J., Werner, I.A., Niederberger, M., Mechanistic aspects of molecular formation and crystallization of zinc oxide nanoparticles in benzyl alcohol. *Nanoscale*, 4, 1982, 2012.
82. Cushing, B.L., Kolesnichenko, V.L., O'Connor, C.J., Recent advances in the liquid-phase synthesis of inorganic nanoparticles. *Chem. Rev.*, 104, 3893, 2004.
83. Demazea, G., Solvothermal processes: Definition, key factors governing the involved chemical reactions and new trends. *Z. Naturforsch. B*, 65, 999, 2010.

84. Dinh, C.T., Nguyen, T.D., Kleitz, F., Do, T.O., Shape-controlled synthesis of highly crystalline titania nanocrystals. *ACS Nano*, 3, 3737, 2009.
85. Liang, X., Bai, S., Wang, X., Dai, X., Gao, F., Sun, B., Ning, Z., Ye, Z., Jin, Y., Colloidal metal oxide nanocrystals as charge transporting layers for solution-processed light-emitting diodes and solar cells. *Chem. Soc. Rev.*, 46, 1730, 2017.
86. Li, Z., Stable perovskite solar cells based on WO_3 nanocrystals as hole transport layer. *Chem. Lett.*, 44, 1140, 2015.
87. Krishnapriya, R., Praneetha, S., Vadivel Murugan, A., Energy efficient, microwave-hydro-/solvothermal synthesis of hierarchical flowers and rice-grain like ZnO nanocrystals as photoanodes for high performance dye-sensitized solar cells. *CrystEngComm*, 17, 8353, 2015.
88. Beusen, J., Van Bael, M.K., Van den Rul, H., D'Haen, J., Mullens, J., Preparation of a porous nanocrystalline TiO_2 layer by deposition of hydrothermally synthesized nanoparticles. *J. Eur. Ceram. Soc.*, 27, 4529, 2007.
89. Papadas, I.T., Savva, A., Ioakeimidis, A., Eleftheriou, P., Armatas, G.S., Choulis, S.A., Employing surfactant-assisted hydrothermal synthesis to control $CuGaO_2$ nanoparticle formation and improved carrier selectivity of perovskite solar cells. *Mater. Today Energy*, 8, 57, 2018.
90. Ai, X.D., Yan, L.T., Liu, Y.C., Li, T.X., Dou, S.Y., Dai, C.A., Preparation and influencing factors of ZnO whisker by hydrothermal microemulsion process. *Adv. Mater. Res.*, 528, 193, 2012.
91. Langevin, D., Micelles and microemulsions. *Annu. Rev. Phys. Chem.*, 43, 341, 1992.
92. Capek, I., Preparation of metal nanoparticles in water-in-oil (w/o) microemulsions. *Adv. Colloid Interface Sci.*, 110, 49, 2004.
93. Malik, M.A., Wani, M.Y., Hashim, M.A., Microemulsion method: A novel route to synthesize organic and inorganic nanomaterials: 1st Nano Update. *Arab. J. Chem.*, 5, 397, 2012.
94. Li, X.C., He, G.H., Xiao, G.K., Liu, H.J., Wang, M., Synthesis and morphology control of ZnO nano-structures in microemulsions. *J. Colloid Interface Sci.*, 333, 465, 2009.
95. Yan, L., Uddin, A., Wang, H., ZnOTetrapods: Synthesis and applications in solar cells. *Nanomater. Nanotechnol.*, 5, 19, 2015.
96. Mirzaei, A. and Neri, G., Microwave-assisted synthesis of metal oxide nanostructures for gas sensing application: A review. *Sens. Actuators B Chem.*, 237, 749, 2016.
97. Rao, K.J., Vaidhyanathan, B., Ganguli, M., Ramakrishnan, P.A., Synthesis of inorganic solids using microwaves. *Chem. Mater.*, 11, 882, 1999.
98. Fathy, M., Hassan, H., Hafez, H., Soliman, M., Abulfotuh, F., Kashyout, A.E.H.B., Simple and fast microwave-assisted synthesis methods of nanocrystalline TiO_2 and rGO materials for low-cost metal-free DSSC applications. *ACS Omega*, 7, 16757, 2022.

99. Wang, H.-E., Zheng, L.-X., Liu, C.-P., Luan, C.-Y., Cheng, H., Li, Y.Y., Martinu, L., Zapien, J.A., Bello, I., Rapid microwave synthesis of porous TiO_2 spheres and their applications in dye-sensitized solar cells. *J. Phys. Chem. C*, 115, 10419, 2011.
100. Nehru, L.C. and Sanjeeviraja, C., Rapid synthesis of nanocrystalline SnO_2 by a microwave-assisted combustion method. *J. Adv. Ceram.*, 3, 171, 2014.
101. Sun, H., Sun, L., Sugiura, T., White, M.S., Stadler, P., Sariciftci, N.S., Masuhara, A., Yoshida, T., Microwave-assisted hydrothermal synthesis of structure-controlled ZnO nanocrystals and their properties in dye-sensitized solar cells. *Electrochem.*, 85, 253, 2017.
102. Gao, X., Li, X., Yu, W., Structural and morphological evolution of ZnO cluster film prepared by the ultrasonic irradiation assisted solution route. *Thin Solid Films*, 484, 160, 2005.
103. Ohayon, E. and Gedanken, A., The application of ultrasound radiation to the synthesis of nanocrystalline metal oxide in non-aqueous solvent. *Ultrason. Sonochem.*, 17, 173, 2010.
104. Meybodi, S.M., Hosseini, S.A., Rezaee, M., Sadrnezhaad, S.K., Mohammadyani, D., Synthesis of wide band gap nanocrystalline NiO powder via a sonochemical method. *Ultrason. Sonochem.*, 19, 841, 2012.
105. Singh, J., Dutta, T., Kim, K.H., Rawat, M., Samddar, P., Kumar, P., "Green" synthesis of metals and their oxide nanoparticles: Applications for environmental remediation. *J.Nanobiotechnol.*, 16, 1, 2018.
106. Mironyuk, I.F., Soltys, L.M., Tatarchuk, T.R., Savka, K.O., Methods of titanium dioxide synthesis (review). *Phys. Chem. Solid State*, 21, 462, 2020.
107. Bandeira, M., Giovanela, M., Roesch-Ely, M., Devine, D.M., da Silva Crespo, J., Green synthesis of zinc oxide nanoparticles: A review of the synthesis methodology and mechanism of formation. *Sustain. Chem. Pharm.*, 15, 100223, 2020.
108. Gnanasangeetha, D. and Thambavani, S.D., Facile and eco-friendly method for the synthesis of zinc oxide nanoparticles using azadirachta and emblica. *Int. J. Pharm. Sci. Res.*, 5, 2866, 2014.
109. Thamima, M. and Karuppuchamy, S., Biosynthesis of titanium dioxide and zinc oxide nanoparticles from natural sources: A review. *Adv. Sci. Eng. Med.*, 7, 18, 2015.
110. Pal, G., Rai, P., Pandey, A., Green synthesis of nanoparticles: A greener approach for a cleaner future, in: *Green Synth. Charact. Appl. Nanoparticles*, p. 1, 2019.
111. Ahmad, W., Jaiswal, K.K., Soni, S., Green synthesis of titanium dioxide (TiO_2) nanoparticles by using Mentha arvensis leaves extract and its antimicrobial properties. *Inorg. Nano-Met. Chem.*, 50, 1032, 2020.
112. Berber, M., Bulto, V., Kliß, R., Hahn, H., Transparent nanocrystalline ZnO films prepared by spin coating. *Scr. Mater.*, 53, 547, 2005.

113. Verma, A. and Vijayan, N., Sol–gel-derived nanocrystalline aluminium-doped zinc oxide thin films for use as antireflection coatings in silicon solar cells. *J. Mater. Res.*, 28, 2990, 2013.
114. Arora, M., Zargar, R.A., Khan, S.D., EPR spectroscopy of different sol concentration synthesized nanocrystalline-ZnO thin films. *Int. J. Spectrosc.*, 2015, Article ID 431678, 2015.
115. Singh, J., Singh, A., Kumar, Kumar, K. et al., Optical properties of TiO_2 thin film: Via dip coating method. *J. Nano Phys.*, 14, 02019(3pp), 2022.
116. Amri, A., Jiang, Z.-T., Pryor, T., Yin, C.-Y., Xie, Z., Mondinos, N., Optical and mechanical characterization of novel cobalt-based metal oxide thin films synthesized using sol–gel dip-coating method. *Surf. Coat. Technol.*, 207, 367, 2012.
117. Mahan, J.E., *Physical vapor deposition of thin films*, 1st ed., p. 336, Wiley-VCH, New York, NY, USA, 2000.
118. Savale, P.A., Physical vapor deposition (PVD) methods for synthesis of thin films: A comparative study. *Arch. Appl. Sci. Res.*, 8, 1, 2016.
119. Kaufman, H.R., Cuomo, J.J., Harper, J.M.E., Technology and applications of broad-beam ion sources used in sputtering part I, ion source technology. *J. Vac. Sci. Technol.*, 21, 725, 1982.
120. Raut, H.K., Ganesh, V.A., Nair, A.S., Ramakrishna, S., Anti-reflective coatings: A critical, in-depth review. *Energy Environ. Sci.*, 4, 3779, 2011.
121. Fancey, K.S., A coating thickness uniformity model for physical vapor-deposition systems: Overview. *Surf. Coat. Technol.*, 71, 16, 1995.
122. Martin, P.M., Deposition technologies: An overview, in: *Handbook of Deposition Technologies for Films and Coatings, Science, Application and Technology*, p. 1, Elsevier, Amesterdam, Netherlands, 2010.
123. Yan, Y., Zhang, Y., Meng, G., Zhang, L., Synthesis of ZnO nanocrystals with novel hierarchical structures via atmosphere pressure physical vapor deposition method. *J. Cryst. Growth*, 294, 184, 2006.
124. Jimenez-Cadena, G., Comini, E., Ferroni, M., Vomiero, A., Sberveglieri, G., Synthesis of different ZnO nanostructures by modified PVD process and potential use for 1dye-sensitized solar cells. *Mater. Chem. Phys.*, 124, 694, 2010.
125. Choopuna, S., Tabata, H., Kawai, T., Self-assembly ZnO nanorods by pulsed laser deposition under argon atmosphere. *J. Cryst. Growth*, 274, 167, 2005.
126. Fusi, M., Russo, V., Casari, C.S., Li Bassi, A., Bottani, C.E., Titanium oxide anostructured films by reactive pulsed laser deposition. *Appl. Surf. Sci.*, 255, 5334, 2009.
127. Yu, S., Ding, L., Xue, C., Chen, L., Zhang, W.F., Transparent conducting Sb-doped SnO_2 thin films grown by pulsed laser deposition. *J. Non-Cryst. Solids*, 358, 3137, 2012.
128. Farhad, S.F.U., Cherns, D., Smith, J.A., Fox, N.A., Fermín, D.J., Pulsed laser deposition of single phase n- and p-type Cu_2O thin films with low resistivity. *Mater. Des.*, 193, 108848, 2020.

129. Aydin, E., Troughton, J., De Bastiani, M., Ugur, E., Sajjad, M., Alzahrani, A., Neophytou, M., Schwingenschlögl, U., Laquai, F., Baran, D., De Wolf, S., Room-temperature sputtered nanocrystalline nickel oxide as hole transport layer for p–i–n perovskite solar cells. *ACS Appl. Energy Mater.*, 1, 6227, 2018.
130. Chen, C., Cheng, Y., Dai, Q., Song, H., Radio frequency magnetron sputtering deposition of TiO_2 thin films and their perovskite solar cell applications. *Sci. Rep.*, 5, 17684, 2015.
131. Abe, S., InSb-added TiO_2 nanocomposite films by RF sputtering. *Nanoscale Res. Lett.*, 8, 269, 2013.
132. Abd El-Moulaa, A.A., Raaifb, M., El-Hossary, F.M., Optical properties of nanocrystalline/amorphous TiO_2 thin film deposited by rf plasma magnetron sputtering. *Acta Phys. Pol. A*, 137, 1068, 2020.
133. Subramanyam, T.K., Uthanna, S., Naidu, B.S., Preparation and characterization of CdO films deposited by dc magnetron reactive sputtering. *Mater. Lett.*, 35, 214, 1998.
134. Wisz, G., Sawicka-Chudy, P., Sibinski, M., Płoch, D., Bester, M., Cholewa, M., Wozny, J., Yavorskyi, R., Nykyruy, L., Ruszała, M., TiO_2/CuO/Cu_2O photovoltaic nanostructures prepared by DC reactive magnetron sputtering. *Nanomaterials*, 12, 1328, 2022.
135. Wisz, G., Sawicka-Chudya, P., Sibińskib, M., Starowiczc, Z., Płocha, D., Góralc, A., Bestera, M., Cholewaa, M., Woźnyb, J., Sosna-Głębskab, A., Solar cells based on copper oxide and titanium dioxide prepared by reactive direct-current magnetron sputtering. *Opto-Electron. Rev.*, 29, 97, 2021.
136. Shukao, A., Haider, A., Takano, I., Electrical and optical properties of copper oxide thin films prepared by DC magnetron sputtering. *J. Vac. Sci. Technol. B*, 38, 012803, 2020.
137. Lange, F.F., Chemical solution routes to single-crystal thin films. *Science*, 273, 903, 1996.
138. Lokhande, C.D., Chemical deposition of metal chalcogenide thin films. *Mater. Chem. Phys.*, 28, 1, 1991.
139. Lokhande, C.D., More, A.M., Gunjakar, J.L., Microstructure dependent performance of chemically deposited nanocrystalline metal oxide thin films. *J. Alloys Compd.*, 486, 570, 2009.
140. Mane, R.S., Lee, W.J., Pathan, H.M., Han, S.-H., Nanocrystalline TiO_2/ZnO thin films: Fabrication and application to dye-sensitized solar cells. *J. Phys. Chem. B*, 109, 24254, 2005.
141. Sagadevan, S. and Podder, J., Optical and electrical properties of nanocrystalline SnO_2 thin films synthesized by chemical bath deposition method. *Soft Nanosci. Lett.*, 5, 55, 2015.
142. Ebrahimiasl, S., Yunus, W.M.Z.W., Kassim, A., Zainal, Z., Synthesis of nanocrystalline SnO_x (x = 1–2) thin film using a chemical bath deposition method with improved deposition time, temperature and pH. *Sensors*, 11, 9207, 2011.

143. Khan, A.F., Mehmood, M., Aslam, M., Ashraf, M., Characteristics of electron beam evaporated nanocrystalline SnO_2 thin films annealed in air. *Appl. Surf. Sci.*, 256, 2252, 2010.
144. Ali, F., Pham, N.D., Fan, L., Tiong, V., Ostrikov, K., Bell, J., Wang, H., Tesfamichael, T., Low hysteresis perovskite solar cells using e-beam evaporated WO_{3-x} thin film as electron transport layer. *ACS Appl. Energy Mater.*, 2, 5456, 2019.
145. Hossain, M.I., Zakaria, Y., Zikri, A., Samara, A., Aissa, B., El-Mellouhi, F., Hasan, N.S., A. Belaidi, A., Mahmood, A., Mansour, S., E-beam evaporated hydrophobic metal oxide thin films as carrier transport materials for large scale perovskite solar cells. *Mater. Technol.*, 37, 248, 2022.
146. Yushkov, Y.G., Tyunkov, A.V., Zolotukhin, D.B., Oks, E.M., Electron beam evaporation of boron for ion-plasma coating synthesis at fore vacuum pressures. *J. Appl. Phys.*, 120, 233302, 2016.
147. Mergel, D., Buschendorf, D., Eggert, S., Grammes, R., Samset, B., Density and refractive index of TiO_2 films prepared by reactive evaporation. *Thin Solid Films*, 371, 218, 2000.
148. Mattox, D.M., *Handbook of physical vapor deposition (PVD) processing*, 2nd Ed., Elsevier/William Andrew, Amsterdam, Netherlands, 2010.
149. Mattox, D.M., Atomistic film growth and some growth-related film properties, in: *Handbook of Physical Vapor Deposition (PVD) Processing*, 1998.
150. Jin, M., Feng, J., De-heng, Z., Hong-lei, M., Shuying, L., Optical and electronic properties of transparent conducting ZnO and ZnO-Al films prepared by evaporating method. *Thin Solid Films*, 357, 98, 1999.
151. Wisitsoraat, A., Pimtara, I., Phokharatkul, D., Jaruwon- grangsee, K., Tuantranont, A., Zinc oxide nanopolypods synthesized by thermal evaporation of carbon nanotubes and zinc oxide mixed powder. *Curr. Nanosci.*, 6, 1, 2010.
152. Suhail, H., Issam, M., Rao, M., Characterization and gas sensitivity of cadmium oxide thin films prepared by thermal evaporation technique. *Int. J. Thin Film Sci. Technol.*, 1, 1, 2012.
153. Kaito, C. and Saito, Y., Structure and crystallization process of a thin film prepared by vacuum evaporation of SnO_2 powder. *J. Cryst. Growth*, 79, 403, 1986.
154. Shah, A.H., Zargar, R.A., Arora, M., Sundar, P.B., Fabrication of pulsed laser-deposited Cr-doped zinc oxide thin films: Structural, morphological, and optical studies. *J. Mater. Sci.: Mater. Electron.*, 31, 21193–21202, 2020.
155. Franco Arias, L.M., Kleiman, A., Vega, D., Fazio, M., Halac, E., Márquez, A., Enhancement of rutile phase formation in TiO_2 films deposited on stainless steel substrates with a vacuum arc. *Thin Solid Films*, 638, 269, 2017.
156. Rao, M.C. and Hussain, O.M., Growth and characterization of vacuum evaporated WO_3 thin films for electrochromic device application. *Res. J. Chem. Sci.*, 1, 92, 2011.

157. Pandey, R.K., Sahu, S.N., Chandra, S., *Handbook of semiconductor electrodeposition*, Marcel Dekker, Inc., New York, 1996.
158. Lokhande, C.D. and Pawar, S.H., Electrodeposition of thin film semiconductors. *Phys. Status Solidi (a)*, 111, 17, 1989.
159. Cruickshank, C., Tay, S.E.R., Illy, B.N., Campo, R.D., Schumann, S., Jones, T.S., Heutz, S., McLachlan, M.A., McComb, D.W., Riley, D.J., Ryan, M.P., Electrodeposition of ZnO nanostructures on molecular thin films. *Chem. Mater.*, 23, 3863, 2011.
160. Tazangi, F.E. and Rahmati, A., Characteristics of electrodeposited ZnO nanostructures: Size and shape dependence. *Armen. J. Phys.*, 9, 20, 2016.
161. Guo, L., Tang, Y., Chiang, F.-K., Ma, L., Chen, J., Density-controlled growth and passivation of ZnO nanorod arrays by electrodeposition. *Thin Solid Films*, 638, 426, 2017.
162. Gong, D., Grimes, C.A., Varghese, O.K., Hu, W., Singh, R.S., Chen, Z., Dicky, E.C., Titanium oxide nanotube arrays prepared by anodic oxidation. *J. Mater. Res.*, 16, 3331, 2001.
163. Zhang, Q., Ma, L., Shao, M., Huang, J., Ding, M., Deng, X., Wei, X., Xu, X., Anodic oxidation synthesis of one-dimensional TiO_2 nanostructures for photocatalytic and field emission properties. *J. Nanomater.*, 2014, 831752, 2014.
164. Wessels, K., Maekawa, M., Rathousky, J., Oekermann, T., One-step electrodeposition of TiO_2/dye hybrid films. *Thin Solid Films*, 515, 6497, 2007.
165. Lin, J., Liu, X., Zhu, S., Chen, X., Anodic nanostructures for solar cell applications, in: *Green Nanotechnology - Overview and Further Prospects*, Larramendy, M.L. and Soloneski, S. (Eds.), pp. 147–171, 2016, Chapter 6.
166. Yang, D.-J., Park, H., Cho, S.-J., Kim, H.-G., Choi, W.-Y., TiO_2-nanotube-based dye-sensitized solar cells fabricated by an efficient anodic oxidation for high surface area. *J. Phys. Chem. Solids*, 69, 1272, 2008.
167. Wang, B., Cao, B., Wang, C., Zhang, Y., Yao, H., Wang, Y., The optical and electrical performance of CuO synthesized by anodic oxidation based on copper foam. *Materials*, 13, 5411, 2020.
168. Zargar, R.A., Fabrication and improved response of ZnO-CdO composite flms under diferent laser irradiation dose. *Sci. Rep.*, 12, 10096, 2022.
169. Zargar, R.A., Chackrabarti, Malik, M.H., Sol-gel syringe spray coating: A novel approach for Rietveld, optical and electrical analysis of CdO film for optoelectronic applications. *Phys. Open*, 7, 10069, 2021.
170. Zargar, R.A., Kumar, K. *et al.*, Structural, optical, photoluminescence, and EPR behaviour of novel ZnO· 80CdO· 20O thick films: An effect of different sintering temperatures. *J. Lumin.*, 245, 118769, 2022.
171. Zargar, R.A. *et al.*, ZnCdO thick film films: Screen printed TiO2 film: A candidate for photovoltaic applications. *Mater. Res. Express*, 7, 065904, 2020.
172. Zargar, R.A., ZnCdO thick film: A material for energy conversion devices. *Mater. Res. Express*, 6, 9, 095909, 2019.

173. Zargar, R.A., Arora, M., Bhat, R.A., Study of nanosized copper-doped ZnO dilute magnetic semiconductor thick films for spintronic device applications. *Appl. Phys. A*, 124, 1–9, 2018.
174. Zargar, R.A., Arora, M., Alshahrani, T., Shkir, M., Screen printed novel ZnO/MWCNTs nanocomposite thick films. *Ceram. Int.*, 47, 6084–6093, 2020.

2

Experimental Realization of Zinc Oxide: A Comparison Between Nano and Micro-Film

Rayees Ahmad Zargar[1*], Shabir Ahmad Bhat[2], Muzaffar Iqbal Khan[1†], Majahid Ul Islam[1], Imran Ahmed[1] and Mohd Shkir[3]

[1]Department of Physics, BGSB University, Rajouri (J&K), Jammu, India
[2]Department of Chemistry, Government Degree College Women Anantnag (J&K), Srinagar, India
[3]Department of Physics, Faculty of Science, King Khalid University Abha, Saudi Arabia

Abstract

Because of its outstanding commercial applications, ZnO-based material synthesis is gaining prominence as compared to conventional production methods. In this chapter, we present a nanotechnology, basics characteristics of ZnO and comparison of ZnO micro- and nano-films fabricated by low-cost sol-gel syringe spray technique. The various techniques like XRD, SEM, UV-visible, PL and two-probe methods were used for structural, optical and electrical conduction mechanism. Zinc oxide synthesis, characterization, and application are discussed. Thin and thick films have found use in a variety of industries, ranging from straightforward coatings for wear and corrosion protection to more sophisticated applications like antireflective coatings, microelectronics, and photovoltaics.

Keywords: ZnO, coated films, structural, optical and electrical nature

2.1 Introduction

Development of material science is responsible for the advancement of technology and eminence of manhood life. This question started in 1959, on the inauguration of modern age of nanotechnology, by the lecture of

---

*Corresponding author: rayeesphy12@gmail.com
†Corresponding author: muzaffariqbalkhan786@gmail.com

Richard Feynman's "There's Plenty of Room at the Bottom". In 1974, word "Nano-technology" had been given by Norio Taniguchi in 1974. Nano is derived from the Greek word "nanos", mean "dwarf" and technology is derived from two Greek words "teknikos" and "logos" have meanings of "art, artifice, to weave, build or join" and "to utter the inner expressions" respectively. This is a point of fact; the word nanotechnology means full control over matter [1]. Nanotechnology gives us guarantee that what we can define accurately we can create. Basics of nanotechnology are making use of nanoscience efficiently to manufacture products (high-tech goods). Nanotechnology is the making, practice, and understanding of tools, machines and skills to answer a problem or execute a definite task at the nanoscale. Or simply we can say, nanotechnology is the action to operate materials at very tiny scales – at the level of atoms and molecules [2]. Rules of classical physics cannot be applied to material shaving size less than 100 nanometers and initiates the exhibition of exclusive and astonishing properties like quantum confinement and many others.

At room temperature gold turn into liquids, silver starts showing anti-microbial properties, inactive materials like become catalysts, flammability is taken by stable materials. These novel properties unlocked the doors for many interesting and thrilling fields having large number of applications. Like Industrial and other revolutions, nanotechnology is a big revolution of 20th century. Begins just basic technologies with partial usage but blowout speedily into new-fangled applications and facilitate new markets and industries clues creative destruction and major economic revolutions. Nanotechnology finds applications in many fields; some major applications are discussed here:

> ➢ In medical science, make available new choices for drug delivery and drug therapies ensuring precise delivery at particular affected site inside the body.
> ➢ In electronics, it is used to fabricate the transistors and the chips.
> ➢ In computers, transistors fabricated by carbon nanotubes are used instead of old silicon transistors which reduced the power consumption and size of our processor.

2.2 Approaches to Nanotechnology

In order to discover unique physical properties, phenomena and comprehend the applications of nanostructures and nanomaterials, fabrication and processing of nanomaterials is the first corner stone in nanotechnology. We are saying about nanoscale, a nanometer is ten millionth of one millimeter. Without a doubt it allows us to work with both the 'bottom up' or the 'top down' approaches to synthesize nanomaterials, i.e. either to assemble atoms together or to dis-assemble (break or dissociate) bulk solids into finer pieces awaiting till they leftover only with few number of atoms.

Manufacturing at the nanoscale proceeds by one of the two main routes (Figure 2.1):

(i) **Top-down method:** This method begins with a pattern generated on a larger scale, then reduced to nanoscale. The top-down methodology frequently uses the old-style workspace or micro manufacture methods where outside well-ordered tools are used to cut, mill, and shape materials into the desired shape and order [3]. Stone grinding to make a statue is a good example.

(ii) **Bottom-up method:** This method starts with atoms or molecules and builds up to nanostructures. Bottom-up approaches make complex assemblies by a collection of smaller components. Or simply bottom-up refers to methods where devices 'create themselves' or self-assembly [4]. Quantum dots are created throughout epitaxial growth and also the formation of nanoparticles by the colloidal dispersion is a good example of bottom-up approach.

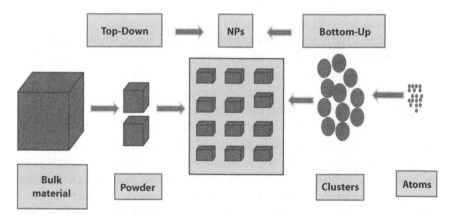

Figure 2.1 Approaches to nanotechnology.

2.3 Wide Band Semiconductors

Wide band gap semiconductors are the semiconductors having larger band gap contrary to the conventional semiconductors. Conventional semiconductors band gap in between 1 and 1.5 eV, however, the band gap energy range from 2 to 4eV for the wide band gap semiconductors. These materials usually belong to III-IV and II-VI groups also named as semiconductors beyond silicon. Silicon carbide SiC (3.3 eV), zinc oxide ZnO (34 eV), cadmium sulphide, CdS (3.4 eV), gallium nitride GaN (3.4 eV), etc. are the few examples of the wide band semiconductors with their respective band gaps [5–7].

Forbidden energy in between valence and conduction band is called as energy band gap. This energy gap in semiconductors provides them a nice characteristic of partial conductivity and capable semiconductors devices on or off to perform the particular function. Superior conductivity with an enhanced switching property about ten times more than silicon can be provided by wide band gap semiconductors. WBG materials have good agreement with power electronics in making smaller, faster, durable and reliable devices [7].

2.4 Zinc Oxide (ZnO)

Zinc oxide has molecular formula ZnO. It is an important and most useful compound which is inorganic. It is a white solid in appearance. Synonym of Zinc Oxide is Oxozinc. It is composed of 80.34% Zinc and 19.6% Oxygen. Zinc oxide (ZnO) is pretty and prominent semiconductor material, usually named as an II-VI semiconductor because zinc and oxygen belong to the second and sixth groups of the periodic table, respectively and mostly shows n-type conductivity as a heavy-duty [8].

ZnO is founded in the crust layer of earth in the form of a mineral named zincite; on the other hand, maximum amount of ZnO produced unnaturally and used commercially. ZnO is an attention-grabbing material from quite a few points of outlook [8]. It is one of the few oxides that show quantum confinement effects. Though, the direct relation in between size and optical absorption is one of the appreciated tools to investigate the growth and synthesis of zinc oxide. Furthermore, nanoparticles of zinc oxide propose significant potential as beginning material for such applications and other commitments such as UV-protection films and chemical sensors [9].

2.4.1 Crystal Structure of ZnO

Metal oxides, particularly zinc oxide, are crystalline materials having atoms grouped in repeating order. The elements contained in this structure determine the material's properties. As an illustration, although graphite's soft, layered structure makes for excellent pencils, silicon's crystal structure enables its widespread use in the semiconductor industry. Crystalline metal oxides, which comprise repeating units of oxygen and metals, are one type of crystalline material essential for a variety of applications, from battery technology to electrolysis of water (i.e., splitting H_2O into its component hydrogen and oxygen). Although their number and the extent of their helpful properties are unclear, researchers believe that there are several crystalline metal oxides that may be useful.

Zinc oxide can be grown in three forms: hexagonal wurtzite, cubic zinc blende, and the cubic rock salt which is observed rarely (Figure 2.2). The wurtzite structure is most stable and thus most common at normal conditions. Zinc blende can be grown on the cubic substrate to stabilize. The zinc blende form can be made stable by growing ZnO on the cubic substrates. Zinc and oxide are tetrahedral for these two circumstances. The rock salt NaCl-type structure can be grown at relatively high pressures ~10 GPa.

Lattices of hexagonal and zinc blende Zinc oxide does not have any reflection symmetry (reflection of a crystal relatively at any given point does not transform it into itself). Piezoelectricity in hexagonal and zinc blende and pyro-electricity in case of hexagonal ZnO are the consequences of these symmetries.

Point group 6 mm (Hermann-Mauguin notation) or C6v and the space group is C_{6v} for the hexagonal structure. The lattice constants are $a = 3.25$ Å and $c = 5.2$ Å; their ratio $c/a \sim 1.60$ is adjacent to the ideal value for hexagonal cell $c/a = 1.633$ [10]. Bonding in zinc oxide is ionic in nature describes sturdy piezoelectricity.

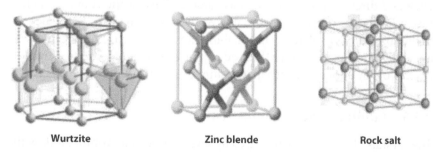

Wurtzite **Zinc blende** **Rock salt**

Figure 2.2 Crystal structures of ZnO.

Zinc and oxygen planes tolerate electric charge both negative and positive correspondingly because of their ionicity. By the reformation of these planes in the absolute materials to preserve electrical neutrality. Zinc oxide has even atomic surfaces, stable and will not be reassembled. All the three crystal structures of zinc oxide are given below in the figures showing their atomic structures.

2.5 Properties of Zinc Oxide

Zinc oxide is one of the most important wide band gap semiconductors. It has good properties which tend to more applications. Some of the important properties are discussed below.

2.5.1 Mechanical Properties

Zinc oxide is a lenient material having hardness of 4.5 on the Mohs scale. Elastic constants of ZnO are smaller than those of III-V semiconductors, such as GaN [11]. Features which make zinc oxide beneficial for the ceramics are greater heat capacity and conductivity, squat thermal expansion and high melting temperature.

In the midst of the semiconductors which have tetrahedral bounding, ZnO is listed with the uppermost piezoelectric tensor, or at least analogous to GaN and AlN. It becomes technically important material because of this property for piezoelectrical applications. However, there is requirement of large electromechanical coupling.

2.5.2 Electronic Properties

Zinc oxide is colorless and has see-through characteristics because of its wide and direct band gap which has value 3.37 eV at the room temperature. With larger band gap material have additional characteristics such as higher breakdown voltages, sustainability for large electric field, can operate at higher temperature [12]. Band gap of ZnO can be modified from 3 eV to 4 eV by alloying it with magnesium oxide or cadmium oxide.

2.5.3 Luminescence Characteristics

The crystals of ZnO can be prepared by several procedures or methods with the size ranging from the value, tens of nanometers to the few centimeters and own n-type conductivity. Two luminescence bands are shown by the various forms of zinc oxide such as single crystals, thin films and threads, nanocrystals, needles, etc., one is short-wavelength band also known as edge luminescence because it has location close to edge of absorption belong to the crystal and second one is long-wavelength all-out lies in the green spectral range. The edge luminescence has an excitonic nature, has maximum value 3.35 eV for the band gap and the decay time is ~0.7 ns. In spite of massive research on the green luminescence, its nature As far as the green luminescence is concerned, despite a vast number of investigations, its nature is up till now not in the knowledge.

2.5.4 Optical Band Gap

Energy gap in between the two bands, valence and conduction band is known as energy gap for specific material. Band gap of zinc oxide film depends up on the concentration of the carriers. It has value of 3.37 eV with the carrier concentration 10^{18} - 10^{20} per centimeter cube. Whereas the band gap calculated by the local density approximation (LDA) is found to be 3.77 eV (Figure 2.3). The quantum confinement effect in grains to confine the electrons within grains produced by the potential barriers offer radical change in the value of band gap [13].

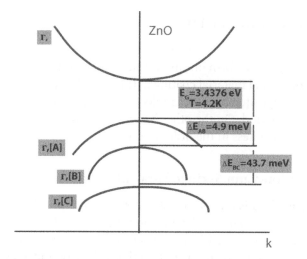

Figure 2.3 Band structure of zinc oxide [14].

2.6 Thin Film Deposition Techniques

Synthesis of nanomaterials is the first corner of nanotechnology which is done by the use of one of the two approaches: top-down or bottom-up. With the most valued properties, preparation of metal oxide nanoparticles is an anew recognized research field draws too much attention to deal with them. The Metal oxides like ZnO thin films have wide-ranging practical applications. Properties of the thin films are reliant on the mode and rate of deposition, on the material and temperature of substrate, and on the related pressure etc.

Various physical processes like thermal evaporation or by bombing ions or electrons in the form of energetic beam to evaporate the atoms or molecules of material to be deposited. Instead one or more chemically reactive species of vapor are made to incident for the deposition of film on substrate. To evade the film from uncleanness of ambient atmosphere, physical deposition is done in a vacuum environ whereas, an inactive carrier gas is exploited by the chemical deposition.

Sol–gel, chemical bath deposition, spray pyrolysis, spin coating, dip coating, solution-cum syringe method etc. are the several procedures be there for the preparation of ZnO thin nano films. Among the different methods, the solution-cum syringe method approach appears to be one of the easiest and most promising methods. Here we discuss some of important deposition techniques [15].

2.6.1 Thin and Thick Film

Over the past decade, semiconductor devices are mostly based on advancement of coating in the form of thin or thick film technology. The thin film is a two-dimensional material coating of thickness in the order of nanometer range and deposited by either molecule-by-molecule or atom-by-atom condensation method. On the other hand, thick film coating is deposition of particles of thickness in the order of micron [16].

2.6.2 Solution-Cum Syringe Spray Method

The syringe method is one of easiest and most economical method comes under the category of chemical deposition techniques. It is one of the novel techniques, not used ever. We used this procedure to deposit thin film on the glass substrate because of easiness of this technique and apparatus used for the deposition is a medical syringe. The choice of the particular method depends on several factors like material to be deposited, nature of substrate, required film thickness, structure of the film, application of the film etc.

Working of this method is discussed in the following steps [17]:

- Solution to be deposited is prepared by any suitable method.
- Substrate on which film is deposited is cleaned by the distil water.
- Solution prepared is stirred well and filled in the syringe.
- Now solution is poured on the substrate by slightly pressing the syringe plunger.
- With the use of syringe needle, solution is spread out on whole substrate.
- At last, film deposited is dried by giving heat-treatment with the muffle furnace.

2.7 Procedure of Experimental Work

The experimental work begins (Figure 2.4) by solving the required amount of zinc acetate in minimum amount of distilled water to form solution 1. Solution 2 is also prepared by dissolving the required amount of sodium hydroxide in a minimum amount of distilled water. Solution 2 is then added to solution 1 dropwise with continuous stirring on magnetic stirrer. Continue the stirring process for three hours and then allowed to settle down overnight. After that solution obtained is filtered by the filter with precise porosity. White gel is dried firstly by giving sun exposure then in oven at 100°C for one hour to get the ZnO powder. Almost half of the powder obtained is kept separately for bulk film preparation and the remaining powder is grinded to reduce the size of particles by the use of mortar and pastel about one hour. Afterward both bulk and nano powder are dissolved in the distilled water and stirred with the magnetic bead for one hour. Then filtered and washed to eliminate the contaminations. Finally, both forms of powders are dried by exposure to sunlight.

Next task is to prepare bulk and nano films which are prepared by using a novel method named as solution-cum syringe spray method not used before. This method is used because of its easiness and also for more economic. To accomplish this bulk and nano powder is dissolved in the 15ml of distilled water separately with the addition of ethylene glycol in small amount as a binder. Now both the bulk and nano ZnO solution are ready. After that the glass substrate on which films are to be deposited are cleaned well with the distilled water and then dried up in front of heat blower. Bulk ZnO solution prepared to cast the films is filled in the medical syringe by simply plunging out the plunger of syringe from its barrel. By the use of syringe solution is sprayed and spread out on the glass substrate as shown in Figure 2.4. After the deposition of film glass substrate is kept in the muffle furnace (Figure 2.4) to dry at 80°C for an hour. Thus, bulk film sample is prepared. Same procedure is followed to cast the nano film. Three samples of eachbulk and nano films are prepared. Samples of both are shown below in Figure 2.4.

2.8 Calculation of Thickness of Thin ZnO Films

The thickness of ZnO bulk and ZnO nano thin films can be calculated by the gravimetric weight difference method regarding deposited weight of a copper oxide film on the glass substrate, per unit area (g/cm^2). The thickness was calculated using formula [17],

$$T = \frac{M}{\rho A}$$

where 'T' is film thickness, 'M' is the mass of the film material in g., 'A' is the area of the film in cm^2 and ρ is the density of the film material.

2.9 Structural Analysis

Enormously important properties of the thin films are the crystallographic properties which prominently effect on the physical properties of the film such as electrical conductivity, electrical resistivity, mobility of charge carriers, optical band gap, and how the atoms are organized. Chemical stoichiometry, procedure followed conditions followed during deposition are basis which decide structural features of the film. These structural features play a vital role in the research and development of devices. Grain size, crystalline phases exist in material and desired orientation of the films are analyzed by the use of a most important and super method of characterization named X-Ray Diffraction (XRD) depicted in Figure 2.5 along with most intense peak of bulk and nano films.

Experimental Realization of Zinc Oxide 55

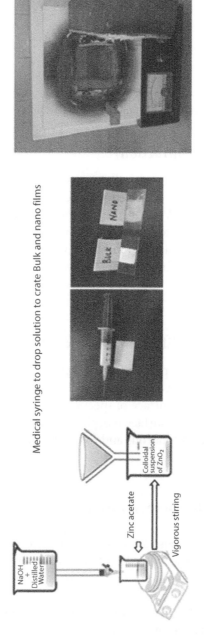

Figure 2.4 Experimental setup for coated films [17].

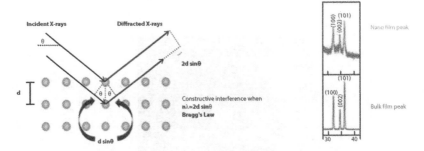

Figure 2.5 XRD representation of ZnO films with comparison of peaks.

2.9.1 XRD (X-Ray Diffraction)

Nineteenth century going to end, still it is a tough job to study the matter at the atomic scale. Discovery of electromagnetic waves having wavelength in between gamma and UV-rays founded a way to investigate the matter atom by atom. X-rays have wavelength in the range of 0.01 to 10 nm comparable to the atomic distance in matter used to study the crystalline structure by the use of diffraction phenomenon. The various structural parameters were calculated and are presented in the Table 2.1 by using different equations [17].

2.9.2 SEM (Scanning Electron Microscope)

Scanning electron microscope is a characterization technique for the compositional analysis (Figure 2.6). It uses intensive and focused beam of electrons to produce signals at surface of specimens which have to be studied. Below figures give complete surface morphology for nano and micro films, both films show dense with porous and rough surface along with agglomerations of particles.

2.10 Optical Characterization

Optical characterization is also a category of characterization used to study the optical properties of the materials. Or, we can say an influential method to characterize the semiconductor materials to gain the information about material.

Various parameters like band gap, recombination mechanisms, impurity levels, constituents and their proportion in the materials, defect types

Table 2.1 Different structural parameters.

Sample	Plane (hkl)	Peak angle (θ)	FWHM (β)	Grain size (nm)	Inter-planar spacing (d) A°	Dislocation density (δ) lines/m	Strain (ε)
Micro-ZnO	(101)	18.23	0.00314	48.57	2.461	439×10^{-15}	0.000784
Nano-ZnO	(101)	18.34	0.00505	28.66	2.447	1217.44×10^{-15}	0.0011021

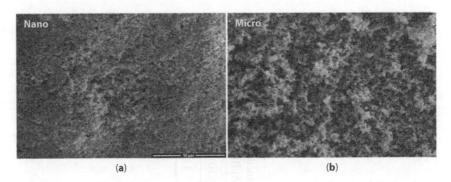

Figure 2.6 Scanning electron microscope pictures for nano- and microfilm [17].

and many other valuable information regarding the material is provided by the optical characterization. We have talked over on the most significant optical characterization techniques, UV-spectroscopy and PL spectroscopy here below [18].

2.10.1 UV Spectroscopy

Ultraviolet spectroscopy is one of leading optical characterization technique used to study the optical properties of the materials. It is named this because it uses UV-rays of em spectrum having wavelength in the range 700-300 nm. In this spectroscopy, beam of UV-light is fragmented into two halves: one half is focused through the cuvette (made of quartz crystal and clear to UV-rays) contains solution of the sample to be analyzed; and second one is passed through reference cell having solvent. Figure 2.7 shows absorbance spectra of nano and micro films with absorption peak positions at 363nm and 387nm respectively. The band gap has been calculated from famous Einstein equation [17].

$$E = h\nu = \frac{hc}{\lambda_p} = \frac{1240}{\lambda_p}$$

where h is Planck's constant, c is the velocity of light, and λ_p is the excitonic absorption edge. The bandgap energy calculated to be ~3.21 eV and ~3.40 eV for micro- and nano-ZnO films, respectively. This blue shift confirms occurrence of quantum confinement from micro- to nano-ZnO films.

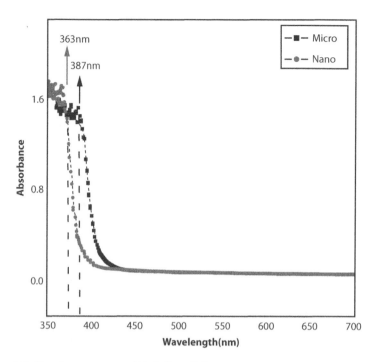

Figure 2.7 Absorbance spectra of ZnO films [17].

2.10.2 Photoluminescence (PL) Spectroscopy

Photoluminescence spectroscopy is also a type of optical characterization helps to study the optical properties of the materials. The word 'Photoluminescence' is composed of two words 'photo' and 'luminescence' collectively have the meaning 'production of light from a sample when light energy is incident on it.

PL spectroscopy monitors the light emitted from atoms or molecules of sample after the absorption of photons [18]. The substances which show the photoluminescence can be studied under this spectroscopy. Either organic and inorganic materials of any size and in any phase solid, liquid, or gas can be characterized by the use of PL spectroscopy. UV rays and visible light of electromagnetic spectrum are used to characterize the material in this technique. Intensity, wavelength, emission stability, and bandwidth of emission peak are four parameters of emitted light to be determined to study the sample. Properties like band gap, impurities, concentration, crystallinity, strain and surface behavior of materials are examined by this characterization technique. Figure 2.8 shows pl spectra of ZnO nano and

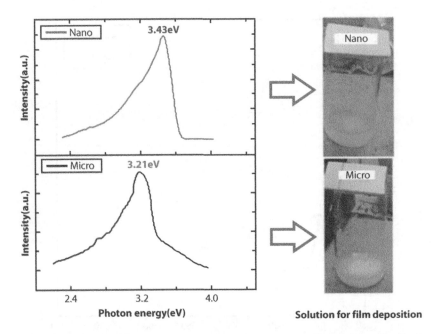

Figure 2.8 Pl spectra along with color representation of nano and micro ZnO films [17].

micro films with clear blue shift along with color that confirms quantum confinement effect.

2.11 Electrical Characterization

Like other sets of characterization, electrical characterization is also an utmost essential method used to study the electrical properties of the materials. Elements which regulate the performance of the device can be understood after determining the electrical properties of semiconductor materials. For the most part, in the case of solar cells, study of electrical parameters like carrier mobility, concentration of charge carriers and electron transport mechanism is very important to deal with the efficiency of the device [15].

2.11.1 Resistivity by Two-Probe Method

Determining resistivity helps us to study the electrical properties of metal films. Contact methods like two-probe, four-probe, and the spreading resistance are commonly used to measure the resistivity. The two probe method is not a complex method with respect to procedure and mostly used applied for high resistive thin films.

The conductivity (σ) of the film is determined by using the standard four probe method. The electric current is restricted to two dimensional (2D) surfaces; the resistance of a rectangular shaped conductor film is easily derived by multiplying the sheet resistivity with film length and divided by film width. The length and width of film acquire same SI unit, it means the film sheet resistivity and (real) resistance are the same and the following equations [19]: The sheet resistance (R_S) was estimated via:

$$R_S = 4.532 \frac{V}{I}$$

By the above formula the conductivity of both the films shows semiconductor nature, also micro films show higher conductivity due to band gap difference and grain boundary mechanism as shown in Figure 2.9 [15–17].

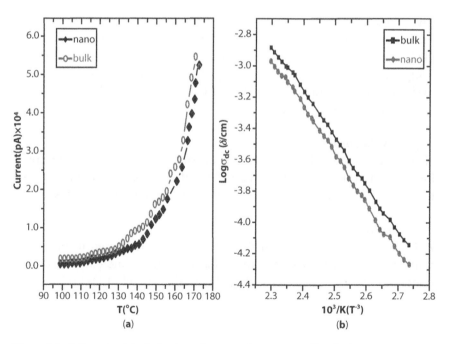

Figure 2.9 DC conductivity behaviors of nano and micro ZnO films [17].

2.12 Applications of Zinc Oxide

The applications of zinc oxide powder are numerous, and the principal ones are summarized below:

(i) Zinc Oxide Nano-Rod Sensor

Nanorod sensors made up of zinc oxide are devices for the detection of electric current which go through the nanowires of zinc oxide because of absorbing gas molecules [20]. Sensitivity of the sensor is improved by the addition of Pd gives the impression to dissociate hydrogen molecules. The sensor detects hydrogen concentrations at room temperature at the rate of ten portions per million, while no reply comes back to oxygen.

(i) Food Additive

Zinc oxide is an essential nutrient and used as an additive in numerous products of food like breakfast cereals etc. to preserve their taste, look and some other qualities. Another compound of zinc named zinc sulphate is also used to perform the same task. A little amount of ZnO is contained in the packed food products, not wished-for the purpose of nutrient. With the contamination of dioxin it is used in pork exports.

(iii) Pigment

As the opaque feature of zinc lies in between lithopone and titanium dioxide it is used as a pigment in the paints. Coating of papers is also done by using this. Chinese white used in the pigments of artists' has superior status in zinc white grades. Zinc oxide is also used as an ingredient in makeup mineral and in oil painting.

(iv) UV Absorber

Nano-sized zinc oxide and titanium dioxide are used in the sunscreen lotions as a protection against the UVA and UVB ultraviolet rays. It also has use in UV-blocking sunglasses during the welding and also in space.

(v) Electronics

ZnO has wide band gap with the value 3.37 eV. It is used to make the transistors as it has good switching characteristics. Zinc oxide has most important use in the manufacturing of nano-lasers and LEDs [21].

2.13 Conclusions and Future Work

According to the experimental work done, the following concepts and technological difficulties emerged: It was discovered that the physical characteristics of the films were improved by the deposition of ZnO films onto glass substrates. To assist in the creation of optoelectronic devices, measure the optical characteristics of ZnO films. The production of ZnO films with various particle sizes aids in the future improvement of electrical and sensing properties.

ZnO has enormous application possibilities in the future. Encouragement as reviewed in this study achieved. There are still significant concerns that require further research. Self-organized development is a characteristic of ZnO. This enables the growth of high-quality crystal ZnO material on any surface. ZnO is appealing for many applications due to its property and the good other features. Last but not least, ZnO is readily available, easy to prepare, and has a variety of uses. Due to their adaptability to provide customized properties, thin and thick films have found use in a variety of industries, ranging from straightforward coatings for wear and corrosion protection to more sophisticated applications like antireflective coatings, microelectronics, and photovoltaics.

References

1. Feynman, R.P., Leighton, R.B., Sands, M., *Feynman lectures on physics*, vol. 1–3, Addison Wesley, New York, 1963.
2. Feynman, R.P., *Surely you're joking, M. Feynman!' Adventures of a curious character*, W. W. Norton, New York, 1985.
3. Zaragr, R.A., Micro size developed TiO2-CdO composite film: Exhibits diode characteristics, *Mater. Lett.*, 335, 15, 133813, 2023.
4. Cai, J., Ruffieux, P., Jaafar, R., Bieri, M., Braun, T., Blankenburg, S., Muoth, M., Seitsonen, A.P., Saleh, M., Feng, X., Müllen, K., Fasel, R., Atomically precise bottom-up fabrication of graphene nanoribbons. *Nature*, 466, 470, 2010.
5. Zargar, R.A., Arora, M., Hafiz, A.K., Investigation of physical properties of screen printed nanosized ZnO films for optoelectronic applications *Eur. Phys. J. Appl. Phys.*, 70, 10403, 2015.
6. Singh, J., Singh, A., Kumar, Kumar, K. et al., Optical properties of TiO_2 thin film: Via dip coating method. *J. Nano Phys.*, 14, 02019(3pp), 2022.
7. Zargar, R.A., Chackrabarti, Malik, M.H., Sol-gel syringe spray coating: A novel approach for Rietveld, optical and electrical analysis of CdO Film for optoelectronic applications. *Phys. Open*, 7, 10069, 2021.

8. Zargar, R.A., Kumar, K. et al., Structural, optical, photoluminescence, and EPR behaviour of novel Zn0·80Cd0·20O thick films: An effect of different sintering temperatures. *J. Lumin.*, 245, 118769, 2022.
9. Zargar, R.A. et al., ZnCdO thick film films: Screen printed TiO2 film: A candidate for photovoltaic applications. *Mater. Res. Express*, 7, 065904, 2020.
10. Zargar, R.A., Arora, M., Bhat, R.A., Study of nanosized copper-doped ZnO dilute magnetic semiconductor thick films for spintronic device applications. *Appl. Phys. A*, 124, 1–9, 2018.
11. Banerjee, P., Lee, W.J., Bae, K.R., Lee, S.B., Rubloff, G.W., Structural, electrical, and optical properties of atomic layer deposition Al-doped ZnO films. *J. Appl. Phys.*, 108, 043504, 2010.
12. Adachi, H. and Wasa, K., *Handbook of sputtering technology*, 2nd edn., pp. 3–39, William Andrew Publishing, Oxford, UK, 2012.
13. Manzoor, U., Islam, M., Tabassam, L., Rahman, S., Quantum confinement effect in ZnO nanoparticles synthesized by co-precipitate method. *Physica E: Low-Dimens. Syst. Nanostruct.*, 4, 1, 1669–1672, 2009.
14. Alivov, I.A. et al., Fabrication and characterization of n-ZnO/p-AlGaN heterojunctionlight-emitting diodes on 6H-SiC substrates. *Appl. Phys. Lett.*, 38, 23, 4720, 2020.
15. Zargar, R.A., ZnCdO thick film: A material for energy conversion devices. *Mater. Res. Express*, 6, 9, 095909, 2019.
16. Shabnam, M., Zayed, M., Hamdy, H., Nanostructured ZnO thin films for self-cleaning applications. *RSC Adv.*, 7, 617–631, 2017.
17. Zargar, R.A. et al., A comparative study of micro- and nano-ZnO films fabricated by sol-gel syringe spray method. *Int. J. Ceram. Eng. Sci.*, 2, 169–176, 2020.
18. Zargar, R.A., Arora, M., Alshahrani, T., Shkir, M., Screen printed novel ZnO/MWCNTs nanocomposite thick films. *Ceram. Int.*, 47, 6084–6093, 2020.
19. Hasiah, S., Ibrahim, K., Senin, H.B., Halim, K.B.K., Electrical conductivity of chlorophyll with polythiophene thin film on indium tin oxide as P-N heterojunction solar cell. *J. Phys. Sci.*, 19, 77–92, 2008.
20. Juarez, B.H., Garcia, P.D., Golmayo, D., Blanco, A., Lopz, C., ZnO inverse opals by chemical vapor deposition. *Adv. Mater.*, 17, 2761–5, 2005.
21. Maruska, H.P. and Tietjen, J.J., The preparation and properties of vapor-deposited single-crystalline GaN. *Appl. Phys. Lett.*, 15, 327, 1969.

3

Luminescent Nanocrystalline Metal Oxides: Synthesis, Applications, and Future Challenges

Chandni Puri[1,2], Balwinder Kaur[3], Santosh Singh Golia[1], Rayees Ahmad Zargar[4] and Manju Arora[1*]

[1]*CSIR-National Physical Laboratory, Dr. K.S. Krishnan Marg, New Delhi, India*
[2]*Academy of Scientific and Innovative Research (AcSIR), Ghaziabad, India*
[3]*Govt Degree College, R.S. Pura, Jammu, Jammu and Kashmir, India*
[4]*Department of Physics, BGSBU, Rajouri-(J&K), India*

Abstract

Luminescent metal oxides have drawn significant attention of the research community owing to their wide range of applications in the field of solid-state lighting, displays, gas sensors, and biomedical etc. The basic concepts of luminescence, types of luminescence, up-conversion, down-conversion and persistent luminescence mechanisms, role of metal oxide nanoparticles in luminescence, synthesis of size and shape controlled luminescent metal oxides nanoparticles (LMONPs) by different chemical methods and their characterization are briefly described in this chapter. The highly efficient and long-life energy saving light emitting diodes to overcome lighting crisis, mesoporous structure and their nanocomposites with organic and inorganic compounds provides better gas sensing activity and biocompatibility of magnetic/non-magnetic luminescent metal oxide nanoparticles encourage their usage in biomedical applications are also concisely discussed. The future challenges in this field are also cited along with proposed solutions to check the toxicity of nanoparticles and standard quality documentation for the global acceptance of synthesized LMNOPs size, shape, and morphology by using a particular method, precursors and solvents used, synthesis parameters for the particular device application.

---

*Corresponding author: manjuarorain@gmail.com

Rayees Ahmad Zargar (ed.) Metal Oxide Nanocomposite Thin Films for Optoelectronic Device Applications, (65–100) © 2023 Scrivener Publishing LLC

Keywords: Luminescence, phosphorescence, fluorescence, luminescent metal oxides, nanoparticles, synthesis

3.1 Introduction

In the present era, the luminescent materials are widely used in various applications especially in the field energy saving LED bulbs/tubes, mobile displays, LED TV screens, biomedical instruments display etc. [1–5]. The light is produced either from incandescence or luminescence process. Incandescence process utilizes heat energy to produce light like natural sun and stars natural resources of light, heating of metals/materials at high temperatures to glow and conventional filament bulbs etc. Various luminescent and phosphors materials have been used for generating cold light at ambient and lower temperatures. In luminescence process, the electron of an atom absorbs energy to excite from its ground state to higher excited state and then returns back to ground state by releasing energy in the form of light in the visible, ultra-violet or infrared region. In the two steps cold emission process, the wavelength of emitted light is characteristic of a luminescent material and not of the incident radiation. In nature this luminescence phenomenon is seen in glow-worms, fireflies, some sea bacteria and in deep-sea animals. In present scenario, this phenomenon has been extensively exploited by the researchers all over the world in the different fields of science and Industrial Applications [3]. The emission of light takes place at a particular characteristics time (τ_c) after absorption of the radiation, this τ_c sub-divides the luminescence process into temperature independent fluorescence and temperature dependent short as well as long period phosphorescence processes. The schematic of

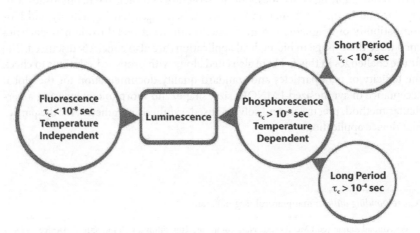

Figure 3.1 Schematic of processes in luminescence phenomena.

processes involved in luminescence phenomena along with characteristics time is presented in the following Figure 3.1.

The temperature dependent phosphorescence phenomenon is subdivided into two parts: (i) short period also known as Thermoluminescence (TL), in which the emission time has duration τ_c less than 10^{-4} s and (ii) long-period phosphorescence takes characteristic time (τ_c) more than 10^{-4} s, while fluorescence process is temperature independent with τ_c less than 10^{-8} s. Many organic and inorganic materials are reported having band gap energy in the range of 2–4 eV which exhibit luminescent property. The conventional wide band gap insulating and semiconducting materials of value more than 3 eV withstand very fast switching frequencies, high power and high temperatures with very high carrier density and their mobility in semiconductors [3–7]. In this category, silicon carbide and gallium nitride are extensively used in LEDs, diode lasers and photovoltaic cell [4, 6, 8]. But these materials suffer with disadvantage of their chemical sensitivity for oxidation reactions which restricts their usage in device fabrication.

In inorganic materials the fluorescence emission of light occurs as long as the materials are excited with photons and the phosphorescence emission of light takes place when excitation light is switched off. While in organic materials; the fluorescence radiative transition occurs between the two states of same spin multiplicity and in phosphorescence, it happens between the states of different spin multiplicity. The other alternatives of luminescence are up-conversion luminescence (UC), down-conversion (DC) and persistence luminescence (PERL). In UC, the low energy photon is converted to high energy photon by the absorption of two photons from near infrared (NIR) to Ultraviolet (UV), NIR to Visible (Vis) and Vis to UV regions i.e. Anti-Stokes shift [9–11]. This process is used for the chemical reaction activation, for example in drug delivery for releasing drug and sensitization of silicon camera for infrared light. On the other hand, in DC, the high energy photon releases excessive energy to form low energy photon/particles to depict Stokes shift [10, 12, 13]. When a high energy photon is converted into two low energy photons by DC, then this process is known as quantum cutting luminescence (QCL) [12, 13]. By exploiting this process one can develop more efficient solar cells. In PERL, the emission of light sustains for longer period even after switching off excitation source [14, 15].

3.2 Different Types of Luminescence

Luminescence types are categorized on the basis of excitation sources/centers/phenomena responsible for exhibiting characteristic

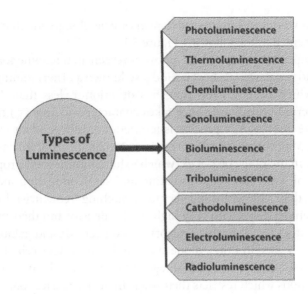

Figure 3.2 Different types of luminescence.

luminescence property in the materials and presented in the following schematic Figure 3.2.

3.2.1 Photoluminescence

The photoluminescence (PL) process gives information about the band gap energy, recombination phenomena, surface effects, vacancy, deep level as well as shallow level defects and impurities present in wide band semiconducting oxides. The material absorbs photons from electromagnetic radiation and then re-emits photons. In this process, when the material is excited by the photons of energy more than the band gap energy of absorbing material, the electrons are excited from ground state or valence band (VB) to a higher energy level or conduction band (CB) and then they return back to a lower energy level by photons emission. In this photoexcitation process, the excessive energy of excited electrons is lost before reaching to rest state at the lowest energy level of the CB. The electron energy comes back to that of the VB level. This released of energy by electrons is transformed into a luminescent photon which is emitted by the material. It means that the energy of the emitted photon is equivalent to the band gap energy (Eg) of material. Hence, this process of photon excitation with subsequent photon emission is known as photoluminescence. This process exhibits resonant radiation, fluorescence and phosphorescence phenomena by the materials on photons absorption and emission. The fast resonant radiation process involves the absorption of photons by the material and then emits back with same energy without any delay in about 10 ns time period. In the fluorescent materials, internal energy transitions take place in the material after photon absorption and before the photon energy emission process. Though, this is also a fast process completes in 10^{-8}-10^{-4} s duration, but some of the absorbed photon energy is lost due to internal energy level transitions present within materials band structure. Hence, the emitted light photons have lesser energy as compared to the absorbed ones. On the other hand, in phosphorescent materials with different spin states, the electronic transitions occurring via absorbed photons energy follows intersystem energy states crossing to move to a higher multiplicity a triplet spin state. When the photon energy is stuck in the triplet state, then transition back to the lower singlet energy states is not allowed based on quantum mechanics principle. This slows down process of radiative transition process to singlet state that may take few minutes to hours for completion. That's why the phosphorescence is rare and the molecule in triplet state prefers intersystem crossing to ground state before the initiation of phosphorescence phenomena. PL spectroscopy technique has been used for exploring the recombination phenomena, evaluation of band gap energy, qualitative and quantitative estimation of impurities, defects in semiconducting materials.

3.2.2 Thermoluminescence

Thermoluminescence (TL) a thermally stimulated luminescence is observed by the emission of light from the thermal recombination of formerly trapped electrons and holes at defects or impurities sites present in the solid materials [16–18]. The electrons and holes are produced in the material on exposing them to suitable light from electromagnetic or ionizing radiation of wavelength more than band gap energy of solid material. The inherent defects and impurities energy levels within band gap region of materials trap separated electrons and holes which drastically reduce the recombination process. The recombination of such trapped electrons or holes is initiated by increasing the temperature of material to raise thermal energy to escape from the trapped sites. If their recombination takes place by a radiative process then, on increasing the temperature, the sample will begin to emit light till all the traps vacates. So in TL, the thermal excitation is involved in activating energy release, imparted to the material from another source of excitation. This technique has been used in archaeological and geological investigations for radiation dating and dosimetry.

3.2.3 Chemiluminescence

In chemiluminescence (CL), the emitted light is obtained as final chemical reaction product whose wavelength lies in UV-Vis or infrared region of electromagnetic spectrum. During chemiluminescence reaction, the electronically excited intermediate or final product of chemical reaction shows luminescence or transfer to another luminescent center/molecule/ion etc. [19–21]. The CL reaction proceeds by two ways (i) direct reaction in which two chemicals e.g. substrate and an oxidant in presence of catalyst react to form a product. Out of that, small fraction of the product attains an electronically excited state and that relax to the ground state with emission of a photon and (ii) in indirect CL process, the energy is transferred by the excited species to a fluorophore (or fluorochrome) for excitation to emit light. In this process, though the material is not CL active but by donating excessive energy to a fluorophore for its excitation and returning back to its ground state with photon emission. For a chemical reaction to produce light, it should meet some essential requirements. It means that CL reaction is exothermic whose excessive energy has been used for electron transitions to excited state with favorable pathway for some of the product and that slowly releases energy non-radiatively in comparison to the rest of the product.

3.2.4 Sonoluminescence

Sonoluminescence (SL) is resulted from the excitation by ultrasonic waves. The liquid on exposure to sound waves of sufficient pressure forms bubbles which are vibrating violently. This is acoustic cavitation process, the fundamental step of sonochemistry and sonoluminescence (SL) [22–24]. In this wide parameter space, bubbles are moved into highly nonlinear oscillations analyzed from slow volume growth with subsequent compression and collapse. At around maximum collapsing stage, the chemical reactions start and light is emitted. The bubble retains its maximum size only for about 0.003% of one period. This spatiotemporal region shows the extremity range of cavitation. Earlier SL studies are based on the light generated from a cloud of cavitating bubbles formed in an ultrasonic field with a frequency in 20 kHz to 2 MHz region. Now, this phenomenon is described in the following manner that when a small gas bubble is acoustically suspended and intermittently driven in a liquid solution at ultrasonic frequencies, causes bubble collapse, cavitation, and light emission [25–27]. The energy of a sound wave in a fluid can concentrate of about 12 orders of magnitude to create flashes of light of life ~50 picoseconds. The origin of flashes from hot spots is formed inside bubbles by nucleation, expansion, and bursting mechanism on passing the sound wave.

3.2.5 Bioluminescence

Bioluminescence (BL) a "living light" deals with the generation and emission of light by living organisms by chemical reactions. Bioluminescent organisms are present on the surface and deep inside sea beds [28]. In the darkness of the sea, the sunlight is not able to reach there. Many living organisms residing there emit light as a product of chemical luminogenic reactions. Since, these enzymatic reactions are going on with luminescent pigment in the presence of proteins and calcium or magnesium cofactors within living organism, that's why this phenomenon is known as bioluminescence (BL). Fireflies, some fungi, marine animals, and some bacteria are some of the examples of this family and they utilize BL reactions for hunting prey, warning for security, mate attraction, camouflage and for surrounding illumination etc. Green and blue BL are widely observed, though some species emit a red glow also.

3.2.6 Triboluminescence

In triboluminescence (TbL), light is produced by the materials when subjected to mechanical stimulation in the form of rubbing, grinding/crushing, ball milling, stretching or compressing [29–33]. The purpose of this action is to dissociate chemical bonds in the material to emit light without using any external source. TbL obtained by grinding or rubbing materials comes under fracture TbL category. While the plastic TbL, and elastic TbL are resulted from the stretching or compressing impact on materials. The extensive work has been going on the synthesis and characterization of better performance new and existing organic as well as inorganic TbL materials for various applications. Based on such investigations, TbL materials are now widely used in the fields of mechanical engineering, energy, biological *tracking*, lighting, detectors and display devices. The huge variety of organic crystals (aliphatic and aromatic), organometallic (transition metal and rare earth metal complexes), and inorganic (host and luminescent dopant) compounds is available showing characteristic TbL phenomenon.

3.2.7 Cathodoluminescence

Cathodoluminescence (CL) is an optical and electromagnetic phenomenon in which electrons strike a luminescent material e.g., phosphor, cause the emission of photons, which may have wavelengths in the visible spectrum [34]. For example, the electron gun as a source of electron beam scans the phosphor-coated inner surface of the old cathode ray tube (CRT) based television, computer, scanning electron microscope (SEM) or transmission electron microscope (TEM) screens to produce light. Cathodoluminescence is basically the inverse of the photoelectric effect, in which electron emission occurs by photons irradiation. In CL, the electron beam has been used as excitation source which generates small de Broglie wavelengths to overcome Abbe's diffraction limit for better imaging resolution which arises from photon-intervened excitation. In materials characterization, SEM and TEM are vital techniques have been used to scan the images which give information about the structure, surface morphology, particle size and shape. These details are further used for revealing the composition, optical and electronic properties of material at the micro and nano levels.

3.2.8 Electroluminescence

Electroluminescence (EL) is an optical phenomenon which is involved in the production of photons by the radiative recombination of holes and electrons on passing an electric current through a material. It has two categories: (i) low-field EL and (ii) high-field EL. In low-field EL, the introduction of minority charge carriers takes place, and while in high-field EL, most of the charge carriers acquire high optical energies [35, 36]. Low-field EL is generally found in p-n junction semiconductors e.g. LEDs where electron hole recombination dominates. They are separated from each other by doping semiconducting material with suitable dopants to form p-n junction before following recombination process on subjecting electric current of few volts per cm. On the other hand, in high-field EL displays are fabricated by using phosphor or organic luminescent materials in powder and thin films form. On applying strong electric field of ~ 106V/cm, the electrons are escalated with very high energies in the material for the excitation of luminescent centers present in them. These excited states of the luminescent centers on relaxation emit EL emission. High-field EL devices can work in AC and DC operating systems.

3.2.9 Radioluminescence

Radioluminescence (RL) is an outcome of the interactions between the material and ionizing radiation particles, cosmic rays or photons like α or β particles and X-rays or γ- rays and appears as burst of emitted light [37, 38].

Radio luminescent materials are also known as scintillators. The excitation process induced by ionizing radiation is similar to cathodoluminescence phenomenon. X- and γ-ray excitations are actuated by the loss of photons energy on interaction with matter. The high energy photons release energy actuated by photoionization reaction. The higher atomic number materials with more number of electrons absorb energy more efficiently as compared to lower atomic number ones. The charged particles and secondary electrons similar to gas discharge emerging from γ-rays, when hitting the substrate surface generate excited species. The penetration depth of charged species depends upon their radiation power. In this process, the electrons are generally trapped in lattice sites. The depth of the trap sites decides that the electron can recombine swiftly or trapped at sites for longer time, while their release leads a delayed recombination luminescence [39, 40]. Radioluminescence phenomena are extensively exploited in many radiation sensors.

3.3 Luminescence Mechanism in Nanomaterials

In luminescent molecules and materials, excitons are the energy source whose conversion between a spin-singlet exciton and two spin-triplet excitons has been utilized in DC or UC. In molecular DC by singlet fission process, one singlet exciton forms two triplet excitons [41] and in molecular UC using triplet fusion or triplet–triplet annihilation process, two spin-triplet excitons produce a singlet exciton [42]. The energy of these excitons is retrieved either by conversion to a photon by the emission from singlet exciton/from an intermediate to access the triplet excitons or by transporting it to another material for the excitation of electronic transitions. In some special cases, where UC is thermally activated for slow fluorescence, the exciton energy conversion takes place between triplet and singlet exciton states. Inorganic materials are extensively studied for UC and DC. The matter of concern in these explorations is the production of multiple exciton generation (MEG) in nanomaterials, and such one high-energy (hot) excitation in an inorganic semiconductor leads to multiple band edge excitations [43]. Rare-earth dopant ions in DC store energy released from well-defined atomic states. Hence, these states are used for splitting the energy from higher-energy excitations [44]. The f-electron electronic configuration of lanthanide ions doped materials is also actively used in UC for biological applications [45]. It means that Up- and Down-Conversion of energy in molecules and materials helps in better understanding the mechanism behind luminescence phenomena, for efficient usage in different applications and to provide direction and challenges in the development of new advanced and existing materials to meet future demand.

3.4 Luminescent Nanomaterials Characteristic Properties

The plasmonic nanoparticles, inorganic compounds and carbon-based quantum dots, photonic crystals, rare earth/transition metal ions doped metal oxide nanoparticles, organic-inorganic nanocomposites, organic dyes are some examples of optically active/luminescent materials. Plasmonic transitions in metallic nanoparticles are the root cause of their unique optical and photocatalytic properties [46, 47]. The nanophotonic characteristics have been exploited in photonic. The quantum confinement effects in semiconductors are used for the optimization of bandgap energy which shifts emission peak in the optical absorption spectrum [48, 49]. Now-a-days luminescent nanomaterials (LNMs) are widely used in biomedical, drug delivery, solar cell, sensors, displays, energy saving lighting and photonic devices owing to their unique optical properties [50, 51]. The properties of pure and doped metal oxide LNMs depends upon the dopant concentration and site symmetry in lattice, particle size, shape and surface morphology, host material, host to dopant energy transfer or vice-versa, luminescence time etc. The quantum confinement in ultrasmall nanoparticles of size less than 5 nm show non-linear optical (NLO) properties [51], while the anisotropy in optical properties of LNMs is observed indifferent size and shaped nanoparticles. Hence, the optical properties of the nanoparticles can be optimized by selecting appropriate synthesis process, reactants, reaction parameters to obtain desired size, shape and structure for the particular application. Relative absorption and scattering of light are nanoparticles size dependent phenomena. The absorption predominates when nanoparticles have sizes <20 nm and scattering of light dominates for nanoparticles of size range >100 nm. The aggregation/agglomeration of nanoparticles into bigger ensembles also increases light scattering. In nanodomain,phosphor particles bring changes in structure, crystallinity, intrinsic optical properties, defect density have long life effects on quantum efficiency (QE) [52, 53]. LNMs effects emission intensity, spectral profile, peak position, excited-state lifetime, excitation intensity etc. The enhanced surface defects in nanostructure causes non-radiative transitions to quench fluorescence. To overcome this effect, luminescent nanoparticles are encapsulated with surfactant molecules/polymeric thin film/fiber to increase the stability and efficiency by reducing non-radiative transitions [54, 55].

3.5 Synthesis and Shape Control Methods for Luminescent Metal Oxide Nanomaterials

The pure and doped nanocrystalline luminescent metal oxides were prepared by the bottom-up chemical and top-down physical routes. Both the methods have own merits and demerits. The top-down methods need costly instrumentation, maintenance and skilled person for operation. The produced nanomaterial is suffered with surface defects, non-uniform shapes and large-scale production for industrial usage. On the other hand, in the bottom-up approach for the synthesis of nanocrystalline metal oxides by chemical methods is economic, simple and easily scalable to large scale production. Especially, wet chemical methods have better control over structure, composition, size, shape and surface morphology of nanoparticles for optimizing their optical, electronic and surface properties under controlled synthesis parameters like selection of reactants, concentration, solvents, heating method, heating/cooling rates, temperature, ligands/surfactants and their nature etc. The as-synthesized optimal size and shaped luminescent metal oxide nanoparticles (LMONPs) exhibit good reproducibility in terms of their quality and yield output. As the metal oxides have distinct intrinsic charge separation property which makes them different from metals. Hence, the selection of suitable synthesis route is very important to achieve desired configuration nanoparticles as per the application demand. The chemical vapor synthesis [56–61], thermal decomposition [62–70], pulsed electron beam evaporation [71–78], microwave-assisted combustion [79–86], hydrothermal/solvothermal [87–96], sol-gel [97–101], coprecipitation [102–108], sonochemical [109–118], continuous flow synthesis [119–130], aerosol pyrolysis [131–139], polyol-mediated [140–147], two-phase routes [148–156], microemulsion [157–166], and Green synthesis [167–175] are some methods used in the synthesis of LMNOPs. The brief details of these methods are discussed in the following subsections.

3.5.1 Chemical Vapor Synthesis Method

CVS is a modified version of chemical vapor deposition (CVD) method in which non-agglomerated, crystalline, high purity LMONPs of size less than 10 nm by thermal pyrolysis of precursor compound are formed via chemical reactions instead of film deposition by the reaction between substrate surface and precursor gaseous entities [56–59]. In CVS method, the particle and grain sizes, agglomeration and crystallinity are governed by the five vital reaction parameters such as (i) reactants used as precursor materials, (ii) the decomposition temperature, (iii) the total process pressure, (iv) the partial pressure of the precursors, and (v) the length of the hot zone. By controlling these parameters, one can optimize nanoparticles size, shape and agglomeration. This process is generally performed at higher reaction temperatures, higher precursor partial pressure, and takes longer duration for nanoparticles formation as compared to CVD for film deposition [60, 61]. The CVS method has been used in the synthesis of pure, poly and monodispersed ZnO nanoparticles synthesis by using organometallic ZnO precursor. CVS is performed at higher process temperatures, higher precursor partial pressure, and longer residence time than chemical vapor deposition (CVD), thus resulting in particle formation [60, 61]. The merit of selecting the molecular organometallic precursors in the CVS method over ionic precursors is their higher vapor pressure and lower decomposition temperatures, which encourages the formation of very small sized particles in initial stage of the nucleation reaction.

3.5.2 Thermal Decomposition Method

In thermal decomposition method, the metal precursor is heated above its decomposition temperature in high boiling solvent. When water and volatile organic solvent are used for the synthesis of metal oxide nanoparticles, they show polydispersity, agglomeration and poor crystallinity. In thermal decomposition method, the as-prepared metal oxide nanoparticles do not require any type of post-synthesis thermal treatment to form monocrystalline nanoparticles with mono-disperse shape and size. The nucleation and growth rate are prime steps in metal oxide nanoparticles production in this method. The particle size and shape are controlled by slightly varying the reaction parameters. During decomposition, the presence of several reagents and their ionic forms undergo complex reactions. The synthesis of LMONPs and luminescent magnetic iron oxide nanoparticles by the thermal decomposition of organometallic precursors in organic solvents with surfactant as capping agent in an inert atmosphere has been reported by the different groups of researchers all over the world. The nanoparticles obtained by this method have almost uniform size, shape, and predominance of mono-dispersity [62–65]. Since, the thermal decomposition reaction is performed under the conditions of high temperatures of the organic liquid and vapor phases with no oxygen. This process is widely used in the preparation of luminescent magnetic nanoparticles which at first form a non-magnetic Wüstite phase which on exposing to surrounding environment oxygen form the magnetic magnetite/maghemite phases of ferrites/iron oxides [66]. The control of particle size and shape can be controlled by the ligand assisted capping process during the period of nanoparticle core growth after nucleation process [62, 67, 68]. V.K. LaMer and R.H. Dinegar [69] and D.V. Talapin *et al.* [70] revealed the reason behind the low polydispersity is the effective separation of the nucleating particle during growth process. This uniform growth rate of the nanoparticles enhances monodisperse nanoparticle population with the size sensitive to the molar ratio of precursor to ligand.

3.5.3 Pulsed Electron Beam Evaporation Method

The synthesis of metal oxides and complex metal oxides nanoparticles of about 10 nm size and a narrow particle size distribution with high purity is still a challenging problem for researchers. The weak agglomeration among nanoparticles leads to achieve compact material with high density when in bulk nanosized materials are produced, and form uniform mixtures of nanopowder [71–78]. A narrow particle size distribution and small dimensions of particles permits the maximum realization of unique properties of the as synthesized nanoparticles having high activity, low sintering temperature, high mutual solubility of substances, large sorption ability, etc. The preparation of such nanopowders is carried out by the evaporation-condensation method, which involves the heating of the target by various means e.g. resistive heating, low-temperature plasma heating, heating of conducting materials with a high-current pulse, laser beam heating, heating with beams of accelerated electrons and ions, eddy currents, etc. The nanocrystalline metal oxides without any or minimum contamination from the starting target materials in the form of coarse powders, powder mixtures, metals, alloys, mixtures of metals and nonmetals, etc. are derived by heating them with either a laser beam or by an electron beam source. Electron beam accelerators have better electric power conversion efficiency than laser beam [72, 74–78].

The energy is lost generally by the absorption and dissipation of radiation by the plasma and vapors of the evaporated material over the target surface. The energy deposition time i.e. the pulse length should be less than the time of considerable expansion of heated vapors of the material to the gas medium surrounding the target. Only expansion to a low-pressure gas results high expansion rates to reduce the vapor concentration and formation of small-size particles [74, 75]. That's why, pulsed electron accelerators are preferred for the production of nanoparticles because of the simple adjustment of the energy, pulse length, and the pulse reiteration rate. The use of a pulsed electron accelerator helps in getting a high-purity product with a high coefficient of conversion of the electric power to the electron energy, low cost, availability and diversity of targets in metals, alloys, powder mixtures, conducting and nonconducting materials. Hence, the as produced NPs with unique and special characteristic properties are used in various applications, like in medical, solid state lighting, optoelectronics, and spintronic devices.

3.5.4 Microwave-Assisted Combustion Method

Microwave-assisted synthesis method for the preparation of luminescent metal oxides nanoparticles is very simple and fast suitable for large scale production with no requirement of high temperature or high pressure [79–86]. The heating of reactant materials via microwave irradiation relies on the property of solvent or reagent microwave radiation absorption ability to transform it into heat. This process is explained by considering the two effects: (i) dipole rotation and (ii) ionic conduction, which means that the reversal of solvent dipoles induces the replacement of charged ions of a solute. Polar reactant materials with a high microwave extinction coefficient are easily excited by the direct absorption of microwaves. The difference in the solvent and reactant dielectric constants causes selective dielectric heating which markedly enhances reaction rates. The metal precursors have large microwave absorption as compared to the solvent, because of that very high effective reaction temperatures are obtained. The swift transfer of energy immediately to the reactants before their relaxation results in spontaneous internal temperature increase. The activation energy decreases as compared to conductive heating to proportionately increase the rate of reaction. Since the reaction is carried out at low temperatures, then the problem pertaining to hotspots or other temperature related defects are restricted. Hence, reaction parameters like temperature, time, and pressure are easily controlled in microwave-assisted synthesis of metal oxide nanoparticles. This process permits the rapid decomposition of the precursors to form highly supersaturated solutions in which nucleation and growth takes place simultaneously to produce the desired stoichiometric nanocrystalline metal oxides with monodisperse property as final reaction product. The growth of nanoparticles is effectively prevented by the adsorption of organic surfactant molecules at the surface for their stabilization and surface passivation. This method helps in synthesizing the very good quality highly crystalline, small size luminescent non-magnetic and magnetic metal oxide nano particles of desired morphology in short time interval. The different research groups have used microwave-assisted method for the synthesis of pure and mixed LMONPs for different device applications [80–86].

3.5.5 Hydrothermal/Solvothermal Method

In hydrothermal/solvothermal synthesis of LMONPs, the chemical reaction is performed in water/solvent in a sealed vessel/autoclave cell to bring the temperature of water/solvent near to their critical points by heating simultaneously at autogenic pressure [87–96]. This process is known as hydrothermal when water is used as solvent and solvothermal when organic solvents are used for carrying out chemical reaction by using suitable precursors, mineralizers and additives. The precursor reactants are taken in solution, gel or suspension form. Mineralizers are basically organic or inorganic additives used in high concentrations to maintain pH of solution. While organic or inorganic additives are added in low concentration to encourage particles dispersion by capping them with controlled size and morphology. The high temperature and high pressure of water/solvent for materials which are insoluble at temperature and pressure i.e. <100°C and <1 atmosphere respectively. As the ionic product (K_w) of reaction attains maximum value in 250–300°C range, that's why the hydrothermal synthesis is mostly carried out at a temperature less than 300°C. The critical temperature and pressure of water are 374°C and 22.1 MPa respectively. The dielectric constant and solubility of solvent for many compounds varies vividly under supercritical conditions. Under supercritical conditions, the dielectric constant of water/solvent should be below 10; then the notable chemical reaction rates are governed by electrostatic theory. The supercritical water favors the chemical reaction to proceed for metal oxide nanoparticle formation due to enhanced reaction rate and massive supersaturation as per the nucleation theory by solubility lowering. These methods have many advantages like large scale production of nanomaterials at low, high yield with well-controlled size and shape of metal oxide nanoparticles. Sometime, the hydrothermal and solvothermal methods are associated with other chemical methods like microwave-assisted, sol-gel, sonochemical and magnetic fields for the synthesis of good quality reproducible nanomaterials. The solvothermal process produces smaller and narrow size distribution metal oxides nanoparticles at low temperature as compared to hydrothermal method. Extensive work has been reported on the preparation of many nanostructured metal oxides by hydrothermal and solvothermal processes [89–96]. The Y_2O_3:Eu^{3+} microspheres prepared by solvothermal method exhibit a strong red emission at 610 nm pertaining to $^5D_0 \rightarrow {}^7F_2$ transition of Eu^{3+} under UV excitation wavelength 259 nm and low-voltage electron beams excitation (1-5 kV) and finds usage fluorescent lamps and field emission display applications [91].

3.5.6 Sol-Gel Method

The wet chemical sol-gel method is extensively used for the synthesis high-quality LMONPs and their mixed oxide composites owing to the significant control over the size, shape, texture and surface properties of the grown nanoparticles [97–101]. The sol–gel method follows (i) hydrolysis, (ii) condensation/polycondensation, (iii) aging, (iv) drying and (v) thermal decomposition reaction steps to form metal oxide nanoparticles. At first, the precursors mostly metal alkoxides solutions are prepared in water or alcohol/organic solvent to undergo hydrolysis to form metal hydroxide [97–99]. When water is used as reaction medium, this process is termed as aqueous sol–gel method; and on using organic solvent, then this is called as nonaqueous sol–gel method. In addition to water and organic solvent, sometimes addition of an acid or a base is also used for the hydrolysis of the precursors. The gel formation is sensitive to amount of water used. The higher water content results in the higher ratio of bridging to nonbridging oxygen groups which in turn form more polymerized and branched structure in the second condensation step. In condensation step: the water/alcohol is eliminated and metal oxide (M-O) bonds are formed. The polymeric network consists of hydroxyl (-OH), bridging (-OH) group between two metal (M-(OH)-M), oxo (-O-), bridging oxo (M-O-M) groups which form colloidal solution. The condensation or polycondensation increases the viscosity of solvent by developing a porous structure and maintaining liquid phase known as gel. The size and the cross-linking within the colloidal particles vary with the alkoxide used as precursor and pH of the solution [100, 101]. While in aging process, the polycondensation goes on within the localized solution and consequently reprecipitation of the gel network. During this process, the porosity decreases, with the enhancement in thickness between colloidal particles. This induces the structure and properties variations in the gel. The fourth drying process is very complex because water and organic molecules are removed to form gel by different means like areal drying, thermal drying, supercritical drying, or by deep freeze-drying methods. Heating the porous gel at high temperature leads to densification and the pores removal from gel to produce xerogel. In supercritical drying aerogels are obtained with very high pore volume, surface area and almost similar gel network. In freeze drying, the solvents are frozen to form cryogel. In drying process, the relative humidity (RH) plays vital role for the stability and performance of nanomaterials. The thermal decomposition of dried gel is carried out by thermal treatment/calcination process to remove the left residues and water molecules. The calcination temperature is a very important parameter in controlling the pore size and the density of LMONPs.

3.5.7 Chemical Co-Precipitation Method

The chemical co-precipitation method deals with the precipitation of salt from a homogeneous liquid by varying the reaction parameters like reactants and their concentrations, solvents, acidic and basic chemicals used for precipitation, temperature, pH of solution, and complexing agents/surfactants. During precipitation, the nucleation and stabilization processes are going on either simultaneously or sequentially. The tiny and stable new phase nanoparticles nucleation begins which stabilize themselves by growing or agglomeration process. The kinetics of nucleation and particle growth reaction process is controlled by the released anions and cations in homogeneous solution to develop monodisperse nanoparticles. By the optimization of reactants concentration and pH of reacting solution, one can easily obtain the uniform and narrow size distribution nanoparticles [102–108]. For the synthesis of metal oxides, in co-precipitation reaction, at first the metal hydroxides precipitates by the adding NaOH, NH_4OH, urea, etc. to a reactant's aqueous solution. The metal oxides are derived by heat treatment/calcination to the as obtained dried metal hydroxide precipitates at low temperatures to small nanosized particle. When two or more metal salts are used, then the stoichiometric composition of the precipitate is based on the differences in solubility between the components and the chemical reaction at the time of co-precipitation. The heterogenous coprecipitate is obtained by the slow precipitation rate or poor mixing of reaction solutions. The co-precipitate hydrothermal treatments help in transforming amorphous precipitates to crystalline materials with better thermal stability and surface activity. This method is also used in the synthesis of LMONPs.

3.5.8 Sonochemical Method

In sonochemical method, the ultrasound wave is not interacting with precursor chemicals at molecular level. But in the 20 kHz to 1 GHz frequency range, they generate acoustic cavitation bubble with the concentrated ultrasonic energy [24–27, 109–118]. This cavitation process takes place in liquids on exposing them with high to moderate range intensity and minimum power of ultrasonic waves [112]. The liquid expands by the sound wave field which causes rapid growth at the weak sites of the liquid having dissolved gases. They act as cavitation nuclei for vapor or gas-filled cavities or microbubbles. The compression of bubbles during cavitation process is faster than thermal transport process and generates localized hot-spots. The compression of bubbles during cavitation process leads to the huge concentration of energy within the small volume of the collapsed bubble. The bubbles are continuously growing during the expansion/compression cycles of the ultrasound till they acquire a critical diameter having transient temperature about 5000 K and reactor temperature is close to ambient temperature. At such high temperature, the hot spots of size less than 1 mm and life ~ 1 ms produces the product by the conversion of surface energy into chemical energy [113]. While excessive kinetic energy of liquid is dissipated in the form of heat. In this fast process, the reaction is controlled at the atomic level of the surfaces at the time scale in ms range. This method is used for the synthesis of different LMONPs [114–118].

3.5.9 Continuous Flow Method

Recently, the continuous flow method is reported for the preparation of wide range of nanomaterials via chemical reactions at high temperature and pressure with marked decrease in the reactor size [119–123]. The small dimensional reactor helps in swift mixing and large interfacial area enhances the mass transfer and high heat transfer rates throughout the reactor because of its high surface to volume ratio [124, 125] and ability to reduce the back mixing. The most important aspect of this method is the automated electromechanical programming which overall manages the continuous flow reaction process by precisely regulating the synthesis parameters e.g. flow rate, dimensions of reactor, temperature, concentration, and mixing for the chemical reactions to form the stable and reproducible nanoparticles [126–130]. The control over nanoparticles nucleation and growth kinetics is derived by effectively reducing the variations in temperature and concentration for monodisperse particle preparation. These features make continuous flow synthesis, a suitable alternative method for the production of various types of noble metals, their colloidal nanocrystals [126–128], metal oxide [129] and quantum dots [130]. The continuous flow technology capabilities are also explored in conjunction with hydrothermal, solvothermal, microwave-assisted techniques for the chemical synthesis of monodisperse, stable, better quality small size and different shaped LMONPs for various applications. The continuous flow method can also be adopted for large scale production of good quality stable nanoparticles.

3.5.10 Aerosol Pyrolysis Method

The aerosol-based synthesis method produces precise morphology and chemical composition nanomaterials by either atomization or aero-to-particle conversion processes. The liquid-feed aerosol formation leads to generate solid particles by both gas-to-particle and droplet-to-particle conversion processes which relies upon process parameters [131, 132]. In a gas-to-particle conversion process, fine droplets with the support and metal precursors evaporate quickly upon mixing with the hot combustion gases to produce precursor vapors which nucleate and grow to the desired product composition in the gas phase, or after particle nucleation and growth process. Hence, in droplet-to-particle conversion case, the reaction occurs in the liquid droplet phase, and each droplet is converted to a single solid particle. While, in gas-to-particle, droplet-to-particle, or a combination of the two states depends upon the relative rates of reaction and evaporation process. In atomization process, a solution of desired composition is atomized into fine droplets via nebulization or electrohydrodynamic method which subsequently crystallizes into nanoparticles on the evaporation of the solvent via nucleation and growth in the condensation step [133]. The hydraulic pressure, pneumatic, ultrasonic nozzle, or rotary disk atomizers, are generally used for producing different sized fine droplets [134]. While aerosol-to-particle preparation is carried out by the spark discharge, flame, furnace or plasma techniques. The electrospray, ultrasound, and pneumatic atomization processes are used for the generation of aerosol particles by atomizing solvents of particular chemical composition [135, 136]. In this method, the nanoparticle formation is based upon the evaporation of the solvent in the atomized droplets leads to the crystallization of solid nanomaterials by condensation and coagulation processes. This method is also employed in the synthesis of LMONPs and their thin films [137–139].

3.5.11 Polyol-Mediated Methods

The polyol-mediated method has been also used for the synthesis of nanocrystalline metal oxides of different shaped particles having size in the range of 30–300 nm [140–144]. This process involves preparation of metal precursor in glycol or diethylene or tetraethyleneglycol as polyol with specific amount of water. Mostly the metal precursors are the metal acetate, alcoholate or halogenides. Since the polarity of the dielthylene glycol as polyol is more than 30, that's why most of the inorganic compounds are easily dissolved in it [141–144]. The boiling points of glycol, diethylene glycol and tetraethylene glycol are 197°C, 246°C and 314°C respectively. The solution is rapidly heated to the boiling point of polyol for the nucleation and growth of well-crystallized LMONPs [145–147] which governs the luminescence, color, electron transport, electrical conductivity, magnetic properties of the nanomaterials. The complexation of the metal oxide nanoparticle with polyol molecule restricts the particle growth, and prevents agglomeration of nanoparticles due to the limited availability of active centers at the surface of nanoparticles [142, 143]. The low molecular weight polyols also play the role of weak stabilizers and they can be removed from the particle surface by the variations in experimental conditions. In the polyol-mediated method, one can easily optimize the solvent exchange and controlled precipitation of nanomaterials. Hence, this process is best suited for the large-scale production of precise size with narrow size distribution, shape, non-agglomerated metal oxide nanoparticles. This method has been used for the synthesis of ZnO nanorods and nanocomposites [145], rare earth doped CeO_2 [146] and different metal tungstate nanoparticles [147].

3.5.12 Two-Phase Method

The synthesis of monodisperse nanocrystals is in demand from both fundamental and applications point of view for researchers. As it is well known that characteristic electrical, optical, and magnetic properties are very sensitive to the size and size distribution of nanomaterials. The synthesis of monodisperse nanocrystals is carried out by two ways (i) single-phase [148, 149] and (ii) two-phase methods [150–156]. In single-phase method, the nucleation and growth processes are separated at first by rapidly mixing two precursors at high temperature for burst-nucleation and then on decreasing temperature, the growth process of nanoparticles slows down in a homogeneous medium. While in the two-phase method, the chemical reaction takes place between two molecular precursors which are separated in the organic and aqueous phases. The nucleation and growth process for nanoparticles formation takes place only at the liquid–liquid interface. In two-phase method, the nucleation step stays for longer time as compared to single phase process and submerges with the growth process. The nanoparticles formed by this method have very narrow size distributions. The synthesis of nanocrystalline metal oxides by two-phase method uses metal alkoxides or metal stearate as metal precursors, and tert-butylamine as activation agent to lower the reaction temperature for precursor hydrolysis and stabilized the metal oxide nanocrystals by capping them with oleic acid as surfactant and dissolved in toluene. The desired shape nanocrystals synthesized by two phase method are obtained by selecting the appropriate concentration of monomer, surfactants and reaction temperature. The very good quality, narrow distribution luminescent metal oxides and colloidal metal oxides synthesized via two phase method of different size and shape are reported by different workers [152–156].

3.5.13 Microemulsion Method

The thermodynamically stable microemulsion method has been used for the synthesis of metal oxide nanoparticles in which optically isotropic solution of two immiscible liquids like oil in water (aqueous) or water in oil (non-aqueous). They have microdomains in one or both liquids to create micro-heterogenous medium. The reactant distribution in the droplets and dynamics involved during inter-droplet exchange of reactants controls the nucleation, growth and agglomeration process during nanoparticles formation [157–161]. These microdomains develop cage-like effect to form microcavities stabilized by capping with the surfactant molecules [161]. When the micelles are formed in aqueous medium, then the hydrophobic end of hydrocarbon chains of the surfactant molecules are directed toward the micelle, and the hydrophilic functional groups of the surfactants are oriented around the surrounding of aqueous medium. While in reverse micelles which are developed in non-aqueous medium have hydrophilic end toward the core of micelles, and the hydrophobic organic chain end are projected outward in the surrounding medium [162]. The stability of the emulsions is controlled by the particle size, particle-particle interaction, particle–water interaction and particle-oil interaction [158]. The merit of the microemulsion method is that the ratio of surfactant to water can be used in the optimization of nanoparticle size. The size of nanoparticles is nearly similar to the water droplet in the reverse microemulsion. This method has been used in the synthesis of different LMONPs [163–166].

3.5.14 Green Synthesis Method

The low-cost green synthesis method is widely used in the synthesis of nanometals and nano sized metal oxides by using different parts of plants, plant extracts or microorganisms or fungi or algae etc. [167–170]. The degradation of any organic compounds by the green synthesis in which plant extracts polyphenols are involved for the redox chemical reaction with precursor material to produce metal oxide nanoparticles. This method is categorized under the bottom-up approach of nanomaterials synthesis in which the reactants undergo oxidation or reduction reactions initiated by the intermediate reactive oxygen species formed during reaction. The very fast green method is also used for large scale production of metal oxide nanoparticles at room temperature electrochemically by the anodization of metals and alloys in KCl electrolyte [167, 168]. The as-grown crystalline nanoparticles and their colloidal suspensions in ethanol or in water are very stable for many days. The size of the synthesized NPs can be readily tuned by changing the anodizing voltage. This process follows twelve green chemistry rules. These are the development and designing of nanoparticles by using non-hazardous chemicals, renewable material, eco-friendly solvents, and degradable reaction waste/undesired by-products. For the synthesis of nanomaterials by using a green chemistry route, the following three basic very important requirements to be fulfilled for the successful completion of reaction to obtain desired composition final product. These are non-toxic solvents, reducing/oxidizing agent, and surfactants for the stabilization of nanoparticles. The selection of appropriate surfactant is another important parameter used for controlling the size, shape morphology and passivation of the prepared nanoparticles surface by capping their surface for different device applications [170–175]. Plants/its parts extracts mainly leaf extracts are used for extracting more stable nanoparticles than that obtained from microorganisms because of their slow rate of synthesis and presence of high contamination.

3.6 Characterization of Nanocrystalline Luminescent Metal Oxides

The structure, morphology, size and shape of nanoparticle, composition, optical, electrical and magnetic properties of the synthesized LMONPs by different methods are characterized by powder X-ray diffraction (XRD), Fourier Transform Infrared (FTIR), Raman, X-ray Photoelectron spectroscopy (XPS), SEM with EDS accessory, TEM/HRTEM, UV-Vis, Photoluminescence (PL), Electron Paramagnetic Resonance (EPR), Vibration Sample Magnetometer (VSM) techniques. These studies help in confirming the formation of desired structure, morphology and chemical composition of the prepared nanoparticles. These techniques also are also used in refining the synthesis parameters, selection of appropriate precursor and synthesis process. By optimizing these factors, one can achieve LMONPs as per the device demand.

3.7 Applications of Nanocrystalline Luminescent Metal Oxides

Nanocrystalline luminescent metal oxides, their nanocomposites and combination with quantum dots have drawn considerable attention of the researchers owing to their wide range of applications in the field of solid-state lighting devices, displays, sensors, biomedical devices [12, 176–178]. In this section, the brief details are discussed. The mesoporous nanocrystalline metal-oxide film was used as an electron-injecting cathode in organic light emitting diodes (OLEDs) having hybrid inorganic/organic three-component coating which consists of an inorganic nanocrystalline TiO_2 film as an electron transport material, a thin layer of an organic electroluminescent conjugated polymer film of poly(9,9-dioctylfluorene-co-benzothiadiazole) (F8BT) and in pores poly(9,9-dioctylfluorene) (PFO) hole-transporting conjugated polymer. This three-component assembly allows the luminescence and the electron- and hole-charge-transport processes in isolation and tuned individually. This process showed more flexibility in device designing with better efficiency as compared to conventional OLEDs [179]. The work has been reported on the application of nanocrystalline and sub-micron rare earth Dysprosium ions doped Gadolinium oxide phosphors in white light LEDs [180] and lanthanide-doped lanthanum hafnate nanoparticles multicolor phosphors for warm white lighting and scintillators [181]. The low cost and very stable nanocrystalline luminescent metal oxides are also used in the development of resistive semiconductor gas sensors in which light is used as activation in place of thermal heating gases detection. The photoelectric and optical properties of nanocrystalline oxides SnO_2, TiO_2, ZnO, In_2O_3 and WO_3 have been exploited as sensing material for semiconductor gas sensor application [182, 183]. The sensitivity of gas sensor is tested by UV and visible light activation. In UV light activated sensor, they used of metal oxides, chemically modified noble metals and their oxides nanoparticles, and the nanocomposite materials of metal oxides. While in visible light activated sensors, the photo- and gas sensitivity of wide band gap energy metal oxides are improved by doping, dyes as spectral sensitizers in narrow band gap semiconductors in nanoparticle and quantum dots form and incorporation of plasmon nanoparticles. Another important application of some biocompatible luminescent metal oxides like TiO_2, ZnO, CuO, Fe_2O_3, Fe_3O_4, and SiO_2, etc. is in the field of biomedical. They can be used in internal tissue therapy, drug delivery, contrast agent in magnetic resonance imaging, anti-diabetic, dentistry, cancer and tumor treatment, biosensors, bacterial/fungal/viral treatment, clinical diagnosis, fingerprint analysis and biological assays etc. [184–187].

3.8 Conclusion and Future Aspects of Nanocrystalline Luminescent Metal Oxides

LMONPs exhibits unique characteristics properties as compared to their bulk counterparts which encouraged in-depth investigations of the existing and new advanced materials to make them stable and sensitive for multidimensional applications. In this book chapter, the introduction of luminescence, types of luminesce, luminescence mechanism, role of LMONPs, synthesis by different methods, characterization by different analytical techniques and applications have been briefly described. As this is a very fast field, the authors have tried their best to make it simple and informative with conceptual details. As we know, that toxicity of nanoparticles, large scale production for industrial usage, clinical trials for biomedical applications, environmental pollution and degradation/mineralization of hazardous chemicals are some major challenges in this field. The translation of materials prepared in laboratory to pilot plant is really a very big problem for researchers to maintain size, shape and morphology of nanoparticle with minimum degree of agglomeration. In biomedical application, the reactive oxygen species formation has been used for anticancer, antibacterial/antimicrobial treatments, drug-delivery and high contrast imaging. For this one need quality control standard documents exhibiting the permissible limit and uncertainty value. Such systematic documentation approved by International and National apex bodies for in vitro and in vivo protocols should be used in future to reveal specific NPs response along the limitations imposed by their toxicity for a particular standardized method of synthesis and testing guidelines. This concept will open new doors for research specific in nanomaterials synthesis by different chemical and physical methods, characterization techniques and applicable for different types of applications as per device requirement.

References

1. Medhi, R., Marquez, M.D., Lee, T.R., Visible-light-active doped metal oxide nanoparticles: Review of their synthesis, properties, and applications. *ACS Appl. Nano Mater.*, 3, 6156, 2020.
2. Sun, S.-K., Wang, H.-F., Yan, X.-P., Engineering persistent luminescence nanoparticles for biological applications: From biosensing/bioimaging totheranostics. *Acc. Chem. Res.*, 51, 5, 1131, 2018.
3. Vollath, D., Szabo, D.V., Schlabach, S., Oxide/polymer nanocomposites as new luminescent materials. *J. Nanopart. Res.*, 6, 181, 2004.
4. Zaragr, R.A., Micro size developed TiO_2-CdO composite film: Exhibits diode characteristics. *Mater. Lett.*, 335, 15, 133813, 2023.
5. Arunadevi, N. and Kirubavathy, S.J., Chapter 2 - Metal oxides:Advanced inorganic materials, in: *Inorganic Anticorrosive Materials Past, Present and Future Perspectives*, p. 21, 2022.
6. Kumari, A. and Mahata, M.K., Chapter 12 - Upconversion nanoparticles for sensing applications, in: *Upconversion Nanophosphors*, p. 311, Micro and Nano Technologies, Elsevier Publishers, Netherlands, 2022.
7. Mialon, G., Türkcan, S., Alexandrou, A., Gacoin, T., Boilot, J.-P., New insights into size effects in luminescent oxide nanocrystals. *J. Phys. Chem. C*, 113, 18699, 2009.
8. Zargar, R.A., ZnCdO thick film: A material for energy conversion devices. *Mater. Res. Express*, 6, 9, 095909, 2019.
9. Duan, C., Liang, L., Li, L., Zhang, R., Xu, Z.P., Recent progress in upconversion luminescence nanomaterials for biomedical applications. *J. Mater. Chem. B*, 6, 192, 2018.
10. He, Luo, X., Liu, X., Li, Y., Wu, K., Visible-to-ultraviolet upconversion efficiency above 10% sensitized by quantum-confined perovskite nanocrystals. *J. Phys. Chem. Lett.*, 10, 5036, 2019.
11. Hisamitsu, S., Miyano, J., Okumura, K., Hui, J.K.H., Yanai, N., Kimizuka, N., Visible to UV photon upconversion in nanostructured chromophoric ionic liquids. *ChemistryOpen*, 9, 14, 2020.
12. Liu, Y., Ai, K., Lu, L., Designing lanthanide-doped nanocrystals with both up- and down-conversion luminescence for anti-counterfeiting. *Nanoscale*, 3, 4804, 2011.
13. Valenta, J., Repko, A., Greben, M., Nižnanský, D., Absolute up- and down-conversion luminescence efficiency in hexagonal Na(Lu/Y/Gd)F4: Yb, Er/Tm/Ho with optimized chemical composition. *AIP Adv.*, 8, 075226, 2018.
14. Srivastava, B.B., Kuang, A., Mao, Y., Persistent luminescent sub-10 nm Cr doped $ZnGa_2O_4$ nanoparticles by a biphasic synthesis route. *Chem. Commun.*, 51, 7372, 2015.
15. Kabe, R. and Adachi, C., Organic long persistent luminescence. *Nature*, 550, 384, 2017.

16. Mahesh, K., Weng, P.S., Furetta, C., *Thermoluminescence in solids and its applications*, Nuclear Technology Publishing, Ashford UK, 1989.
17. Vij, D.R., *Thermoluminescentmaterials*, PTR Prentice Hall, Englewood Cliffs NJ, 1993.
18. Abass, M.R., Diab, H.M., Abou-Mesalam, M.M., New improved thermoluminescence magnesium silicate material for clinical dosimetry. *Silicon*, 14, 2555, 2022.
19. Nieman, T.A., *Luminescence techniques in chemical and biochemical analysis*, W.R.G. Baeyens, D. De Keukeleire, K. Korkidis, (Eds.), Marcel Dekker, New York, 1991, Chapter 17.
20. Nakashima, K. and Imai, K., *Molecular luminescence spectroscopy*, Part 3, S.G. Schulman, (Ed.), Wiley, New York, 1993, Chapter 1.
21. García-Campaña, A.M. and Bayens, W.R.G. (Eds.), *Chemiluminescence in analytical chemistry*, Marcel Dekker, New York, 2001.
22. Suslick, K.S., Didenko, Y., Fang, M.M., Hyeon, T., Kolbeck, K.J., McNamara III, W.B., Mdleleni, M.M., Wong, M., Acoustic cavitation and its chemical consequences. *Philos. Trans. R. Soc. Lond. Ser. A*, 357, 335, 1999.
23. Crum, L.A.A., Mason, T.J., Reisse, J., Suslick, K.S., *Sonochemistry and sonoluminescence*, AIP Conf. Proc., p. 524, Kluwer Acad, Dordrecht, 1999.
24. Suslick, K.S., Sonochemistry. *Science*, 247, 1439, 1990.
25. Suslick, K.S. and Flannigan, D.J., Inside a collapsing bubble: Sonoluminescence and the conditions during cavitation. *Annu. Rev. Phys. Chem.*, 59, 659, 2008.
26. Didenko, Y.T., McNamara III, W.B., Suslick, K.S., Molecular emission from single-bubble sonoluminescence. *Nature*, 407, 877, 2000.
27. Smart, A.G., Inside a sonoluminescing microbubble, hints of a dense plasma. *Phys. Today*, 65, 18, 2012.
28. Herring, P.J. and Widder, E.A., Bioluminescence, in: *Encyclopedia of Ocean Sciences*, J.H. Steele, S.A. Thorpe, K.K. Turekian, (Eds.), p. 308, Academic Press, San Diego, USA, 2001.
29. Walton, A.J., Triboluminescence. *Adv. Phys.*, 26, 887, 1977.
30. Xie, Y. and Li, Z., Triboluminescence: Recalling interest and new aspects. *Chem.*, 4, 943, 2018.
31. Hsu, C.W., Ly, K.T., Lee, W.-K., Wu, C.-C., Wu, L.-C., Lee, J.-J., Lin, T.-C., Liu, S.-H., Chou, P.-T., Lee, G.-H., Chi, Y., Triboluminescence and metal phosphor for organic light-emitting diodes: Functional Pt(II) complexes with both 2-pyridylimidazol-2-ylidene and bipyrazolate chelates. *ACS Appl. Mater. Interfaces*, 8, 49, 33888, 2016.
32. Smith, C.J., Griffin, S.R., Eakins, G.S., Deng, F., White, J.K., Thirunahari, S., Ramakrishnan, S., Sangupta, A., Zhang, S., Novak, J., Liu, Z., Rhodes, T., Simpson, G.J., Triboluminescence from pharmaceutical formulations. *Anal. Chem.*, 90, 6893, 2018.
33. Dengfeng, P., Bing, C., Feng, W., Recent advances in doped mechanoluminescent phosphors. *ChemPlusChem*, 80, 1209, 2015.

34. Hu, M., Application of lanthanide-doped luminescence nanoparticles in imaging and therapeutics, Chapter-8, in: *Photo Nanotechnology for Therapeutics and Imaging*, p. 205, 2020.
35. Williams, F., High-field electroluminescence. *J. Lumin.*, 23, 1, 1981.
36. Dean, P.J., Comparisons and contrasts between light emitting diodes and high field electroluminescent devices. *J. Lumin.*, 23, 17, 1981.
37. Klein, J., Sun, C., Pratx, G., Radioluminescence in biomedicine: Physics, applications, and models. *Phys. Med. Biol.*, 64, 04TR01 (47 pp.), 2018.
38. Jacobsohn, L.G., Bennett, B.L., Muenchausen, R.E., Martin, M.S., Shao, L., McDaniel, F.D., Doyle, B.L., [AIP Application of accelerators in research and industry: Twentieth international conference - Fort Worth (Texas) (10–15 August 2008)] Radioluminescence investigation of ion-irradiated phosphors. *AIP Conf. Proc.*, 1099, 977, 2009.
39. Guidelli, E.J., Baffa, O., Clarke, D.R., Enhanced UV emission from silver/ZnO and gold/ZnO core-shell nanoparticles: Photoluminescence, radioluminescence, and optically stimulated luminescence. *Sci. Rep.*, 5, 14004, 11, 2015.
40. İlhan, M. and Keskin, İ.Ç., Photoluminescence, radioluminescence and thermoluminescence properties of Eu3+ doped cadmium tantalate phosphor. *Dalton Trans.*, 47, 13939, 2018.
41. Smith, M.B. and Michl, J., Singlet fission. *Chem. Rev.*, 110, 6891, 2010.
42. Sternlicht, H., Nieman, G.C., Robinson, G.W., Triplet-triplet annihilation and delayed fluorescence in molecular aggregates. *J. Chem. Phys.*, 38, 1326, 1963.
43. Schaller, R.D. and Klimov, V.I., High efficiency carrier multiplication in PbSe nanocrystals: Implications for solar energy conversion. *Phys. Rev. Lett.*, 92, 186601, 2004.
44. Kroupa, D.M., Roh, J.Y., Milstein, T.J., Creutz, S.E., Gamelin, D.R., Quantum-cutting ytterbium-doped $CsPb(Cl_{1-x}Br_x)_3$ perovskite thin films with photoluminescence quantum yields over 190%. *ACS Energy Lett.*, 3, 2390, 2018.
45. Zhou, J., Liu, Q., Feng, W., Sun, Y., Li, F., Up-conversion luminescent materials: Advances and applications. *Chem. Rev.*, 115, 395, 2015.
46. Garcia, M.A., Surface plasmons in metallic nanoparticles: Fundamentals and applications. *J. Phys. D Appl. Phys.*, 44, 283001, 2011.
47. Zhang, Q., Tan, Y.N., Xie, J., Lee, J.Y., Colloidal synthesis of plasmonic metallic nanoparticles. *Plasmonics*, 4, 9, 2009.
48. Zargar, R.A., Arora, M., Alshahrani, T., Shkir, M., Screen printed novel ZnO/MWCNTs nanocomposite thick films. *Ceram. Int.*, 47, 6084–6093, 2020.
49. Wen, G.W., Lin, J.Y., Jiang, H.X., Chen, Z., Quantum-confined stark effects in semiconductor quantum dots. *Phys. Rev. B*, 52, 5913, 1995.
50. Zargar, R.A., Fabrication and improved response of ZnO-CdO composite films under diferent laser irradiation dose. *Sci. Rep.*, 12, 10096, 2022.
51. Kumbhakar, P., Ray, S.S., Stepanov, A.L., Optical properties of nanoparticles and nanocomposites. *J. Nanomater.*, 2014, 181365, 2014.

52. Abrams, B.L. and Holloway, P.H., Role of surface of the surface in luminescent processes. *Chem. Rev.*, 104, 5783, 2004.
53. Yen, W.M. and Hajime, Y. (Eds.), *Practical applications of phosphors*, CRC Press, Boca Raton, 2018.
54. Gupta, S.K., Hernandez, C., Zuniga, J.P., Lozano, K.L., Mao, Y., Luminescent PVDF nanocomposite films and fibers encapsulated with $La_2Hf_2O_7$:Eu^{3+} nanoparticles. *SN Appl. Sci.*, 2, 616, 2020.
55. Hernandez, C., Gupta, S.K., Zuniga, J.P., Vidal, J., Galvan, R., Mao, Y., Lozano, K., Performance evaluation of Ce^{3+} doped flexible PVDF fibers for efficient optical pressure sensors. *Sens. Actuators A Phys.*, 298, 111595, 2019.
56. Klein, S., Winterer, M., Hahn, H., Reduced-pressure chemical vapor synthesis of nanocrystalline silicon carbide powders. *Chem. Vap. Deposition*, 4, 143, 1998.
57. Polarz, S., Roy, A., Merz, M., Halm, S., Schröder, D., Schneider, L., Bacher, G., Kruis, F.E., Driess, M., Chemical vapor synthesis of size-selected zinc oxide nanoparticles. *Small*, 1, 540, 2005.
58. Jin, W., Lee, I.-K., Kompch, A., Dörfler, U., Winterer, M., Chemical vapor synthesis and characterization of chromium doped zinc oxide nanoparticles. *J. Eur. Ceram. Soc.*, 27, 4333, 2007.
59. Bacsa, R., Kihn, Y., Verelst, M., Dexpert, J., Bacsa, W., Serp, P., Large scale synthesis of zinc oxide nanorods by homogeneous chemical vapour deposition and their characterisation. *Surf. Coat. Technol.*, 201, 9200, 2007.
60. Vallejos, S., Maggio, F.D., Shujah, T., Blackman, C., Chemical vapour deposition of gas sensitive metal oxides. *Chemosensors*, 4, 4 (18 pp.), 2016.
61. Wang, G.Z., Wang, Y., Yau, M.Y., To, C.Y., Deng, C.J., Ng, D.H.L., Synthesis of ZnO hexagonal columnar pins by chemical vapor deposition. *Mater. Lett.*, 59, 3870, 2005.
62. Park, J., Lee, E., Hwang, N.-M., Kang, M., Kim Sung, C., Hwang, Y., Park, J.-G., Noh, H.-J., Kim, J.-Y., Park, J.-H., Hyeon, T., One-nanometer-scale size-controlled synthesis of monodisperse magnetic iron oxide nanoparticles. *Angew. Chem. Int. Ed.*, 44, 2873, 2005.
63. Sun, S. and Zeng, H., Size-controlled synthesis of magnetite nanoparticles. *J. Am. Chem. Soc.*, 124, 8204, 2002.
64. Yu, W.W., Falkner, J.C., Yavuz, C.T., Colvin, V.L., Synthesis of monodisperse iron oxide nanocrystals by thermal decomposition of iron carboxylate salts. *Chem. Commun.*, 2004, 2306, 2004.
65. Xu, Z., Shen, C., Hou, Y., Gao, H., Sun, S., Oleylamine as both reducing agent and stabilizer in a facile synthesis of magnetite nanoparticles. *Chem. Mater.*, 21, 1778, 2009.
66. Unni, M., Uhl, A.M., Savliwala, S., Savitzky, B.H., Dhavalikar, R., Garraud, N., Arnold, D.P., Kourkoutis, L.F., Andrew, J.S., Rinaldi, C., Thermal decomposition synthesis of iron oxide nanoparticles with diminished magnetic dead layer by controlled addition of oxygen. *ACS Nano*, 11, 2284, 2017.

67. Park, J., Joo, J., Kwon, S.G., Jang, Y., Hyeon, T., Synthesis of monodisperse spherical nanocrystals. *Angew. Chem., Int. Ed.*, 46, 4630, 2007.
68. Kovalenko, M.V., Bodnarchuk, M.I., Lechner, R.T., Hesser, G., Schaeffler, F., Heiss, W., Fatty acid salts as stabilizers in size- and shape-controlled nanocrystal synthesis: The case of inverse spinel iron oxide. *J. Am. Chem. Soc.*, 129, 6352, 2007.
69. LaMer, V.K. and Dinegar, R.H., Theory, production and mechanism of formation of monodispersed hydrosols. *J. Am. Chem. Soc.*, 72, 4847, 1950.
70. Talapin, D.V., Rogach, A.L., Haase, M., Weller, H., Evolution of an ensemble of nanoparticles in a colloidal solution: Theoretical study. *J. Phys. Chem. B*, 105, 12278, 2001.
71. Blair, M.W., Jacobsohn, L.G., Tornga, S.C., Ugurlu, O., Bennett, B.L., Yukihara, E.G., Ross, E., Muenchausen nanophosphor aluminum oxide: Luminescence response of a potential dosimetric material. *J. Lumin.*, 130, 825, 2010.
72. Il'ves, V.G., Kamenetskikh, A.S., Kotov, Y.A., Medvedev, A.I., Sokovnin, S.Y., Production of nanopowders of metal oxides by evaporation in the pulsed electron beam. *Russ. J. Non Ferr. Met.*, 51, 197, 2010.
73. Il'ves, V.G. and Sokovnin, S.Y., Production and studies of properties of nanopowders on the basis of CeO_2. *Nanotechnol. Russ.*, 7, 213, 2012.
74. Il'ves, V.G., Sokovnin, S.Y., Uporov, S.A., Zuev, M.G., Properties of the amorphous-nanocrystalline Gd2O3 powder prepared by pulsed electron beam evaporation. *Phys. Solid State*, 55, 1262, 2013.
75. Sokovnin, S.Y. and Il'ves, V.G., *Use of a pulsed electron beam for obtaining nanopowders of some metal oxides*, p. 316, Publishing house of Ural Branch of the Russian Academy of Sciences, Yekaterinburg, 2012.
76. Sokovnin, S.Y., Murzakaev, A.M., Il'ves, V.G., Spirina, A.V., Research of nanopowders properties pure and doped metal oxides produced by a method of pulsed electron beam evaporation. *Russ. Phys. J. Spec. Issue High Curr. Electron.*, 55, 394, 2012.
77. Sokovnin, S.Y., Il'ves, V.G., Medvedev, A.I., Murzakaev, A.M., Investigation of properties of ZnO-Zn-Cu nanopowders obtained by pulsed electron evaporation. *Inorg. Mater. Appl. Res.*, 4, 410, 2013.
78. Sokovnin, S.Y., Il'ves, V.G., Surdo, A.I., Mil'man, I.I., Vlasov, M.I., Effect of iron doping on the properties of nanopowders and coatings on the basis of Al_2O_3 produced by pulsed electron beam evaporation. *Nanotechnol. Russ.*, 8, 466, 2013.
79. Wang, H.Q. and Thomas, T., Monodisperse upconverting nanocrystals by microwave-assisted synthesis. *ACS Nano*, 3, 3804, 2009.
80. Abdelsayed, V., Aljarash, A., El-Shall, M.S., Al Othman, Z.A., Alghamdi, A.H., Microwave synthesis of bimetallic nanoalloys and co-oxidation on ceria-supported nanoalloys. *Chem. Mater.*, 21, 2825, 2009.
81. Liang, J., Deng, Z., Jiang, X., Li, F., Li, Y., Photoluminescence of tetragonal ZrO_2 nanoparticles synthesized by microwave irradiation. *Inorg. Chem.*, 41, 3602, 2002.

82. Panda, A.B., Glaspell, G., El-Shall, M.S., Microwave synthesis and optical properties of uniform nanorods and nanoplates of rare earth oxides. *J. Phys. Chem. C*, 111, 1861, 2007.
83. Ding, K.L., Miao, Z.J., Liu, Z.M., Zhang, Z.F., Han, B.X., An, G.M., Miao, S.D., Xie, Y., Facile synthesis of high quality TiO2 nanocrystals in ionic liquid via a microwave-assisted process. *J. Am. Chem. Soc.*, 129, 6362, 2007.
84. Wilson, G.J., Will, G.D., Frost, R.L., Montgomery, S.A., Efficient microwave hydrothermal preparation of nanocrystalline anatase TiO_2 colloids. *J. Mater. Chem.*, 12, 1787, 2002.
85. Williams, M.J., Sánchez, E., Aluri, E.R., Douglas, F.J., MacLaren, D.A., Collins, O.M., Cussen, E.J., Budge, J.D., Sanders, L.C., Michaelis, M., Smales, C.M., Cinatl, J., Lorrio, S., Krueger, D., de Rosales, R.T.M., Corr, S.A., Microwave-assisted synthesis of highly crystalline, multifunctional iron oxide nanocomposites for imaging applications. *RSC Adv.*, 6, 83520, 2016.
86. Karthik, K., Dhanuskodi, S., Gobinath, C., Prabukumar, S., Sivaramakrishnan, S., Multifunctional properties of microwave assisted CdO–NiO–ZnO mixed metal oxide nanocomposite: Enhanced photocatalytic and antibacterial activities. *J. Mater. Sci.: Mater. Electron.*, 29, 13 pp, 2018.
87. Byrappa, K. and Yoshimura, M., *Handbook of hydrothermal technology*, William Andrew, Norwich, 2001.
88. Hayashi, H. and Hakuta, Y., Hydrothermal synthesis of metal oxide nanoparticles in supercritical water. *Materials*, 3, 3794, 2010.
89. Wang, X. and Li, Y., Selected-control hydrothermal synthesis of α- and $β-MnO_2$ single crystal nanowires. *J. Am. Chem. Soc.*, 124, 2880, 2002.
90. Tan, B. and Wu, Y., Dye-sensitized solar cells based on anatase TiO_2 nanoparticle/ nanowire composites. *J. Phys. Chem. B*, 110, 15932, 2006.
91. Yang, J., Quan, Z., Kong, D., Liu, X., Lin, J., Y_2O_3: Eu^{3+} microspheres: Solvothermal synthesis and luminescence properties. *Cryst. Growth Des.*, 7, 730, 2007.
92. Liu, B. and Zeng, H.C., Hydrothermal synthesis of ZnO nanorods in the diameter regime of 50 nm. *J. Am. Chem. Soc.*, 125, 4430, 2003.
93. Lu, F., Cai, W.P., Zhang, Y.G., ZnO hierarchical micro/nanoarchitectures: Solvothermal synthesis and structurally enhanced photocatalytic performance. *Adv. Funct. Mater.*, 18, 1047, 2008.
94. Lu, C.L., Lv, J.G., Xu, L., Guo, X.F., Hou, W.H., Hu, Y., Huang, H., Crystalline nanotubes of γ-AlOOH and $γ-Al_2O_3$: Hydrothermal synthesis, formation mechanism and catalytic performance. *Nanotechnology*, 20, 215604 (9pp), 2009.
95. Ge, S., Shi, X., Sun, K., Li, C., Uher, C., Baker, J.R., Banaszak Holl, M.M., Orr, B.G., Facile hydrothermal synthesis of iron oxide nanoparticles with tunable magnetic properties. *J. Phys. Chem. C*, 113, 13593, 2009.
96. Ruixue, W., Youzhi, X., Mengzhou, X., Hollow iron oxide nanomaterials: Synthesis, functionalization, and biomedical applications. *J. Mater. Chem. B*, 9, 1965, 2021.

97. Hench, L.L. and West, J.K., The sol-gel process. *Chem. Rev.*, 90, 33, 1990.
98. Niederberger, M. and Pinna, N., *Metal oxide nanoparticles in organic solvents: Synthesis, formation, assembly and application*, Springer, New York, 2009.
99. Taimur, A., Sang, I.S., Jeong, O.K., Synthesis and characterization of fluoro- and chloro-bimetallic alkoxides as precursors for luminescent metal oxide materials via sol-gel technique. *Chin. J. Chem.*, 25, 998, 2007.
100. Velardi, L., Scrimieri, L., Serra, A., Manno, D., Calcagnile, L., The synergistic role of pH and calcination temperature in sol–gel titanium dioxide powders. *Appl. Phys. A*, 125, 735 (7pp.), 2019.
101. Isley, S.L. and Penn, R.L., Titanium dioxide nanoparticles: Effect of sol-gel pH on phase composition, particle size, and particle growth mechanism. *J. Phys. Chem. C*, 112, 4469, 2008.
102. Massart, R., Preparation of aqueous magnetic liquids in alkaline and acidic media. *IEEE Trans. Magn.*, 17, 1247, 1981.
103. Kim, D.K., Zhang, Y., Voit, W., Rao, K.V., Muhammed, M., Synthesis and characterization of surfactant-coated superparamagnetic monodispersed iron oxide nanoparticles. *J. Magn. Magn. Mater.*, 225, 30, 2001.
104. Costenaro, D., Carniato, F., Gatti, G., Marchesea, L., Bisio, C., Preparation of luminescent ZnO nanoparticles modified with aminopropyltriethoxy silane for optoelectronic applications. *New J. Chem.*, 37, 2103, 2013.
105. Liying, H.E., Yumin, S.U., Jiang, L., Shikao, S.H.I., Recent advances of cerium oxide nanoparticles in synthesis, luminescence and biomedical studies: A review. *J. Rare Earths*, 33, 791, 2015.
106. Janjua, M.R.S.A., Synthesis of Co_3O_4 nano aggregates by co-precipitation method and its catalytic and fuel additive applications. *Open Chem.*, 17, 865, 2019.
107. Hui, B.H. and Salimi, M.N., Production of iron oxide nanoparticles by co-precipitation method with optimization studies of processing temperature, pH and stirring rate. *2020 IOP Conf. Ser.: Mater. Sci. Eng.*, vol. 743, p. 012036, 2021.
108. Rajakumar, G., Mao, L., Bao, T., Wen, W., Wang, S., Gomathi, T., Gnanasundaram, N., Rebezov, M., Shariati, M.A., Chung, I.-M., Thiruvengadam, M., Zhang, X., Yttrium oxide nanoparticle synthesis: An overview of methods of preparation and biomedical applications. *Appl. Sci.*, 11, 2172 (24 pp.), 2021.
109. Price, G., *Current trends in sonochemistry*, G.J. Price (Ed.), p. 87, Royal Society of Chemistry, Cambridge, 1992.
110. Leighton, T.G., *The acoustic bubble*, Academic Press, London, 1994, [2 [3]].
111. Young, F.R., *Cavitation*, McGraw-Hill, London, 1989.
112. Henglein, A., Contributions to various aspects of cavitation chemistry. *Adv. Sonochem.*, 3, 17, 1993.
113. Rae, J., Ashokkumar, M., Eulaerts, O., Sonntag, C.V., Reisse, J., Grieser, F., Estimation of ultrasound induced cavitation bubble temperatures in aqueous solutions. *Ultrason. Sonochem.*, 12, 325, 2005.

114. Wang, C., Cheng, H., Sun, Y., Lin, Q., Zhang, C., Rapid sonochemical synthesis of luminescent and paramagnetic copper nanoclusters for bimodal bioimaging. *ChemNanoMat*, 1, 1, 27, 2015.
115. Xiong, H.-M., Shchukin, D.G., Möhwald, H., Xu, Y., Xia, Y.-Y., Sonochemical synthesis of highly luminescent zinc oxide nanoparticles doped with magnesium(II). *Angew. Chem. Int. Ed.*, 48, 2727, 2009.
116. Yin, L.X., Wang, Y.Q., Pang, G.S., Koltypin, Y., Gedanken, A., Sonochemical synthesis of cerium oxide nanoparticles-effect of additives and quantum size effect. *J. Colloid Interface Sci.*, 246, 78, 2002.
117. Reaz, M., Haque, A., Cornelison, D.M., Wanekaya, A., Delong, R., Ghosh, K., Magneto-luminescent zinc/iron oxide core-shell nanoparticles with tunable magnetic properties. *Physica E: Low-Dimens. Syst. Nanostruct.*, 123, 114090, 2020.
118. Shchukin, D.G. and Möhwald, H., Self-repairing coatings containing active nanoreservoirs. *Small*, 3, 926, 2007.
119. Zhang, L. and Xia, Y.N., Scaling up the production of colloidal nanocrystals: Should we increase or decrease the reaction volume. *Adv. Mater.*, 26, 2600, 2014.
120. Volk, A.A., Epps, R.W., Abolhasani, M., Accelerated development of colloidal nanomaterials enabled by modular microfluidic reactors: Toward autonomous robotic experimentation. *Adv. Mater.*, 33, 2004495, 2021.
121. Roberts, E.J., Karadaghi, L.R., Wang, L., Malmstadt, N., Brutchey, R.L., Continuous flow methods of fabricating catalytically active metal nanoparticles. *ACS Appl. Mater. Interfaces*, 11, 27479, 2019.
122. Breen, C.P., Nambiar, A.M.K., Jamison, T., Jensen, K., Ready set, flow! Automated continuous synthesis and optimization. *Trends Chem.*, 3, 373, 2021.
123. Shen, J., Shafiq, M., Ma, M., Chen, H.R., Synthesis and surface engineering of inorganic nanomaterials based on microfluidic technology. *Nanomaterials*, 10, 1177, 2020.
124. Köhler, J.M., Li, S., Knauer, A., Why is micro segmented flow particularly promising for the synthesis of nanomaterials? *Chem. Eng. Technol.*, 36, 887, 2013.
125. Hartman, R.L., McMullen, J.P., Jensen, K.F., Deciding whether to go with the flow: Evaluating the merits of flow reactors for synthesis. *Angew. Chem. Int. Ed.*, 50, 7502, 2011.
126. Zhang, L. and Xia, Y., Scaling up the production of colloidal nanocrystals: Should we increase or decrease the reaction volume? *Adv. Mater.*, 26, 2600, 2014.
127. Duraiswamy, S. and Khan, S.A., Droplet-based microfluidic synthesis of anisotropic metal nanocrystals. *Small*, 5, 2828–2834, 2009.
128. Zhang, L., Niu, G.D., Lu, N., Wang, J.G., Tong, L.M., Wang, L.D., Kim, M.J., Xia, Y.N., Continuous and scalable production of well-controlled noble-metal nanocrystals in milliliter-sized droplet reactors. *Nano Lett.*, 14, 6626, 2014.

129. Jiao, M.X., Jing, L.H., Wei, X.J., Liu, C.Y., Luo, X.L., Gao, M.Y., The Yin and Yang of coordinating co-solvents in the size-tuning of Fe_3O_4 nanocrystals through flow synthesis. *Nanoscale*, 9, 18609, 2017.
130. Abdel-Latif, K., Bateni, F., Crouse, S., Abolhasani, M., Flow synthesis of metal halide perovskite quantum dots: From rapid parameter space mapping to AI-guided modular manufacturing. *Matter*, 3, 1053, 2020.
131. Koirala, R., Pratsinis, S.E., Baiker, A., Synthesis of catalytic materials in flames: Opportunities and challenges. *Chem. Soc. Rev.*, 45, 3053, 2016.
132. Schimmoeller, B., Pratsinis, S.E., Baiker, A., Flame aerosol synthesis of metal oxide catalysts with unprecedented structural and catalytic properties. *ChemCatChem*, 3, 1234, 2011.
133. Graves, B., Engelke, S., Jo, C., Baldovi, H.G., De La Verpilliere, J., De Volder, M., Boies, A., Plasma production of nanomaterials for energy storage: Continuous gas-phase synthesis of metal oxide CNT materials via a microwave plasma. *Nanoscale*, 12, 5196, 2020.
134. Yu Vasilyev, A., Domrina, E.S., Kaufman, S.V., Maiorova, A.I., Classification of atomization devices. *J. Phys. Conf. Ser.*, 1359, 012131, 2019.
135. Wang, W.-N., Purwanto, A., Lenggoro, I.W., Okuyama, K., Chang, H., Jang, H.D., Investigation on the correlations between droplet and particle size distribution in ultrasonic spray pyrolysis. *Ind. Eng. Chem. Res.*, 47, 1650, 2008.
136. Kim, H.-U., Kulkarni, A., Ha, S., Shin, D., Kim, T., Note: Electric field assisted megasonic atomization for size-controlled nanoparticles. *Rev. Sci. Instrum.*, 88, 076106, 2017.
137. González-Carreño, T., Morales, M.P., Serna, C.J., Barium ferrite nanoparticles prepared directly by aerosol pyrolysis. *Mater. Lett.*, 43, 97, 2020.
138. Basak, S., Rane, K.S., Biswas, P., Hydrazine-assisted, low-temperature aerosol pyrolysis method to synthesize γ-Fe_2O_3. *Chem. Mater.*, 20, 4906, 2008.
139. Nie, P., Xu, G., Jiang, J., Dou, H., Wu, Y., Zhang, Y., Wang, J., Shi, M., Fu, R., Zhang, X., Aerosol-spray pyrolysis toward preparation of nanostructured materials for batteries and supercapacitors. *Small Methods*, 2017, 1700272 (24 pp.), 017.
140. Lide, D.R. (Ed.), *Handbook of chemistry and physics*, p. 84, CRC Press, Boca Raton, FL, 2003.
141. Feldmann, C. and Jungk, H.-O., Polyol-mediated preparation of nanoscale oxide particles. *Angew. Chem. Int. Ed.*, 40, 359, 2001.
142. Feldmann, C., Polyol-mediated synthesis of nanoscale functional materials. *Adv. Funct. Mater.*, 3, 101, 2003.
143. Feldmann, C., Polyol-mediated synthesis of nanoscale functional materials. *Solid State Sci.*, 7, 868, 2005.
144. Ammar, S. and Fievet, F., Polyol synthesis: A versatile wet-chemistry route for the design and production of functional inorganic nanoparticles. *Nanomaterials*, 10, 1217 (8 pp.), 2020.

145. Shah, A.H., Zargar, R.A., Arora, M., Sundar, P.B., Fabrication of pulsed laser-deposited Cr-doped zinc oxide thin films: Structural, morphological, and optical studies. *J. Mater. Sci.: Mater. Electron.*, 31, 21193–21202, 2020.
146. Ivanova, O.S., Gasymova, G.A., Shcherbakov, A.B., Ivanov, V.K., Baranchikov, A.E., Gil', D.O., Tretyakov, Y.D., Polyol-mediated synthesis of nanocrystalline ceria doped with neodymium, europium, gadolinium, and ytterbium. *Dokl. Chem.*, 443, 82, 2012.
147. Ungelenk, J., Speldrich, M., Dronskowski, R., Feldmann, C., Polyol-mediated low-temperature synthesis of crystalline tungstate nanoparticles MWO_4 (M = Mn, Fe, Co, Ni, Cu, Zn). *Solid State Sci.*, 31, 62, 2014.
148. LaMer, V.K. and Dinegar, R.H., Theory, production and mechanism of formation of monodispersed hydrosols. *J. Am. Chem. Soc.*, 72, 4847, 1950.
149. Park, J., An, K., Hwang, Y., Park, J.-G., Noh, H.-J., Kim, J.-Y., Park, J.-H., Hwang, N.-M., Hyeon, T., Ultra-large-scale syntheses of monodisperse nanocrystals. *Nanomaterials*, 3, 891, 2004.
150. Pan, D., Wang, Q., Jiang, S., Ji, X., An, L., Controlled synthesis of monodisperse nanocrystals by a two-phase approach without the separation of nucleation and growth. *J. Mater. Chem.*, 19, 1063, 2009.
151. Kockmann, A., Hesselbach, J., Zellmer, S., Kwade, A., Garnweitner, G., Facile surface tailoring of metal oxide nanoparticles via a two-step modification approach. *RSC Adv.*, 5, 60993, 2015.
152. Nguyen, T.-D. and Do, T.-O., General two-phase routes to synthesize colloidal metal oxide nanocrystals: Simple synthesis and ordered self-assembly structures. *J. Phys. Chem. C*, 113, 11204, 2009.
153. Pan, D., Zhao, N., Wang, Q., Jiang, S., Ji, X., An, L., Facile synthesis and characterization of luminescent TiO_2 nanocrystals. *Adv. Mater.*, 17, 1991, 2005.
154. Zhao, N., Pan, D., Nie, W., Ji, X., Two-phase synthesis of shape-controlled colloidal zirconia nanocrystals and their characterization. *J. Am. Chem. Soc.*, 128, 10118, 2006.
155. Nguyen, T.-D., Mrabet, D., Do, T.-O., Controlled self-assembly of Sm_2O_3 nanoparticles into nanorods: Simple and large scale synthesis using bulk Sm_2O_3 powders. *J. Phys. Chem. C*, 112, 15226, 2008.
156. Yulizar, Y., Abdullah, I., Surya, R.M., Parwati, N., Apriandanu, D.O.B., Two-phase synthesis of $NiCo_2O_4$ nanoparticles using Bryophyllumpinnatum (Lam) Oken leaf extract with superior catalytic reduction of 2,4,6-trinitrophenol. *Mater. Lett.*, 311, 131465, 2022.
157. Malik, M.A., Wani, M.Y., Hashim, M.A., Microemulsion method: A novel route to synthesize organic and inorganic nanomaterials. *Arab. J. Chem.*, 5, 397, 2012.
158. Richard, B., Lemyre, J.-L., Ritcey, A.M., Nanoparticle size control in microemulsion synthesis. *Langmuir*, 33, 19, 4748, 2017.
159. Eastoe, J., Hollamby, M.J., Hudson, L., Recent advances in nanoparticle synthesis with reversed micelles. *Adv. Colloid Interface Sci.*, 128–130, 5, 2006.

160. López-Quintela, M.A., Tojo, C., Blanco, M.C., García, R.L., Leis, J.R., Microemulsion dynamics and reactions in microemulsions. *Curr. Opin. Colloid Interface Sci.*, 9, 264, 2004.
161. Sunaina, Sethi, V., Mehta, S.K., Ganguli, A.K., Vaidya, S., Understanding the role of co-surfactants in microemulsions on the growth of copper oxalate using SAXS. *Phys. Chem. Chem. Phys.*, 21, 336, 2019.
162. Sharma, S., Yadav, N., Chowdhury, P.K., Ganguli, A.K., Controlling the microstructure of reverse micelles and their templating effect on shaping nanostructures. *J. Phys. Chem. B*, 119, 11295, 2015.
163. Adithya, G.T., Rangabhashiyam, S., Sivasankari, C., Lanthanum-iron binary oxide nanoparticles: As cost-effective fluoride adsorbent and oxygen gas sensor. *Microchem. J.*, 148, 364, 2019.
164. Salvador, M., Gutiérrez, G., Noriega, S., Moyano, A., Blanco-López, M.C., Matos, M., Microemulsion synthesis of superparamagnetic nanoparticles for bioapplications. *Int. J. Mol. Sci.*, 22, 427, 2021, https://doi.org/10.3390/ijms22010427.
165. Tiseanu, C., Parvulescu, V.I., Boutonet, M., Cojocaru, B., Primus, P.A., Teodorescu, C.M., Solans, C., Dominguez, M.S., Surface versus volume effects in luminescent ceria nanocrystals synthesized by an oil-in-water microemulsion method. *Phys. Chem. Chem. Phys.*, 13, 17135, 2011.
166. Pemartin-Biernath, K., Vela-González, A.V., Moreno-Trejo, M.B., Leyva-Porras, C., Castañeda-Reyna, I.E., Juárez-Ramírez, I., Solans, C., Sánchez-Domínguez, M., Synthesis of mixed Cu/Ce oxide nanoparticles by the oil-in-water microemulsion reaction method. *Materials*, 9, 480, 2016.
167. Ali, G., Park, Y.J., Kim, J.W., Cho, S.O., A green, general and ultra-fast route for the synthesis of diverse metal-oxide nanoparticles with controllable sizes and enhanced catalytic activity. *ACS Appl. Nano Mater.*, 1, 6112, 2018.
168. Singh, J., Dutta, T., Kim, K.H., Rawat, M., Samddar, P., Kumar, P., "Green" synthesis of metals and their oxide nanoparticles: Applications for environmental remediation. *J. Nanobiotechnol.*, 16, 1, 2018.
169. Sheldon, R.A., Arends, I.W.C.E., Hanefeld, U., *Green chemistry and catalysis*, Wiley-VCH Verlag GmbH & Co. KGaA, Weinheim, Germany, 2007.
170. Anastas, P.T. and Warner, J.C., *Green chemistry: Theory and practice*, Oxford University Press, New York, NY, USA, 1998.
171. Vanathi, P., Rajiv, P., Narendhran, S., Rajeshwari, S., Rahman, P.K.S.M., Venckatesh, R., Biosynthesis and characterization of phyto mediated zinc oxide nanoparticles: A green chemistry approach. *Mater. Lett.*, 134, 13, 2014.
172. Chen, L., Batjikh, I., Hurh, J., Han, Y., Huo, Y., Ali, H., Li, J.F., Rupa, E.J., Ahn, J.C., Mathiyalagan, R., Yang, D.C., Green synthesis of zinc oxide nanoparticles from root extract of Scutellariabaicalensis and its photocatalytic degradation activity using methylene blue. *Optik*, 184, 324, 2019.
173. Sharma, J.K., Akhtar, M.S., Ameen, S., Srivastava, P., Singh, G., Green synthesis of CuO nanoparticles with leaf extract of calotropis gigantea and its dye-sensitized solar cells applications. *J. Alloys Compd.*, 632, 321, 2015.

174. Suresh, S., Ilakiya, R., Kalaiyan, G., Thambidurai, S., Kannan, P., Prabu, K.M., Suresh, N., Jothilakshmi, R., Karthick Kumar, S., Kandasamy, M., Green synthesis of copper oxide nanostructures using cynodondactylon and cyperus rotundus grass extracts for antibacterial applications. *Ceram. Int.*, 46, 12525, 2020.
175. Nasrollahzadeh, M. and Sajadi, S.M., Synthesis and characterization of titanium dioxide nanoparticles using euphorbia heteradenaJaub root extract and evaluation of their stability. *Ceram. Int.*, 41, 14435, 2015.
176. McKittrick, J. and Shea-Rohwer, L.E., Review: Down conversion materials for solid-state lighting. *J. Am. Ceram. Soc.*, 9, 1327, 2014.
177. Han, J.K., Choi, J.I., Piquette, A., Hannah, M., Anc, M., Galvez, M., Talbot, J.B., McKittrick, J., Phosphor development and integration for near-UV LED solid state lighting. *ECS J. Solid State Sci. Technol.*, 2, R3138, 2012.
178. Sternig, A., Bernardi, J., McKenna, K., Diwald, O., Surface-specific visible light luminescence from composite metal oxide nanocrystals. *J. Mater. Sci.*, 50, 8153, 2015.
179. Haque, S.A., Koops, S., Tokmoldin, N., Durrant, J.R., Huang, J., Bradley, D.D.C., Palomares, E., A multilayered polymer light-emitting diode using a nanocrystalline metal-oxide film as a charge-injection electrode. *Adv. Mater.*, 19, 683, 2007.
180. Jayasimhadri, M., Ratnam, B.V., Jang, K., Lee, H.S., Chen, B., Yi, S.S., Jeong, J.H., Moorthy, L.R., Combustion synthesis and luminescent properties of nano and submicrometer-size $Gd_2O_3:Dy^{3+}$ phosphors for white LEDs. *Int. J. Appl. Ceram. Technol.*, 8, 709, 2011.
181. Gupta, S.K., Zuniga, J.P., Abdou, M., Thomas, M.P., Goonatilleke, M.D.A., Guiton, B.S., Mao, Y., Lanthanide-doped lanthanum hafnate nanoparticles as multicolor phosphors for warm white lighting and scintillators. *Chem. Eng. J.*, 379, 122314, 2020.
182. Zargar, R.A., Arora, M., Bhat, R.A., Study of nanosized copper-doped ZnO dilute magnetic semiconductor thick films for spintronic device applications. *Appl. Phys. A*, 124, 1–9, 2018.
183. Zargar, R.A. et al., ZnCdO thick film films: Screen printed TiO2 film: A candidate for photovoltaic applications. *Mater. Res. Express*, 7, 065904, 2020.
184. Nikolova, M.P. and Chavali, M.S., Metal oxide nanoparticles as biomedical materials. *Biomimetics*, 5, 27(47pp.), 2020.
185. Tufani, A., Qureshi, A., Niazi, J.H., Iron oxide nanoparticles based magnetic luminescent quantum dots (MQDs) synthesis and biomedical/biological applications: A review. *Mater. Sci. Eng.: C*, 118, 111545 (21 pp.), 2020.
186. Gautam, M., Kim, J.O., Yong, C.S., Fabrication of aerosol-based nanoparticles and their applications in biomedical fields. *J. Pharm. Invest.*, 51, 361, 2021.
187. Guzman, M., Flores, B., Malet, L., Godet, S., Synthesis and characterization of zinc oxide nanoparticles for application in the detection of fingerprints. *Mater. Sci. Forum*, 916, 232, 2018.

4

Status, Challenges and Bright Future of Nanocomposite Metal Oxide for Optoelectronic Device Applications

Ajay Singh[1]*, Sunil Sambyal[2], Vishal Singh[3], Balwinder Kaur[4] and Archana Sharma[5]

[1]*Department of Physics, GGM Science College, (Constituent College of Cluster University of Jammu) Jammu, India*
[2]*Department of Physics, School of Applied Sciences, Shri Venkateshwara University, Gajraula, UP, India*
[3]*Department of Nano Sciences and Materials, Central University of Jammu, Bagla Rahya Suchani, Jammu, India*
[4]*Department of Physics, Govt. Degree College, R.S. Pura, Jammu, India*
[5]*Department of Physics, University of Jammu, Jammu, J&K (UT), India*

Abstract

Semiconductor metal oxides possess excellent properties which make them promising candidate for optoelectronic and photonic applications. Development of nanotechnology opens lot of possibilities to manufacture a great variety of different nanostructures of metal oxides. By tuning/controlling parameters like size, shape, composition, crystallinity, and band gap, one can tune the metal oxides for the chosen application. Electronic structure, and charge transport mechanism are the properties that make MOs different and fascinating from conventional semiconductors and group III–V compounds. Recently, lot of advances using MOs as a core materials have been achieved including synthesis of new transparent conducting oxides (TCO), p-type and n-type MO semiconductors for transistors, p–n junctions, printing MO electronics and, most importantly, commercialization of amorphous oxide semiconductors for flat panel displays. This chapter focuses on status, challenges, and bright future of nanocomposite metal oxide for optoelectronic device applications. In optoelectronic application of MOs, the focus will be on light emitting devices, photodetectors, solar-cells, and transparent conducting layers. In addition to this, the physics and chemistry of MOs versus other conventional electronic materials will be discussed.

---

*Corresponding author: ajay.dadwal1234@gmail.com

Rayees Ahmad Zargar (ed.) Metal Oxide Nanocomposite Thin Films for Optoelectronic Device Applications, (101–128) © 2023 Scrivener Publishing LLC

Keywords: Semiconductor metal oxide, optoelectronic devices, nanocomposites, chemical technique

Abbreviations

AES	Auger electron spectroscopy
ALD	Atomic layer deposition
AMLCD	Active matrix liquid crystal display
AMOLED	Active matrix organic light emitting diode
AOS	Amorphous nano oxide semiconductor
CB	Conduction band
CBD	Chemical bath deposition
CBM	Conduction band minimum
CTL	Charge transport layer
DC	Down converting
DOS	Density of state
DSSCs	Dye sensitized solar cells
ECD	Electrochemical deposition
EDS	Energy dispersive spectroscopy
EL	Electro luminescence
ETL	Electron transport layer
FTIR	Fourier – transform infra-red
FTO	Fluorine doped tin oxide
HOMO	Highest occupied molecular orbital
H-P-G	Host-passivation-guest
HTL	Hole transport layer
IGZO	Indium gallium zinc oxide
ITO	Indium-tin-oxide
LECBD	Low energy cluster beam deposition
LEDs	Light emitting diode
LUMO	Lowest occupied molecular orbital
MCP	Mechanochemical processing
MO NPs	Metal oxide nanoparticles
MOs	Metal oxides
MOSFET	Metal oxide semiconductor field effect transistor
OLED	Organic light emitting diode
OPV	Organic photovoltaic cells
PCE	Power conversion efficiency
PL	Photo luminescence

PVS	Physical vapor synthesis
RBS	Rutherford backscattering spectroscopy
RL	Radio luminescence
SEM	Scanning electron microscopy
SILAR	Successive ionic layer adsorption and reaction
SIMS	Secondary ion mass spectrometry
SMO	Semiconducting metal oxide
TCE	Transparent conductive electrode
TCO	Transport conducting layer
TEM	Transmission electron microscope
TF	Thin film
TFTs	Thin film transistors
UP	Up conversion
VBM	Valence band maximum
WF	Work function
XPS	X-ray photoelectron spectroscopy
XRD	X-Ray diffraction

4.1 Introduction

The most abundant compounds exist in all three state of matter i.e., in gas, liquid and solid on the earth is Oxides consisting of one oxygen and one other element. A very common example of oxides in liquid state is hydrogen oxide (H_2O) known as water, a universal solvent and a gas oxide is carbon dioxide (CO_2). A familiar example of solid oxide is silicon dioxide (SiO_2) also called silica. In the modern semiconductor industry [1, 2], solid oxides proved to be indispensable after the advent of first MOSFET in the late 1950s. In a metal-oxide-semiconductor structure, silicon oxide acts as an insulating layer which is the core of MOSFET. This insulating layer performs a dielectric property resulting in the formation of channel between oxide and semiconductor which prevent electrons or holes to tunnel from semiconductor to metal gate. Oxides are thought to be an insulator; however, some metal oxides such as ZnO, In_2O_3, SnO2, etc. [3] and complicated oxides like Zn-Sn-O, In-Sn-O [4, 5] are semiconductors. Due to superior electrical performance, good transparency, uniformity and excellent stability, semiconducting metal oxide (SMO) are promising candidates for various electronic devices. Thin film transistors (TFTs) are fabricated by deposited semiconductor materials, dielectric materials and conducting materials in the form of thin film on a certain substrate [6]. TFT is used in display products like liquid crystal display (LCD), organic light-emitting diode (OLED), active matrix liquid crystal display (AMLCD), and active matrix organic light-emitting diode (AMOLED). A variety of semiconducting materials are employed to made TFT which is used in AMLCD and AMOLED. TFTs can be made totally transparent by using semiconducting metal oxides as they possess high mobility and transparency under visible light, for example, several groups used ZnO as n-channel to fabricate highly transparent TFTs [7] which shows mobility larger than 1 cm^2 V^{-1}. s^{-1}.

SMO are non-stoichiometric compounds belong to ionic solid having ionic bonds between positive metallic ions and negative oxygen ions. SMO show superior thermal/chemical stability, tunable energy bands, and high dielectric constants [8] due to incomplete filled 'd' shell. As compared to other semiconductor materials, SMO show more intrinsic defects because ionic bonds are easily broken in comparison to covalent bonds [9]. SMO differ in crystalline structure, energy band levels and other physical aspect because they belong to large group of materials. For instance, ZnO exist in hexagonal wurtzite [10] or cubic zinc blende, In_2O_3 crystallize in cubic structure and SnO_2 exist in rutile tetragonal [3]. There are six point defects in SMO, such as, site of metallic atom (M) replaced with oxygen atom (O), a site of O replaced with M, no atom on a site of M, no atom on a site of O, M or O existing in an interstitial region. Undoped SMO exhibits n-type semiconducting properties due to the presence of oxygen vacancies and metallic interstitial. ZnO shows n-type conductivity due to inherent defects instead of intentional doping [11]. Hence, the origin of n-type semiconducting is controversial in some SMO.

Many physical properties of solids such as resistance, and absorption can be explained on the basis of band theory of solids. The band gap is generally wide (> 3 eV) for most of the SMO. Due to wide-band gap, SMO materials have high breakdown voltage and bear large electric field, which means that the electronic devices made up of SMO materials can be operated at high temperature and power. Wide band gap is a desired characteristic in transparent electronics. Thus, SMOs show high optical transparency in the visible region of the electromagnetic spectrum. For example, the band gap of ZnO, In_2O_3, SnO_2 is about 3.4eV, 3.5 eV, 3.6 eV [3].

In addition to band theory, doping is crucial parameter in the variation of electrical and optical properties of semiconductors. Depending upon the type of dopant used, it will introduce addition energy levels which are close to valence band or conduction band. Doped semiconductors have higher carrier concentration as compared to intrinsic semiconductors which results in improved conductivity. Carrier mobility is the key factor for electronic devices and impurity level has direct impact on the carrier mobility. Thus, SMOs are widely used in electronic devices.

This chapter explores advancement in metallic oxide nanoparticles optoelectronic application such as light emitting devices, photodetectors, solar-cells and transparent conducting layers. The focus will be on the uniqueness and universality of MOs for enhancing the performance of optoelectronic devices. This book chapter elaborates the importance of the problem with brief historical, global scenario with past, present and future perspectives with proposed solutions/advantages/disadvantages of MOs for optoelectronic device application.

4.2 Synthesis of Nanocomposite Metal Oxide by Physical and Chemical Routes

A wide range of techniques are used for the preparation of MO nanoparticles such as chemical techniques, physical techniques, and mechanical procedures. These techniques are shown in Figure 4.1 and discussed below.

4.2.1 Synthesis of Metal Oxides Nanoparticles by Chemical Technique

In this technique, supersaturation of liquid medium containing various reactants is achieved by the precipitation of solid phase (NPs). The supersaturation can be created either by adding the necessary reactants or by cooling of a previously heated solution. Various surfactants can be used in chemical synthesis which control the structural shape and size of the nanoparticles by controlling surface tension and also prevents agglomeration of nanoparticles.

(i) Sol-Gel Method
The method of synthesizing metal oxide nanoparticles using a sol-gel is perhaps the most popular. "Sol" refers to the solution containing mixture of metal oxide distributed in an organic solvent. The procedure can be designated as aqueous or non-aqueous depending on whether water is the solvent. The transformation of sol-gel involves the reaction of hydrolysis, condensation, and polymerization. After polymerization of hydroxide, gel is formed which is subjected to heating to obtain required oxide as shown in Figure 4.2. MOs such as TiO_2, ZnO, MgO, CuO, ZrO_2 [12], SnO_2 and many others having size > 20 nm [13] can be synthesized by sol-gel technique.

106 Metal Oxide Nanocomposite Thin Films

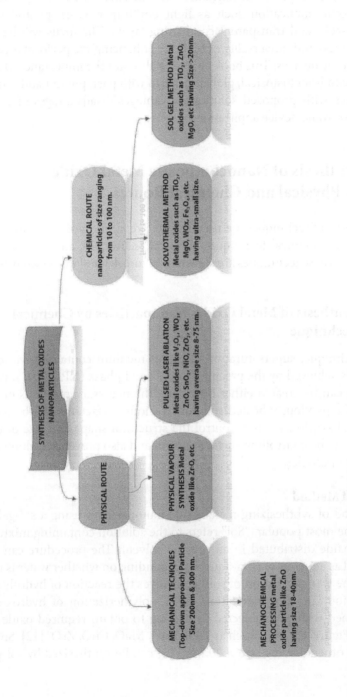

Figure 4.1 Techniques for the synthesis of MOs.

(ii) Solvothermal Method
In this method the starting materials are dispersed in the solvent. The solution is subjected to high temperature and pressure resulting in supersaturation and growth of particle takes place. By varying reactants and thermodynamical parameters such as temperature, pressure, and time, different shape and size of the particles are formed [12]. If solvent used is water, the process is known as "hydrothermal synthesis". MOs such as MgO, TiO_2, Fe_3O_4, WOx, and many others having ultra-small size can be synthesized by solvothermal techniques [14].

4.2.2 Synthesis of Metal Oxides Nanoparticles by Physical Technique

In this technique, a desire shape of a nanostructure is obtained by breaking down a solid precursor into single atoms/molecules. This procedure is expensive since this involves costly vacuum technique and the chamber dimensions put a restriction on the area of deposited layer. The following physical deposition techniques are successfully used for the preparation of MO NPs (nanoparticles):

(i) Physical Vapor Synthesis (PVS)
In this technique, vapor is produced by the application of plasma arc to a solid target which triggers supersaturation reaction. The formation of molecular clusters takes place by cooling down of mixture of resultant atoms and reactive gas which further aggregate in the form of particles [15]. Type of reactant gas used and cooling rate [16] are the parameters which decides the formation of particles. For the formation of ZnO NPs, low energy cluster beam deposition (LECBD) method is used which involves supersonic adiabatic expansion of supersaturated gas [17].

(ii) Pulsed Laser Ablation Method
In this method, the atoms are vaporized from the solid target by using a high-power laser beam. The procedure is carried out in an extremely high or in the presence of gas, such as oxygen. When gas is used, the target might be formed entirely of metal, and as it is being deposited, it oxidizes. MOs such as V_2O_5, WO_3, ZnO, SnO_2 [18], NiO, ZrO_2, Cu/Cu_2O [12] and many others having average size 8-75 nm [18] can be synthesized by this technique.

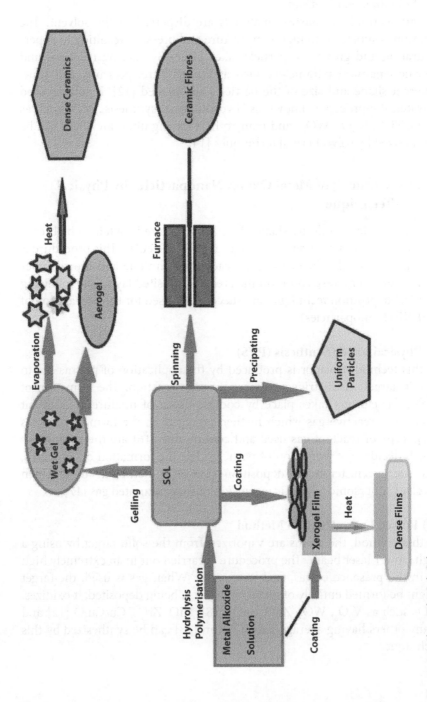

Figure 4.2 Shows the Sol-Gel process.

4.2.3 Synthesis of Metal Oxides by Mechanical Technique

The top-down approach for creating nanostructures is the mechanical techniques. In this technique, the milling of bulk MOs is done to obtain a fine and homogeneous dispersion of particles. The particles produced by this method have dimensions between 200 nm and 300 nm [19, 20]. In order to produce MO NPs of much smaller sizes, mechanochemical processing is often employed.

(i) Mechanochemical Processing (MCP)
In this technique, size reduction is done by using ball mill with a chemical reaction that is initiated during milling [15]. An exchange reaction takes place when the precursors collide in a proper diluent in the ball mill. For further decomposition of compound, heat treatment is applied, for example, $ZnCO_3$ obtained during preparation decomposes to Zinc oxide after heated. Depending upon the milling duration and heating temperature, NPs of ZnO having 18-40 nm in size have been synthesized by using this approach [20].

4.3 Characterization Techniques Used for Metal Oxide Optoelectronics

4.3.1 X-Ray Diffraction (XRD)

To investigate the structural properties of MO materials, XRD is the main technique. With the help of XRD, the information like formation of desired phase of precursor, level of dopant, effect of dopant on phase, peak broadening, crystalline size, and crystallinity of material can be obtained. Fasasi et al. [21] confirm from XRD analysis that after sintering of $BaTiO_3$ and Gd_2O_3, their peaks are not present. To know the efficiency of the preparation technique, XRD analysis is done. To evaluate the level of crystallinity of MOs film deposited on different substrate, XRD analysis is very useful.

4.3.2 Scanning Electron Microscopy (SEM)

To investigate grain size/shape of MOs and of thin films, SEM technique is widely used. Djuriić et al. [22] evaluate sizes and dimensions of nanomorphologies of ZnO by using SEM images and elementary analysis through energy dispersive spectroscopy (EDS). To study the shape and size of multi-layer material such as TCO/blocking layer/electron transport layer [ETL]/hole transport layer (HTL)/thin film of perovskite, SEM is used [23].

4.3.3 Transmission Electron Microscopy (TEM)

To investigate the growth morphology such as columnar, layered, stacked etc. [24], TEM is used. Grain size in thin film depends upon parameters used for their synthesis such as substrate, temperature or speed of pulse. To evaluate grain size, TEM can be useful. To know about the in depth stacking of oxides and lattice deformation, TEM is used. For example, Yu *et al.* [25] evaluate stacking of InO_2 and GaO $(ZnO)_5$ along (0001) direction in $InGaO_3(ZnO)_5$ lattice by employing this technique.

4.3.4 Rutherford Backscattering Spectrometry (RBS)

This technique is used for micro analytical study of MOs. This technique is sensitive to various events (Physical/chemical) occurring at the surface [26]. This technique allows in-depth profiling of MOs up to the resolution of nm. RBS is better technique because RBS does not experience the issue of charging of sample during characterization as is experienced in X-ray photoelectron spectroscopy (XPS), and Secondary ion mass spectrometry (SIMS), RBS not only separate the masses of different elements but also used for compositional depth profiling. This characterization technique is very important as it provides information such as composition, trace defects/impurities, growth mechanism of film, diffusion, etc. which are helpful in understanding optoelectronic applications of MOs.

4.3.5 Fourier-Transform Infra-Red (FTIR)

This technique is used for vibrational probing of MOs. The information regarding chemical, optical constant α (hω) (absorption coefficient) and n (hω) (refractive index), through transmission, adsorption, and reflection measurements can be obtained from this technique. The optical parameters obtained from this characterization is used to calculate band gap by using Tauc plot. In the region $h\omega < E_g$, the Urbach energy can be determined [27]. To ascertain chemical composition of material, FTIR is used. For example, in composite materials, bands at 816 and 775 cm^{-1} are dedicated to the O-WO stretching modes for WO_3 nanoparticles [28] and the bands at 991, 880, and 615 cm^{-1} are assigned to Mo-O, O atoms stretching vibration in the Mo-OMo units and vibration of MoO bonds for MoO_3 nanoparticles respectively [29].

4.3.6 Raman Spectroscopy

For examining the chemical composition, strain, crystal orientation, symmetry, and optical properties of optoelectronic materials, Raman spectroscopy is used which is based on molecular vibration. Raman spectroscopy and FTIR are interchangeable and complementary to each other. The difference between the two techniques is that Raman activity involves two photon transitions, 'a driving and a scattering photon', whereas, FTIR involves single photon transition [30, 31]. To calculate change in lattice and bonds in MOs, Raman spectroscopy is useful. For example, in ZnO/NiO composite, by changing Ni concentration, a change in strain is obtained by observing Raman shift of E_2^{high} of the ZnO lattice [32]. The peak broadening in Raman spectra indicating defects [33], which is very useful measurement in optoelectronic devices.

4.3.7 Luminescence Technique

There are various luminescence methods which utilize the ability of optoelectronic materials to emit light are radio-luminescence (RL), photoluminescence (PL), and electroluminescence (EL). Among them, PL is the most popular characterization techniques. In case of RL, fast particles/ionizing energy bombardment is used [33], in EL, input energy used is an electrical bias [34] and in PL, optical radiation, typically a laser is used. Using RL, various emission features are observed and characterized for HfO_2 nanocrystals [35]. The effect of chemical modification on deposition of ZnO on plastic sheet can be studied using RF technique. To further authenticate and correlate the result, RL can be performed in conjunction with PL [36]. In EL, the excitation is through tunneling process or avalanche [37]. EL is used to study silicon rich oxide/silicon films. Electroluminescence spectroscopy providing information regarding band gap and chromaticity.

In the next section, we will discuss the details of opto-electronic devices based on MOs nanocomposites, their working principle and application.

4.4 Optoelectronic Devices Based on MOs Nanocomposites

4.4.1 Light-Emitting Device

Metal oxides synthesized by solution processed method such as TiO_2 [38], ZnO [39, 40], and ZrO_2 [41] have high stability to oxygen and moisture and are used as electron injection layer (EIL) in OLEDs. Owing to their optoelectronic properties such as mobility and transparency, metal oxides are used in OLEDs. Metal oxides thin film is used in inverted device configuration having Indium-tin oxide (ITO) cathode and metallic anode i.e., Au or Ag. TiO_2 has greater stability, high conductivity, high transparency and conduction band (CB) ~ 3.8 eV which is close to lowest unoccupied molecular orbit (LUMO) level of OLEDs.

For the Protection of conventional device structure from oxygen and moisture, an air stable EIL made up of titanium sub-oxide (TiOx) of approximately 30 nm thick is employed [42]. This layer plays the role of both blocking oxygen and electron injection property.

ZnO film synthesized by solution processed method has CB of approximately 4 eV which is close to the LUMO of light-emitting polymers and is employed as an electron injection layer (EIL) in the inverted OLED [43]. In comparison, OLED having PEDOT: PSS as a hole injection layer (HIL) and barium as an EIL shows less efficiency than ZnO film as an EIL and MoO_3 as HIL. Although, the energy level of ZnO and TiO_2 are similar [38], ZnO shows higher luminance, and efficiency because of superior electron injection property. By using ZnO NPs as an EIL, F8BT-based polymeric light emitting devices (PLEDs) show luminance > 10,000 cd/cm^2 at 7 V [39]. Because of difference in energy barrier of CB of ZnO (4 eV) and LUMO of PLEDs (2-3 eV), various methods are used to increase electron injection from ZnO into the PLED by surface modification of the ZnO with Cs_2CO_3 [44], or barium hydroxide ($Ba(OH)_2$), and others [45]. Surface modification layers of ZnO by an ionic liquid molecule (ILM) [46] such as methyl benzoic acid (BA-CH_3), methoxy benzoic acid (BA-OCH_3) [47], and tetra-n-butyl ammonium tetrafluoborate (TBABF4) [48] remarkably reduces the electron injection barrier between the CB of ZnO and the LUMO level of the light-emitting polymers. The ZnO modified with polyethylene mine (PEI) or and polyethylene mine ethoxylated (PEIE) exhibits reduction of work function ~ 2-3 eV compared with the ZnO without PEI or PEIE. Surface modification procedure is highly recommended for inverted OLEDS.

4.4.2 Photodetector

By combining p-NiO and n-TiO_2 in a metal oxide heterojunction, an entirely transparent photodetector is created. This transparent photodetector operates on its own using the photovoltaic effect and does not require external bias. The carrier collections are controlled by the spontaneously produced electric field, which develop potential between two metal oxide layers. Fluorine-doped tin oxide (FTO) layer served as a back contact, while silver nanowires (Ag NWs) act as the top contact. The photodetector of AgNWs/NiO/TiO$_2$/FTO/Glass was obtained with a comparatively good transparency (44%). The N$_2$-doping procedure was used to execute energy band tuning in order to provide superior performing and broad wavelength photodetection. The transparent photodetector that has been coated with N$_2$ is capable of efficiently detecting of ultraviolet, green, and blue light with high photocurrents of 558 µA^0, 171 µA^0, and 66 µA^0, respectively. The photodetector also attained outstanding responsiveness (136 mA/W), noise-equivalent power (9.2 × 10^{-10} W. Hz$^{-1/2}$), linear dynamic range (34 dB), and detectivity (1.11 × 10^9 Jones). The band energies of metal-oxide having wide-bandgap may be modifying by functional doping, thus increasing their range of detection and enhanced photoelectric devices.

4.4.3 Solar Cell

Due to less expensive and straightforward construction, dye sensitized solar cells (DSSCs) having potential for the replacement of traditional silicon-based solar cells. Other oxide, such as ZnO or SnO$_2$, might be employed in addition to the often-used TiO$_2$ [49–52]. The performance of DSSCs are decided by the nature of material used, morphology [49] and material characteristics (native flaws) [52]. The drawback of DSSCs is that a liquid electrolyte is necessary. As a result, organometallic lead halide perovskite-based solar cells have gained more and more attention [53–55]. Similar to DSSCs, they frequently employ titania, but other metal oxides are of interest as well since they show superior charge transport and may be less expensive than titania.

(i) Cu$_2$O Solar Cells

Copper oxides (Cu$_2$O) are the most widely used MOs compound. The band gap of various phases of copper oxides (CuO$_x$) such as Cu$_4$O$_3$ (paramelaconite), Cu$_2$O (cuprous oxide), and CuO (tenorite or cupric oxide) lies in the range between 1.4 and 2.2 eV [56]. Owing to its low cost, good mobility, wide availability, and high absorption coefficient, Cu$_2$O which is a p-type semiconductor is widely used as a photovoltaic material.

Cu$_2$O has been utilized as an electronic material since 1920. In photoelectrochemical cells, Cu$_2$O is employed as a photocathode because its power conversion efficiency (PCE) was more than 8% in Schottky (heterojunction) solar cells [57]. Heterojunction solar cell based on transparent conducting oxides (TCO) such as ITO/ZnO/Cu$_2$O, shows best efficiency of above 2% [58, 59] but still display lower theoretical efficiency of Cu_2O which is about 19–20%. In homogeneous solar cell, the p-n junction is exclusively made of Cu_2O and has an efficiency of 1.06% [60, 61]. For improving the film photoactivity, n-Cu_2O and p-Cu_2O film deposited by electrodeposition is found to be useful [62]. It was discovered that crystal orientation, grain size and synthesis procedure significantly affected the performance of homojunction Cu_2O solar cells [63].

(ii) Binary Heterojunction Solar Cells

A heterojunction (p-n junction) solar cell in which one of the semiconducting materials has a stronger affinity for electrons and the other semiconducting material has affinity for holes results in the production of local electric field and hence current is produced. A bilayer ZnO oxide [64, 65] is utilized as an oxide window for Cu_2O because ZnO has large direct bandgap ~3.37 eV, strong electron mobility ~120 cm^2 $V^{-1}s^{-1}$, and low-temperature synthesis [66]. Depending upon the deposition techniques used such as atomic layer deposition (ALD) and pulsed layer deposition (PLD), Ga_2O_3/Cu_2O shows enhanced performance of 3.97% and 5.38% as compared to ZnO [67, 68]. MO materials with wide energy bandgap are used in all transparent heterojunction solar cells, however, because of lattice mismatch or other defects, having low performance solar cell.

(iii) Thin-Film Solar Cells

The process of creating a thin-film (TF) solar cell involves depositing photovoltaic material on various substrates. Ferroelectric based thin film solar cells show greater photovoltaic voltage as compared to conventional semiconductors [69]. The first ferroelectric-based thin-film solar cell was reported by Dharmadhikari in 1982 [70]. The most significant ferroelectric material for a variety of applications is barium titanate ($BaTiO_3$ or BTO). Ferroelectric materials, e.g., $Pb(Zr_{1-x}Ti_x)O_3$ and $LiNbO_3$, show photocurrent and voltage which is below the photo-electronics requirements for device application. Thin film made of $BaTiO_3$ (BTO) have been the subject of in-depth research over the years. $BaTiO_3$, although being an environmentally favorable material, has poor thermal stability and loses ferroelectricity at roughly 130 °C due to phase change and strain relaxation. The ferroelectric phase of BTO was stabilized when pores were added to the $BaTiO_3$ and $SrTiO_3$ nanocomposites that induce anisotropic stresses or heterointerfaces [71]. In a different investigation, sol-gel deposition of Nd-doped $BaTiO_3$ thin films is used to reduce the bandgap of BTO material. The average grain size marginally increased when doping with

2% Nd, and it gradually reduced as the Nd concentration increases up to 5%. A shift in bandgap and residual polarization occur at 2% Nd [72]. A deeper understanding is required for improved photoresponse for creating novel materials having higher conductivity, thermal stability, coupled with decreased bandgap and thickness.

(iv) Dye Sensitive Solar cells
The solar cells based on dye and quantum dot sensitized are derived from nanomaterials have sufficient surface area for the adsorption of dye molecules and quantum dots as monolayer to capture maximum number of photon and minimum interface charge recombination. They impart role in optical absorption antenna. The inorganic electrode materials must have exceptional charge mobility and stable long lifetime with light scattering/photon trapping property. From structural point of view, the inorganic nanomaterial has perfect crystalline structure with minimum defects (surface and bulk) and low grain boundaries joining between individual nanostructures. Dye-sensitized solar cells should have large internal surface area, one-dimensional nanostructure solar cell should have optical effects due to antireflection, metal nanoparticles based solar cell should have surface plasmon resonance with nanoparticles agglomeration, and three-dimensional solar cell should have host–passivation–guest (H-P-G) core–shell spherical structures. The efficiency of solar cell can be increased by optimizing porous structure and adsorption statue. The following synthesis methods have been used (i) successive ionic layer adsorption and reaction (SILAR), (ii) chemical bath deposition (CBD), electrochemical deposition (ECD), electrophoresis (EP), and linker-assisted binding, or by combining two or more of these methods. Solar cell having efficiency ~ 4.92 % is achieved in CdS and CdSe quantum dot-sensitized TiO_2 nanocrystalline film passivated by ZnS [73]. An increased efficiency of approximately 5.06 % is achieved in Sb_2S_3-sensitized TiO_2 nanocrystalline film [74]. In dye sensitized solar cells (DSCs), nanowires/nanorods and nanotubes of ZnO have been used with reported values of efficiencies in the range of (1.0 –2.5) % [75] and its value is improved by modifying ZnO nanowires surface with a thin layer of TiO_2 [76]. TiO_2 nanotubes an example of one-dimensional nanostructure in DSC has maximum efficiency 6.9% which can be enhanced to about 7.4% in them modifies nanotubes. Though one-dimensional nanostructures have advantage in electrons conduction as compared to their nanoparticles but the solar efficiencies of such DSCs are much less than that of TiO_2 nanoparticles i.e. about 11–12%.

4.4.4 Charge-Transporting Layers Using Metal Oxide NPs

For the creation of charge transporting layers (CTL), Metal oxides have been employed as bulk and nanostructured materials. This gives the freedom for the development of organic free LED which are more stable and enables the use of higher current densities. Large number of metal oxides find application in charge transport layer such as CuO is used as the hole transporting layer (HTL) [77], nanoparticles of ZrO_2 as the electron transporting layer (ETL) [78], QD LEDs with MoO_3 used as an HTL and TiO_2 used as an ETL [79], and ZnO as the ETL [80]. These layers find application in photodetectors, solar cells, and light generating devices. TiO_2 act as charge carrying layer and its surface area is employed for dye absorption, plays a particularly significant role in organic dye sensitized solar cells (DSSC). The surface area per millimeter of thickness of titania is increased by around 100 times compared to flat films when it has a mesoporous structure, which may be created, for example, by sintering TiO_2 NPs [81]. A bigger active area results in a well packed monolayer of the dye and an improvement in the solar cell's capacity to capture light.

4.4.5 MO NPs as a Medium for Light Conversion

A light converting layer may also be made using the photoluminescence property of NPs. It may be useful in photovoltaics, for example, as typical homojunction solar cells can't exploit the entire sunlight spectrum, with UV and NIR being particularly difficult. As down-converting (DC) agents, ZnO NPs have been proposed because they can absorb UV light and re-emit it in the visible range [82–84]. They can have their luminescence duration and efficiency extended by doping them or combining them with rare earth elements [85].

Additionally, metal oxide nanoparticles absorb low energy near-infrared photons in order to produce visible photon and is used in the up-conversion (UC) process [86]. The MO NPS, for example, ZnO, ZrO_2, Y_2O_3, Lu_2O_3, $BaTiO_3$, and TiO_2 [87] act as a host matrix for rare earth elements as Er, Ho, Nd, and Yb [88] in this application.

4.4.6 Transparent Conducting Oxides (TCO)

According to conventional electro-optical criteria for transparent conductors, the average transmittance must be above 80% in the visible region, and the resistivity must be in the range $10^{-3}\Omega$-cm or less. In order to be acceptable for this application, a semiconductor material must have an energy band gap over 3.0 eV and a carrier concentration of at least 1020 cm^{-3} [89]. These characteristics are shared by a significant number of MOs such as SnO_2, ZnO, In_2O_3, TiO_2, $CdO-In_2O_3$, $CdO-SnO_2$, $MgIn2O4$, $GaInO_3$, $CdO-In_2O_3-SnO_2$, $ZnO-In_2O_3-SnO_2$ etc. Among metal oxides, indium-tin oxide (ITO) is the most popular transparent conductor. The ideal electro-optic characteristics are produced by an atomic ratio of 90/10 between In_2O_3 and SnO_2 [90]. However, this material has some drawbacks, including brittleness [91, 92] and supply scarcity [90].

4.5 Advantages of Pure/Doped Metal Oxides Used in Optoelectronic Device Fabrication

Owing to complex crystal and electronic structures, metal oxides possess optical, optoelectronic, photoelectrochemical, and electrical properties. MOs possess different oxidation state, symmetry, coordination number, and acid-base surface properties which make them fascinating compounds. MOs shows semiconducting, metallic, metallic, insulating, and magnetic behaviors with sudden or continuous transition between these states. Nanodevices can be generated by integrating metal oxide as 0-D quantum dots, 1-D nanowires and nanotubes, thin film coatings [93].

A conventional material such as hydrogenated amorphous silicon (a-Si:H) [94] and thin film transistor using semiconductor are widely used in display backplane electronics but these have limitations for future applications due to their low carrier mobility (~0.5–1.0 cm^{2}V^{-1} s^{-1}), modest flexibility, and optical opacity. MOs on the other hand possess excellent carrier mobilities and high optical transparency and is used in flexible OLED displays and solar cells on plastic substrates.

Metal oxides such as ITO and fluorine-doped tin oxide have superior transparency and conductivity and are used for optoelectronic application. However, they are brittle and hence have less flexibility [95]. To overcome this limitation, thin layer of carbon nanotubes, graphene, silver nanowires, Ti_3C_2 (MXene), and conductive polymers like PEDOT: PSS are used in printed optoelectronic devices [96–100] because these materials have favorable Young's modulus and superior transparency. oxide semiconductors are modified via doping depending on the purposes of device applications, e.g., TFT (Thin Film Transistor) or ITO (Sn-doped In_2O_3) [101].

4.6 Parameters Required for Optoelectronic Devices Applications

The various parameters required for improving the efficiency of metal oxides used for optoelectronic applications are:

(i) Carrier mobility is a major parameter which decides the performance of semiconductor. The conduction band minimum (CBM) and the valence band maximum (VBM) determines the effective masses of electron and holes in MOs. Smaller effective masses mean larger carrier mobilities. Transport in MOs is determined by hybridized $sp\cdot\sigma$-bonding and $sp\cdot\sigma^*$-anti-bonding states [102].

(ii) Band structure and band energy are key factor which determines the performance of MOs. In MOs semiconductor, high conductivity and superior optical transparency seem contradictory [103]. To design p-type MOs, modification of band-structure and band-energy are key strategies. X-ray and photoemission experiments show that greater VBM dispersion is achievable in metal orbitals having energies close to O $2p$ or using more extended anion orbitals, as in $CuGaO_2$ and $SrCu_2O_2$ [104, 105].

(iii) Transparent conductive electrodes (TCE) are another important parameter for determining the performance of organic optoelectronic devices. As properties of TCE depends upon material used and deposition method. ITO is widely used transparent electrode in optoelectronic owing to their excellent optoelectronic properties [106].

(iv) Doping of metal oxides results in the formation of transparent conductive oxides which are used in organic optoelectronic devices. Doping generates interstitial defects, oxygen vacancies which helps the charge carrier transport and results in high conductivity of oxide films [107, 108]. A lot of research is performed to modify the surface of ITO required as TCE in OPV and OLED devices [109–111].

(v) The work function (WF) of an electrode also affects the performance of MOs as it determines the carrier (electron or hole) injection barrier height into the active materials. Several surface treatments have been used to modify the work function for achieving a better device performance in OPV and OLED devices [112–114]. UV-ozone and O_2 plasma treatment are widely used methods for this purpose [115–117].

4.7 Conclusion and Future Perspective of Metal Oxides-Based Optoelectronic Devices

In the past several years, a lot of advancement has been made in optoelectronic devices by capitalizing and implementing unique properties of new metal oxides. The advancement in optoelectronic devices is as follows: (i) in MO the transport phenomena is mainly limited to electrons but the advances in p-type semiconductor opens possibility for p-type TFTs, and oxide CMOS. with this advancement, our understanding about MO changes significantly as to how MO can be used for new functionality. To have better understanding of n- and p-type MO semiconductors, additional fundamental studies are required. (ii) Amorphous nano oxide semiconductor (AOS) shows unique charge transport properties which are not possible in chalcogenides and a-Si:H. (iii) a high performance, flexible, and transparent TFTs, with IGZO (Indium gallium zinc oxide) currently commercialized for OLED, can be achieved by low-temperature film grown on different substrates by various methods. Researchers are exploring MOs to achieve IGZO performance with larger process windows. (iv) various solution-based MO film growth techniques has been developed. For optimal performance, the processing temperature of MOs is higher than organic semiconductor devices. Owing to better transparency and superior stability of MOs, it is imperative that in-depth understanding in these materials will yield high performance as compared to gas-phase growth. Moreover, performance of optoelectronic devices is restricted because of bandgap energy of nano-oxide layer. The loss of energy is due to the low energy of photon or heat loss from high frequency photon. Therefore, multiple excitation generations and band gap tunability is required which has direct effect on photocurrent generation.

Acknowledgement

Dr. Archana Sharma is thankful and acknowledges the financial support provided by the Department of Science and Technology (DST), New Delhi, India, against a financial grant No. SR/WOS-A/PM-3/2018.

References

1. Zargar, R.A., Arora, M., Bhat, R.A., Study of nanosized copper-doped ZnO dilute magnetic semiconductor thick films for spintronic device applications. *Appl. Phys. A*, 124, 1–9, 2018.
2. Lee, M.L., Fitzgerald, E.A., Bulsara, M.T., Currie, M.T., Lochtefeld, A., Strained Si, and Ge channels for high-mobility metal-oxide-semiconductor field-effect transistors. *J. Appl. Phys.*, 97, 1, 011101, 2005.
3. Batzill, M. and Diebold, U., The surface and materials science of tin oxide. *Prog. Surf. Sci.*, 79, 47, 2005.
4. Rajachidambaram, J.S., Sanghavi, S., Nachimuthu, P., Shutthanandan, V., Varga, T., Flynn, B., Thevuthasan, S., Herman, G.S., Characterization of amorphous zinc tin oxide semiconductors. *J. Mater. Res.*, 27, 2309, 2012.
5. Kim, H., Gilmore, C.M., Piqué, A., Horwitz, J.S., Mattoussi, H., Murata, H., Kafafi, Z.H., Chrisey, D.B., Electrical, optical, and structural properties of indium–tin–oxide thin films for organic light-emitting devices. *J. Appl. Phys.*, 86, 6451, 1999.
6. Zargar, R.A., Arora, M., Alshahrani, T., Shkir, M., Screen printed novel ZnO/MWCNT nanocomposite thick films. *Ceram. Int.*, 47, 6084–6093, 2020.
7. Nomura, K., Ohta, H., Ueda, K., Kamiya, T., Hirano, M., Hosono, H., Thin-film transistor fabricated in single-crystalline transparent oxide semiconductor. *Science*, 300, 1269, 2003.
8. Deng, Y., Semiconducting metal oxides for gas sensing, in: *Understanding Semiconducting Metal Oxide Gas Sensors*, Springer, Singapore, 2019.
9. Fisher, E.R., Elkind, J.L., Clemmer, D.E., Georgiadis, R., Loh, S.K., Aristov, N., Sunderlin, L.S., Armentrout, P.B., Reactions of fourth-period metal ions (Ca^+–$Zn+$) with O_2: Metal-oxide ion bond energies. *J. Chem. Phys.*, 93, 2676, 1990.
10. Özgür, Ü., Alivov, Y.I., Liu, C., Teke, A., Reshchikov, M.A., Doğan, S., Avrutin, V., Cho, S.J., Morkoç, H., A comprehensive review of ZnO materials and devices. *J. Appl. Phys.*, 98, 041301, 2005.
11. Look, D.C., Hemsky, J.W., Sizelove, J.R., Residual native shallow donor in ZnO. *Phys. Rev. Lett.*, 82, 2552, 1999.
12. Zargar, R.A., Chackrabarti, Malik, M.H., Sol-gel syringe spray coating: A novel approach for Rietveld, optical and electrical analysis of CdO film for optoelectronic applications. *Phys. Open*, 7, 10069, 2021.
13. Thiagarajan, S., Sanmugam, A., Vikraman, D., Facile methodology of sol-gel synthesis for metal oxide nanostructures, in: *Recent Applications in Sol-Gel Synthesis*, p. 1, 2017.
14. Ren, Y., Liu, Z., Pourpoint, F., Armstrong, A.R., Grey, C.P., Bruce, P.G., Nanoparticulate TiO_2(B): An anode for lithium-ion batteries. *Angew. Chem. Int. Ed.*, 51, 9, 2164, 2012.
15. Espitia, P.J., de Soares, N., Coimbra, J.S., de Andrade, N.J., Cruz, R.S., Medeiros, E.A., Zinc oxide nanoparticles: Synthesis, antimicrobial activity and food packaging applications. *Food Bioprocess Technol.*, 5, 5, 1447, 2012.

16. Casey, P., Nanoparticle technologies and applications, in: *Nanostructure Control of Materials*, p. 1, 2006.
17. Apostoluk, A., Masenelli, B., Tupin, E., Canut, B., Hapiuk, D., Mélinon, P., Delaunay, J.J., Efficient ultraviolet light frequency down-shifting by a thin film of ZnO nanoparticles. *Int. J. Nanosci.*, 11, 04, 1240022, 2012.
18. Huotari, J., Lappalainen, J., Puustinen, J., Baur, T., Alépée, C., Haapalainen, T., Komulainen, S., Pylvänäinen, J., Spetz, A.L., Pulsed laser deposition of metal oxide nanoparticles, agglomerates, and nanotrees for chemical sensors. *Proc. Eng.*, 120, 1158, 2015.
19. Altavilla, C. and Ciliberto, E., *Inorganic nanoparticles: Synthesis, application, and perspectives*, CRC Press, Boca Raton, 2011.
20. Aghababazadeh, R., Mazinani, B., Mirhabibi, A., Tamizifar, M., ZnO nanoparticles synthesised by mechanochemical processing. *J. Phys. Conf. Ser.*, 26, 312, 2006.
21. Zargar, R.A., ZnCdO thick film: A material for energy conversion devices. *Mater. Res. Express*, 6, 9, 095909, 2019.
22. Djuriic, A.B., Ng, A.M.C., Chen, X.Y., ZnO nanostructures for optoelectronics: Material properties and device applications. *Prog. Quantum Electron.*, 34, 4, 191, 2010.
23. Chen, Y., Zhang, L., Zhang, Y., Gao, H., Yan, H., Large-area perovskite solar cells – A review of recent progress and issues. *RSC Adv.*, 8, 19, 10489, 2018.
24. Naghavi, N., Marcel, C., Dupont, L., Rougier, A., Leriche, J.-B., Guéry, C., Structural and physical characterisation of transparent conducting pulsed laser deposited in2o3–zno thin films. *J. Mater. Chem.*, 10, 10, 2315, 2000.
25. Yu, X., Marks, T.J., Facchetti, A., Metal oxides for optoelectronic applications. *Nat. Mater.*, 15, 4, 383, 2016.
26. Perrière, J., Rutherford backscattering spectrometry. *Vacuum*, 9, 429, 1987.
27. Akshay, V.R., Arun, B., Mandal, G., Vasundhara, M., Visible range optical absorption, Urbach energy estimation and paramagnetic response in Cr-doped TiO_2 nanocrystals derived by a sol-gel method. *Phys. Chem. Chem. Phys.*, 21, 24, 12991, 2019.
28. Atta, N.F., Galal, A., El-Ads, E.H., *Perovskite nanomaterials – synthesis, characterization, and applications*, IntechOpen, London, UK, 2016.
29. Dogan, E., Ozkazanc, E., Ozkazanc, H., Multifunctional polyindole/nano-metal-oxide composites: Optoelectronic and charge transport properties. *Synth. Met.*, 256, 116154, 2019.
30. Leng, Y., *Materials characterisation: Introduction to microscopic and spectroscopic methods*, Weiley-VCH, Weinham, 2013.
31. Friedrich, D.M., Two-photon molecular spectroscopy. *J. Chem. Educ.*, 59, 472, 1982.
32. Zargar, R.A., Fabrication and improved response of ZnO-CdO composite flms under diferent laser irradiation dose. *Sci. Rep.*, 12, 10096, 2022.
33. Yi, G.-C., *Semiconductor nanostructures for optoelectronic devices*, Springer, Berlin, Heidelberg, 2012.

34. Jimenez, J. and Tomm, J.W., *Spectroscopic analysis of optoelectronic semiconductors*, Springer International Publishing, Cham, 2016.
35. Villa, I., Lauria, A., Moretti, F., Fasoli, M., Dujardin, C., Niederberger, M., Vedda, A., Radio-luminescence spectral features and fast emission in hafnium dioxide nanocrystals. *Phys. Chem. Chem. Phys.*, 20, 23, 15907, 2018.
36. Zargar, R.A., Kumar, K. et al., Structural, optical, photoluminescence, and EPR behaviour of novel ZnO· 80CdO· 2OO thick films: An effect of different sintering temperatures. *J. Lumin.*, 245, 118769, 2022.
37. Sze, S. and Ng, K.K., *Physics of semiconductors*, Wiley Interscience, New Jersey, 2007.
38. Kabra, D., Song, M.H., Wenger, B., Friend, R.H., Snaith, H.J., High efficiency composite metal oxide-polymer electroluminescent devices: A morphological and material based investigation. *Adv. Mater.*, 20, 3447, 2008.
39. Chiba, T., Pu, Y.J., Hirasawa, M., Masuhara, A., Sasabe, H., Kido, J., Solution-processed inorganic–organic hybrid electron injection layer for polymer light-emitting devices. *ACS Appl. Mater. Interfaces*, 4, 6104, 2012.
40. Chiba, T., Pu, Y.J., Sasabe, H., Kido, J., Yang, Y., Solution-processed organic light-emitting devices with two polymer light-emitting units connected in series by a charge-generation layer. *J. Mater. Chem.*, 22, 22769, 2012.
41. Lone, M.R. et al., Optical properties of ZnO:A simulation study. *J. Nano Electron. Phys.*, 13, 6, 06014, 2021.
42. Lee, K., Kim, J.Y., Park, S.H., Kim, S.H., Cho, S., Heeger, A.J., Air-stable polymer electronic devices. *Adv. Mater.*, 19, 2445, 2007.
43. Bolink, H.J., Coronado, E., Repetto, D., Sessolo, M., Air stable hybrid organic-inorganic light emitting diodes using ZnO as the cathode. *Appl. Phys. Lett.*, 91, 223501, 2007.
44. Park, J.S., Lee, J.M., Hwang, S.K., Lee, S.H., Lee, H.J., Lee, B.R., Park, H.I., Kim, J.S., Yoo, S., Song, M.H., Kim., S.O., A ZnO/N-doped carbon nanotube nanocomposite charge transport layer for high performance optoelectronics. *J. Mater. Chem.*, 22, 12695, 2012.
45. Lu, L.P., Kabra, D., Friend, R.H., Barium hydroxide as an interlayer between zinc oxide and a luminescent conjugated polymer for light-emitting diodes. *Adv. Funct. Mater.*, 22, 4165, 2012.
46. Lee, B.R., Choi, H., SunPark, J., Lee, H.J., Kim, S.O., Kim, J.Y., Song, M.H., Surface modification of metal oxide using ionic liquid molecules in hybrid organic–inorganic optoelectronic devices. *J. Mater. Chem.*, 21, 2051, 2011.
47. Park, J.S., Lee, B.R., Lee, J.M., Kim, J.S., Kim, S.O., Song, M.H., High performance polymer light-emitting diodes with N-type metal oxide/conjugated polyelectrolyte hybrid charge transport layers. *Appl. Phys. Lett.*, 96, 243306, 2010.
48. Youn, H. and Yang, M., Solution processed polymer light-emitting diodes utilizing a ZnO/organic ionic interlayer with Al cathode. *Appl. Phys. Lett.*, 97, 243302, 2010.

49. Sharifi, N., Tajabadi, F., Taghavinia, N., Recent developments in dye-sensitized solar cells. *Chem. Phys. Chem.*, 15, 3902, 2014.
50. Jim, W.Y., Liu, X., Yiu, W.K., Leung, Y.H., Djurišić, A.B., Chan, W.K., Liao, C., Shih, K., Surya, C., The effect of different dopants on theperformance of SnO_2-based dye-sensitized solar cells. *Phys. Status Solidi B*, 252, 3, 553, 2015.
51. Liu, X., Wang, G., Ng, A., Ng, Y.H., Chen, X.Y., Leung, Y.H., Djurišić, A.B., Ng, A.M.C., Chan, W.K., Zinc oxide precursor treatment forimproving dye-sensitized solar cell efficiency. *Phys. Status Solidi B*, 252, 3, 532, 2015.
52. Wong, K.K., Ng, A., Chen, X.Y., Ng, Y.H., Leung, Y.H., Ho, K.H., Djurišić, A.B., Ng, A.M.C., Chan, W.K., Yu, L., Phillips, D.L., Effect of ZnO nanoparticle properties on dye-sensitized solar cell performance. *ACS Appl. Mater. Interfaces*, 4, 1254, 2012.
53. Luo, S. and Daoud, W.A., Recent progress in organic–inorganic halide perovskite solar cells:Mechanisms and material design. *J. Mater. Chem. A*, 3, 8992, 2015.
54. Ren, Z., Ng, A., Shen, Q., Gokkaya, H.C., Wang, J., Yang, L., Yiu, W.-K., Bai, G., Djurišić, A.B., Leung, W. W.-F., Hao, J., Chan, W.K., Surya, C., Thermal assisted oxygen annealing for high efficiency planar CH3NH3PbI3 perovskite solar cells. *Sci. Rep.*, 4, 6752, 1, 2014.
55. Ng, A., Ren, Z., Shen, Q., Efficiency enhancement by defect engineering in perovskite photovoltaic cells prepared using evaporated PbI2/CH3NH3 I multilayers. *J. Mater. Chem. A*, 3, 17, 9223, 2015.
56. Anderson, A.Y., Bouhadana, Y., Barad, H.-N., Kupfer, B., Rosh-Hodesh, E., Aviv, H., Tischler, Y.R., Rühle, S., Zaban, A., Quantum efficiency and bandgap analysis for combinatorial photovoltaics: Sorting activity of Cu–O compounds in all-oxide device libraries. *ACS Comb. Sci.*, 16, 2, 53, 2014.
57. Sullivan, I., Zoellner, B., Maggard, P.A., Copper (I)-based p-type oxides for photoelectrochemical and photovoltaic solar energy conversion. *Chem. Mater.*, 28, 17, 5999, 2016.
58. Georgieva, V., Tanusevski, A., Georgieva, M., Low-cost solar cells based on cuprous oxide, in: *Solar Cells-Thin-Flm Technologies*, InTech, Croatia, 2011.
59. Mittiga, A., Salza, E., Sarto, F., Tucci, M., Vasanthi, R., Heterojunction solar cell with 2% efficiency based on a Cu_2O substrate. *Appl. Phys. Lett.*, 88, 163502, 2006.
60. Olsen, L.C., Addis, F.W., Miller, W., Experimental and theoretical studies of Cu_2O solar cells. *Sol. Cells*, 7, 3, 247, 1982.
61. McShane, C.M. and Choi, K.S., Junction studies on electrochemically fabricated p-n Cu2O homojunction solar cells for efficiency enhancement. *Phys. Chem. Chem. Phys.*, 14, 6112, 2012.
62. Wijesundera, R.P. and Siripala, W., Electrodeposited Cu_2O homojunction solar cells: Fabrication of a cell of high short circuit photocurrent. *Sol. Energy Mater. Sol. Cells*, 157, 881, 2016.
63. Han, K. and Tao, M., Electrochemically deposited p-n homojunction cuprous oxide solar cells. *Sol. Energy Mater. Sol. Cells*, 93, 1, 153, 2009.

64. Minami, T., Miyata, T., Nishi, Y., The relationship between the electrical properties of the n-oxide and p-Cu_2O layers and the photovoltaic properties of Cu_2O-based heterojunction solar cells. *Sol. Energy Mater. Sol. Cells*, 147, 85, 2016.
65. Chen, X., Lin, P., Yan, X., Bai, Z., Yuan, H., Shen, Y., Liu, Y., Zhang, G., Zhang, Z., Zhang, Y., Three-dimensional ordered ZnO/Cu_2O Nano heterojunctions for efficient metal–oxide solar cells. *ACS Appl. Mater. Interfaces*, 7, 5, 3216–23, 2015.
66. Anderson, J. and Van de Walle, C.G., Fundamentals of zinc oxide as a semiconductor. *Rep. Prog. Phys.*, 72, 12, 126501, 2009.
67. Lee, Y.S., Chua, D., Brandt, R.E., Siah, S.C., Li, J.V., Mailoa, J.P., Lee, S.W., Gordon, R.G., Buonassisi, T., Atomic layer deposited gallium oxide buffer layer enables 1.2 V open-circuit voltage in cuprous oxide solar cells. *Adv. Mater.*, 26, 27, 4704, 2014.
68. Musselman, K.P., Wisnet, A., Iza, D.C., Hesse, H.C., Scheu, C., MacManus-Driscoll, J.L., Schmidt-Mende, L., Strong efficiency improvements in ultra-low-cost inorganic nanowire solar cells. *Adv. Mater.*, 22, E254–E258, 35, 2010.
69. Zargar, R.A., Kumar, K., Zakaria, M.M.M., Shkir, M., AlFaify, S., Optical characteristics of ZnO films under different thickness: A MATLAB based computer calculation for photovoltaic applications. *Physica B*, 63, 414634, 2022.
70. Dharmadhikari, V.S. and Grannemann, W.W., Photovoltaic properties of ferroelectric $BaTiO_3$ thin films RF sputter deposited on silicon. *J. Appl. Phys.*, 53, 12, 8988, 1982.
71. Suzuki, N., Osada, M., Billah, M., Bando, Y., Yamauchi, Y., Hossain, S.A., Chemical synthesis of porous barium titanate thin film and thermal stabilization of ferroelectric phase by porosity-induced strain. *J. Vis. Exp.*, 13, 57441, 2018.
72. Jiang, W., Cai, W., Lin, Z., Fu, C., Effects of Nd-doping on optical and photovoltaic properties of barium titanate thin films prepared by sol–gel method. *Mater. Res. Bull.*, 48, 9, 3092, 2013.
73. Hu, L., Huang, Y., Zhang, F., Chen, Q., CuO/Cu_2O composite hollow polyhedrons fabricated from metal–organic framework templates for lithium-ion battery anodes with a long cycling life. *Nanoscale*, 5, 10, 4186, 2013.
74. JA, C., JH, R., SH, I., YH, L., HJ, K., SI, S., MK, N., High-performance nanostructured inorganic-organic heterojunction solar cells. *Nano Lett.*, 10, 7, 2609, 2010.
75. Law, M., Greene, L.E., Johnson, J.C., Saykally, R., Yang, P., Nanowire dye-sensitized solar cells. *Nat. Mater.*, 4, 6, 455, 2005.
76. Matt, L., Lori, E.G., Aleksandra, R., Tevye, K., Jan, L., Peidong, Y., ZnO–Al_2O_3 and ZnO–TiO_2 core–shell nanowire dye-sensitized solar cells. *J. Phys. Chem. B*, 110, 45, 22652, 2006.

77. Ding, T., Yang, X., Bai, L., Zhao, Y., Fong, K.E., Wang, N., Demir, H.V., Sun, X.W., Colloidal quantum-dot leds with a solution-processed copper oxide (CuO) hole injection layer. *Org. Electron.*, 26, 245, 2015.
78. Kim, H.Y., Park, Y.J., Kim, J., Han, C.J., Lee, J., Kim, Y., Greco, T., Ippen, C., Wedel, A., Ju, B.-K., Oh, M.S., Transparent INP quantum dot light-emitting diodes with ZrO_2 electron transport layer and indium zinc oxide top electrode. *Adv. Funct. Mater.*, 26, 20, 3454, 2016.
79. Kwack, Y.-J., Jang, H.-R., Ka, Y., Choi, W.-S., Solution-processed quantum dot leds using molybdenum oxide and titanium oxide as charge transport layers. *J. Nanoelectron. Optoelectron.*, 11, 2, 234, 2016.
80. Portier, J., Hilal, H., Saadeddin, I., Hwang, S., Subramanian, M., Campet, G., Thermodynamic correlations and band gap calculations in metal oxides. *Prog. Solid State Chem.*, 32, 3-4, 207, 2004.
81. Altavilla, C. and Ciliberto, E., *Inorganic nanoparticles*, CRC Press, Boca Raton, 2011.
82. Apostoluk, A., Zhu, Y., Canut, B., Masenelli, B., Delaunay, J.J., Znajdek, K., Sibiński, M., Investigation of luminescent properties of zno nanoparticles for their use as a down-shifting layer on solar cells. *Phys. Status Solidi C*, 10, 10, 1301, 2013.
83. Apostoluk, A., Zhu, Y., Masenelli, B., Delaunay, J.-J., Sibiński, M., Znajdek, K., Focsa, A., Kaliszewska, I., Improvement of the solar cell efficiency by the zno nanoparticle layer via the down-shifting effect. *Microelectron. Eng.*, 127, 51, 2014.
84. Znajdek, K., Sibiński, M., Lisik, Z., Apostoluk, A., Zhu, Y., Masenelli, B., Sędzicki, P., Zinc oxide nanoparticles for improvement of thin film photovoltaic structures' efficiency through down shifting conversion. *Opto-Electron. Rev.*, 25, 2, 99, 2017.
85. Znajdek, K., Szczecińska, N., Sibiński, M., Wiosna-Sałyga, G., Przymęcki, K., Luminescent layers based on rare earth elements for thin-film flexible solar cells applications. *Optik*, 165, 200, 2018.
86. Trupke, T., Green, M.A., Würfel, P., Improving solar cell efficiencies by up-conversion of sub-band-gap light. *J. Appl. Phys.*, 92, 7, 4117, 2002.
87. Kumar, V., Ntwaeaborwa, O.M., Soga, T., Dutta, V., Swart, H.C., Rare earth doped zinc oxide Nanophosphor powder: A future material for solid state lighting and solar cells. *ACS Photonics*, 4, 11, 2613, 2017.
88. Das, R., Khichar, N., Chawla, S., Dual mode luminescence in Rare Earth (Er^{3+}/Ho^{3+}) doped ZnO nanoparticles fabricated by inclusive CO precipitation technique. *J. Mater. Sci.: Mater. Electron.*, 26, 9, 7174, 2015.
89. Minami, T., Transparent conducting oxide semiconductors for transparent electrodes. *Semicond. Sci. Technol.*, 20, 4, S35, 2005.
90. Hecht, D.S., Hu, L., Irvin, G., Emerging transparent electrodes based on thin films of carbon nanotubes, graphene, and metallic nanostructures. *Adv. Mater.*, 23, 13, 1482, 2011.

91. Cairns, D.R., Witte, R.P., Sparacin, D.K., Sachsman, S.M., Paine, D.C., Crawford, G.P., Newton, R.R., Strain-dependent electrical resistance of tin-doped indium oxide on polymer substrates. *Appl. Phys. Lett.*, 76, 11, 1425, 2000.
92. Sibiński, M. and Znajdek, K., Degradation of flexible thin-film solar cells due to a mechanical strain. *Opto-Electron. Rev.*, 25, 1, 33, 2017.
93. Shah, A.H., Zargar, R.A., Arora, M., Sundar, P.B., Fabrication of pulsed laser-deposited Cr-doped zinc oxide thin films: Structural, morphological, and optical studies. *J. Mater. Sci.: Mater. Electron.*, 31, 21193–21202, 2020.
94. Spear, I.W. and Le Comber, P., Substitutional doping of amorphous silicon. *Solid State Commun.*, 17, 1193, 1975.
95. Jin, S.W., Lee, Y.H., Yeom, K.M., Yun, J., Park, H., Jeong, Y.R., Hong, S.Y., Lee, G., Oh, S.Y., Lee, J.H., Noh, J.H., Ha, J.S., Highly durable and flexible transparent electrode for flexible optoelectronic applications. *ACS Appl. Mater. Interfaces*, 10, 36, 30706, 2018.
96. Gao, L., Flexible device applications of 2D semiconductors. *Small*, 13, 35, 1603994, 2017.
97. Kim, C.-L., Jung, C.-W., Oh, Y.-J., Kim, D.-E., A highly flexible transparent conductive electrode based on nanomaterials. *NPG Asia Mater.*, 9, 10, 438, 2017.
98. Liu, J., Zhang, L., Li, C., Highly stable, transparent, and conductive electrode of solution-processed silver nanowire-Mxene for flexible alternating current electroluminescent devices. *Ind. Eng. Chem. Res.*, 58, 47, 21485, 2019.
99. Kim, H. and Alshareef, H.N., MXetronics: MXene-enabled electronic and photonic devices. *ACS Mater. Lett.*, 2, 1, 55, 2020.
100. Zhang, Y.-Z., Wang, Y., Jiang, Q., El-Demellawi, J.K., Kim, H., Alshareef, H.N., MXene printing and patterned coating for device applications. *Adv. Mater.*, 32, 21, 1908486, 2020.
101. Shi, J., Zhang, J., Yang, L., Qu, M., Qi, D.C., Zhang, K.H., Wide bandgap oxide semiconductors: From materials physics to optoelectronic devices. *Adv. Mater.*, 33, 50, 2006230, 2021.
102. Kamiya, T., Nomura, K., Hosono, H., Origins of high mobility and low operation voltage of amorphous oxide TFTs: Electronic structure, electron transport, defects and doping. *J. Disp. Technol.*, 5, 273, 2009.
103. Facchetti, A. and Marks, T.J., *Transparent electronics: From synthesis to application*, Wiley, New York, 2010.
104. Fortunato, E., Barquinha, P., Martins, R., Oxide semiconductor thin-film transistors: A review of recent advances. *Adv. Mater.*, 24, 2945, 2012.
105. Yanagi, H., Kawazoe, H., Kudo, A., Yasukawa, M., Hosono, H., Chemical design and thin film preparation of p-type conductive transparent oxides. *J. Electroceram.*, 4, 407, 2000.
106. Cao, W., Li, J., Chen, H., Xue, J., Transparent electrodes for organic optoelectronic devices: A review. *J. Photonics Energy*, 4, 040990, 2014.
107. Minami, T., Transparent conducting oxide semiconductors for transparent electrodes. *Semicond. Sci. Technol.*, 20, 4, S35, 2005.

108. Irfan, I., Graber, S., So, F., Gao, Y., Interplay of cleaning and de-doping in oxygen plasma treated high work function indium tin oxide (ito). *Org. Electron.*, *13*, 10, 2028, 2012.
109. Xue, J. and Forrest, S.R., Carrier transport in multilayer organic photodetectors. II. Effects of anode preparation. *J. Appl. Phys.*, 95, 4, 1869, 2004.
110. Milliron, D.J., Hill, I.G., Shen, C., Kahn, A., Schwartz, J., Surface oxidation activates indium tin oxide for hole injection. *J. Appl. Phys.*, 87, 1, 572, 2000.
111. Li, C.N., Kwong, C.Y., Djurišić, A.B., Lai, P.T., Chui, P.C., Chan, W.K., Liu, S.Y., Improved performance of OLEDs with ITO surface treatments. *Thin Solid Films*, *477*, 1-2, 57, 2005.
112. Park, Y., Choong, V., Gao, Y., Hsieh, B.R., Tang, C.W., Work function of indium tin oxide transparent conductor measured by photoelectron spectroscopy. *Appl. Phys. Lett.*, *68*, 19, 2699, 1996.
113. Sugiyama, K., Ishii, H., Ouchi, Y., Seki, K., Dependence of indium–tin–oxide work function on surface cleaning method as studied by ultraviolet and X-ray photoemission spectroscopies. *J. Appl. Phys.*, *87*, 1, 295, 2000.
114. Sun, K. and Ouyang, J., Polymer solar cells using chlorinated indium tin oxide electrodes with high work function as the anode. *Sol. Energy Mater. Sol. Cells*, 96, 1, 238, 2012.
115. Hong, Z.R., Liang, C.J., Sun, X.Y., Zeng, X.T., Characterization of organic photovoltaic devices with indium-tin-oxide anode treated by plasma in various gases. *J. Appl. Phys.*, *100*, 9, 093711, 2006.
116. Destruel, P., Bock, H., Séguy, I., Jolinat, P., Oukachmih, M., Bedel-Pereira, E., Influence of indium tin oxide treatment using UV–ozone and argon plasma on the photovoltaic parameters of devices based on organic discotic materials. *Polym. Int.*, 55, 6, 601, 2006.
117. Wu, J., Agrawal, M., Becerril, H.A., Bao, Z., Liu, Z., Chen, Y., Peumans, P., Organic light-emitting diodes on solution-processed graphene transparent electrodes. *ACS Nano*, 4, 1, 43, 2009.

Part II
THIN FILM TECHNOLOGY

5

Semiconductor Metal Oxide Thin Films: An Overview

Krishna Kumari Swain[1*], Pravat Manjari Mishra[2] and Bijay Kumar Behera[3]

[1]*Department of Applied Mechanics, IIT Madras, Chennai, Tamil Nadu, India*
[2]*Department of Environment and Sustainability, CSIR - IMMT, Bhubaneswar, Odisha, India*
[3]*AEBN Division, ICAR-Central Inland Fisheries Research Institute, West Bengal, India*

Abstract

High-throughput fabrication procedures and the selection of materials with acceptable performance characteristics are critical in the development of large-area, low-cost electronics for flat-panel displays, sensor arrays, and flexible circuits. For wide variety of electronics application high charge carrier mobility, high electrical conductivity, large dielectric constants, mechanical flexibility and optical transparency may be required. In this context semiconducting metal oxides (SMO_x) are good choices for selection. Due to their great sensitivity and inexpensive cost, SMO_x are frequently used as the sensitive element in electronic devices. At higher temperature, SMO_xs offer excellent structural stabilities, which helps them to withstand in harsh environmental conditions. When there are modest changes in the environment, SMO_xs quickly change in their electrical characteristics. Because of their quick changes in their electrical properties, they are highly used as in sensing applications. A metal cation and an oxide anion are present in SMO_xs which can also include ternary and quaternary metal oxides. SMO_xs are basically classified in two categories such as n-type and p-type. Despite their high electronic performances, SMO_x suffers difficulties associated with large-area fabrication and its high cost and complexity. To minimize this problem, SMOxs-based next-generation thin films are extensively used in electrical goods due to their elastic and flexible properties. In comparison to thick films, SMOxs thin films have improved surface effects, high specific areas, and high porosity. These characteristics make thin films perfect for a range of uses, such as optoelectronic devices,

---

Corresponding author: swain.krishna1990@gmail.com; https://orcid.org/0000-0001-8505-366X

Rayees Ahmad Zargar (ed.) *Metal Oxide Nanocomposite Thin Films for Optoelectronic Device Applications*, (131–154) © 2023 Scrivener Publishing LLC

catalytic nanomedicine, and particularly chemical sensors. This chapter provides an overview of semiconductor metal oxide based thin films, discusses about the recent developments of SMO_xs thin films and its applications towards sensing field. The first section of the chapter gives an overview of several SMOx based thin films, their properties as well as their synthesis. Also, the chapter discusses about the applications of SMOx based thin films in different fields. Furthermore, the limitation and future developments of SMOx based thin films are addressed in the final part of the chapter.

Keywords: Semiconductors, metal oxides, thin films, sensing mechanisms, nanomaterials

5.1 Introduction

In applications as diverse as radiofrequency identification (RFID) tags, flexible solar cells, transparent flexible displays, conformal sensor arrays, and integration of logic circuits on flexible substrates by high-throughput roll-to-roll processes, large-area, flexible microelectronics holds the promise of lower fabrication costs and new functionality. The scaling up of substrate size and cost-effective fabrication are major concerns in this developing industry, in addition to maximizing circuit performance. When optical transparency and mechanical flexibility are sought, conventional electronics using single crystal silicon wafers are expensive and have limited potential for large-area devices. Amorphous silicon (a-Si:H), which addresses some of these issues, is only partially effective for these new applications due to its performance constraints in terms of mobility, current carrying capacity, and transparency [1]. New materials and fabrication techniques based on scientific knowledge will therefore be needed for all of the material components of thin-film transistors, are comprised of a semiconductor, a gate dielectric and conductors.

Among the essential components utilized in organic semiconductor devices are thin film metal oxides. Since a typical organic semiconductor lacks inherent charge carriers, all charges must be introduced into the device through electrode/organic interfaces, whose energy structure consequently determines how well devices function [2]. The work function of the electrode has a significant impact on the energy barrier at the interface. Because of this, different kinds of thin-film metal oxides can be utilized as a buffer layer to change the electrode's ability to perform its purpose.

5.1.1 Introduction to Semiconducting Metal Oxide

One of the most prevalent compounds on earth is oxide, which may be found everywhere as a gas, liquid, or solid and contains at least one oxygen and one additional element. Hydrogen oxide (H2O), often known as water, which serves as a universal solvent and is essential for life on our blue planet, is a highly common example of a liquid oxide at ambient temperature. People are accustomed to breathing carbon dioxide (CO2), a type of gas oxide that is used in photosynthesis and produced as one of the byproducts of respiration. Solid oxides, in addition to water and gas oxide, are crucial to our daily lives. For instance, silicon oxide (SiO2), also known as silica, is widely used in everyday items, architectural designs, and technological advancements. Quartz, which contains just silicon oxide in an amorphous condition, is the most common type of colorless glass [3]. Quartz is used to produce a variety of semiconductor devices in addition to producing optical fibers for cable television and communication systems due to its exceptional optical qualities and chemical stability. Although metal oxides have traditionally been thought of as insulators, several of them, notably the simple binary oxides ZnO, In_2O_3, SnO_2, and others and the complex oxide Zn-Sn-O and In-Sn-O, actually possess semiconductor characteristics [4].

Semiconducting metal oxide (SMOx), in contrast to molecular oxides, typically takes the form of a solid and is a member of the family of non-stoichiometric compounds, meaning that the proportions of its elemental composition cannot be precisely described by a ratio of small natural numbers [5]. Positive metallic ions and negative oxygen ions are joined by robust ionic bonds in SMOx, a type of ionic solid. The 's' electronic shell of SMOx materials is naturally filled, whereas the 'd' shell is only partially filled; as a result, SMOx materials have superior thermal and chemical durability as well as distinctive features like tunable energy bands, high dielectric constants, and other qualities. And these special qualities give SMOx capability in various applications. Solid oxides have become crucial to the modern semiconductor industry ever since the first metal-oxide semiconductor field-effect transistor (MOSFET) was introduced in the late 1950s [6]. SMOx is a very diverse class of materials, and as a result, they are unique from one another in terms of crystalline structure, innate flaws, energy band level, and other physical characteristics. For instance, in terms of crystal structure, ZnO can be found in cubic zinc blende or hexagonal wurtzite. Humidity, however, makes hexagonal wurtzite more stable and prevalent with the space group P63mc. The wurtzite form of ZnO has lattice constants of a = 3.25 and c = 5.2, with a ratio of c/a that is close to ideal at 1.633 [7]. Contrarily, In_2O_3 crystallizes as a cubic structure with a lattice constant of 10.08 while SnO_2 is a rutile tetragonal structure with lattice constants of 4.737 and 3.185. SMO_x has a variety of distinctive features as a result of these varied crystal forms. The SMOx structure, the heart of the MOSFET, is an insulating layer made of the aforementioned silicon oxide. The device's SMOx structure can be attributed to a planar capacitor, with the semiconductor layer primarily being p-type or n-type silicon and the metal layer possibly being conductive polycrystalline silicon in addition to metals.

5.1.2 Properties of Semiconducting Metal Oxide

Numerous types of defects, such as point, line, plane, and volume flaws, can be found inside materials containing oxides. In actuality, although having periodic crystal formations, all crystalline solids have imperfections and are not perfect. The three basic types of point defects are also known as zero-dimensional defects vacancy defects, interstitial defects, and Frenkel pairs when there are no heteroatoms [8]. In a completely flawless crystal, specific atoms would occupy the unoccupied locations that are known as vacancy defects. Additionally, the nearby atoms of the vacancies won't easily collapse due to the integrity of the crystal structure around them, ensuring their existence. Additionally, after a specific atom enters the vacant space, the emptiness will migrate in the opposite direction and take up residence where the atom had stood. Interstitial defects, on the other hand, manifest themselves with a generally high energy configuration when atoms occur in a location where there are often no atoms present. The final type of defect, known as Frenkel defects, appears to be mixture of vacancy defects and interstitial defects, in which an atom occupies an interstitial site and leaves a vacancy. As was previously said, SMOx is an ionic solid, and as ionic bonds are more easily broken than covalent bonds, SMOx often has more inherent defects than elemental semiconductor materials. There are six traditional point defects for SMOx, including sites where a metallic atom (M) is swapped out for an oxygen atom (O), a site where an O is swapped out for a site where there isn't an atom, and M or O that is present in an interstitial area. Undoped SMOx often demonstrates n-type semiconducting characteristics due to the presence of intrinsic defects such oxygen vacancies and metallic interstitial. The genesis of n-type semiconducting, however, is still up for debate for several SMOxs. For instance, n-type conductivity of ZnO has a different explanation than intrinsic defects in the absence of purposeful doping [9]. Hydrogen is found and described in bulk ZnO using first-principle calculations and experiments. It is assumed to be the plausible culprit because its presence is inevitable in most growth and annealing processes.

5.1.3 Semiconducting Metal Oxide Thin Films

A solid substance is considered to be in thin film form when a single atomic, molecular, or ionic species is physically or ultra-chemically altered to condense onto a solid substrate. In practically all fields today, using thin films to improve the physical and chemical properties of materials is a standard technique. Thin solid films have been developed to perform a wide range of functions in many different sorts of engineering systems. For instance, significant and outstanding advancements in thin film technology have been made, enabling the fast downsizing of electronic devices. The interfaces between materials with various electronic characteristics determine how the confinement of electric charge works in such devices. The demand for thin materials with excellent quality, repeatability, and reliability was another factor that accelerated the development of thin film growing processes and contributed to their major advancement.

Semiconducting metal oxide (SMOx) thin films are typically created in one or more thin layers. Numerous electronic components including transistors, sensors, and photovoltaic devices are frequently used in conjunction with such structures. The production method directly affects the structural, chemical, and physical characteristics of semiconductor thin films, whose thicknesses range from a few nanometers to hundreds of micrometers. Due to their versatility of features and the ease with which they may be produced, SMOx thin films, especially in recent years have emerged as a prospective ideal option in the field of electronic materials [10]. First off, a variety of chemical, electrochemical, and physical deposition processes make it possible to produce semiconductor material at a reasonable cost over vast areas with the necessary geometry and structure. Additionally, it is simple to produce single- or multi-crystalline structures with intricate geometries as well as microstructures of nanocrystalline thin films by altering the production process, temperature, substrate, and other factors according to the method [11]. Different junction types between various SMOx enable the development of industrial applications of electronic materials while enhancing the electrical characteristics of thin films.

5.1.4 Thin Films Deposition Method

For the deposition of thin film, there are several deposition techniques accessible. Because the emphasis is on thin-film deposition methods for generating layers with thicknesses ranging from a few nanometers to roughly 10 micrometers, reducing the number of ways to be evaluated simplifies the classification process. Evaporative methods and chemical reactions that take place in the gaseous and liquid phases are two examples of purely chemical processes that may be used to create thin coatings. Physical-chemical procedures are the umbrella term for a variety of processes based on glow discharges and reactive sputtering that combine physical and chemical interactions. The devices are generated by the combination of thin film materials, their deposition processes, and manufacturing procedures.

These methods can be classified in two different ways such as (a) Physical Vapor Deposition (PVD) and Chemical Vapor Deposition (CVD). While chemical methods rely on physical qualities and form solid films on the substrate, physical methods use deposition techniques like evaporation or sputtering that depend on the evaporation or discharge of a material from a source. Structure-property interactions form the foundation of thin film technologies and are key components of electronic devices. Performance and cost-effectiveness of thin film components are supported by manufacturing methods based on a particular chemical reaction.

5.1.4.1 *Physical Vapor Deposition (PVD) Method*

Owing to their broad bandgaps, recent oxide semiconductor thin films produced using physical vapor deposition (PVD) processes exhibit electrical performance that is similar to poly-Si and outstanding environmental stability. Physical vapor deposition is a method which involves moving atoms from a solid or liquid source to the substrate by evaporation, ion collisions, or sublimation on a target, for quickly and easily creating thin layer materials on a substrate [12]. The PVD operations are performed in a vacuum and typically comprise four phases. During this phase, high energy sources like resistive heating or electron beams vaporize the target first. The vaporized atoms move from the target to the surface of the substrate during the second phase of transportation. The third step comprised of reaction and deposition. The atoms of the substance will interact with the gas in this stage if a gas such as oxygen or nitrogen is present in the solution. However, if the coating is made entirely of the target material and includes no gas at all, then this phase is not a part of the process. The final phase, deposition, is when the substrate's surface is built. The schematic representation of PVD method for thin film fabrication is shown in the Figure 5.1.

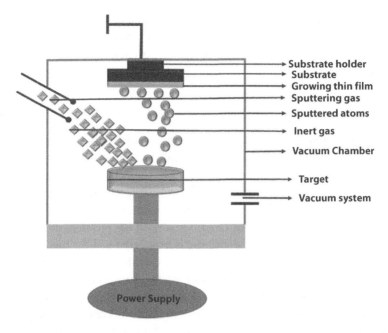

Figure 5.1 Illustrative representation of physical vapor deposition (PVD) method for thin film deposition.

5.1.4.2 Evaporation Methodology

Thermal evaporation or vacuum evaporation, one of the first techniques for forming thin films, is still commonly used in research labs and in industry to deposit metals and metal alloys [13]. The charge, also known as the source material, is heated during the evaporation procedure to produce a vapor plume that is directed toward the substrate. The source material can be a pure metal, an alloy, a mixture, or a combination. As soon as atoms, molecules, and molecular clusters reach the substrate and begin to condense from the vapor phase, a solid film is produced. The substrate takes in the heat from condensation. This method causes localized heating that can have major consequences on a tiny scale. Melting substrates during the early deposition cycles is usual in the development of metal coating methods for small cross section plastic components. Temperature-sensitive substrates can be coated without melting if source-to-substrate separations and deposition rates are controlled. There are several techniques to provide heat to the charge to produce vaporization, including induction heating, electric resistance heating, and electron beam heating. In this context, the deposition of thin films by catholic arc and laser ablation might also be categorized as thermal processes. The laser source is placed outside the evaporation system, and the laser beam is directed on the evaporated material, which is normally in the form of a fine powder, after passing through a window.

5.1.4.3 Thermal Evaporation

The easiest method for producing thin films with a few micrometers (m) or less in thickness is heat evaporation, also known as vacuum deposition [14]. In a vacuum, evaporating and condensing operations make up the thermal evaporation process. Firstly, the cylindrical chamber used in PVD which has a furnace is designed cautiously to maintain a proper flow and have proper deposition, as this chamber is used for physical deposition of material in vapor form. This technique is properly used for metals and the metals to be deposited are placed in inert crucible with a pressure up to 10^{-7} Torr. To evaporate the metal sources, the Tungsten or Electron Beam is used to heat the crucible and the evaporated material gets condensed on to the wafer to form a new layer over it. This process of deposition will be moderate to fast with no damage to the substrate because it is a shear process of condensation. This process can use two types of sources that are resistive and electron beam source.

Diffusion pumps can be utilized as primary pumps to keep chamber pressure below 510-2 millibar and high diffusion pumps to maintain pressure below 510-5 millibar using diffusion lines. Mechanical rotary pumps are used to evacuate the chamber through the roughing line from atmospheric pressure to a level at which these pressures can be maintained. In the electron beam evaporation, the vacuum coating system is made up of these elements.

In the Resistive method to generate a thin layer on the substrate using the resistive process, materials are heated using a resistively heated boat or filament, which is often made of refractory metals like tungsten, molybdenum, and tantalum with or without ceramic coating. First, place substrate in the same chamber as the little amount of coating material in the boat, which is inside the chamber. Then, beginning the vacuum procedure for the chamber in order to lower the pressure and create a very lengthy mean free path for the free atoms or molecules. After that, vaporize the depositing material thermally by-passing high current (10-100A) through the boat, which undergoes resistive heating. The huge vapor is then able to reach the substrate. Eventually, it returned to the solid state and formed thin film.

There are a few issues with thermal evaporation. The impurity layers that form as a result of beneficial substances reacting with the hot boat cause changes in optical characteristics. This approach subsequently has a significant impact on the material's stoichiometry, making the films produced by evaporation unsuitable for optoelectronic applications. Additionally, because the material dissolves at lower temperatures, resistive heating is unable to evaporate materials with high melting points, especially metal oxides like zirconia. This substance will result in an impure layer on the surface if this technique is used. However, electronic beam (e-beam) heating eliminates these issues, making vacuum evaporation technology the preferred method for deposition films. The deposition surface is shielded from fault nucleation and damage due to the low kinetic energy of the material's atoms during thermal evaporation.

5.1.4.4 Molecular Beam Epitaxy

A sophisticated, accurate technique for producing single-crystal epitaxial films in high vacuum is molecular beam epitaxy (MBE) [15]. The films are created on single-crystal substrates by slowly evaporating the elements or molecules that make up the film from separate Knudsen effusion source cells (deep crucibles in furnaces with cooled shrouds) onto substrates and maintained at a temperature that is conducive to chemical reaction, epitaxy, and excess reactant re-evaporation. Small-diameter atomic or molecule beams are produced by the furnaces and are focused on the heated substrate, which is typically silicon or gallium arsenide. Fast shutters are placed between the sources and the substrates. By adjusting these shutters, it is feasible to produce super lattices that have uniformity, lattice match, composition, dopant concentrations, thickness, and interfaces that can be accurately regulated down to the level of atomic layers.

5.1.4.5 Electron Beam Evaporation

In electron beam evaporation, a stream of electrons is focused onto the surface of the material and accelerated through fields that are typically between 5 and 10 kV [16]. The substance melts and evaporates as a result of the electrons' quick loss of energy when they hit the surface. The surface is directly heated by impinging electrons, as opposed to conventional heating processes. Direct heating can be used to evaporate materials from water-cooled crucibles. For evaporating reactive materials, in particular reactive refractory materials, such water-cooled crucibles are necessary in order to almost eliminate interactions with crucible walls. High purity films can be made because crucible materials or their reaction by products are essentially excluded from evaporation [17]. The two types of electron beam weapons are thermionic and plasma electrons. In the first kind, heated refractory metal filaments, rods, or discs are used to thermionically create electrons. The latter form involves the extraction of electron beams from plasma that is confined in a narrow area.

5.1.4.6 Advantages and Disadvantages of PVD Method

PVD has several advantages including: (i) Compared to the substrate material, coatings created through PVD may have better characteristics. (ii) All varieties of inorganic materials and certain varieties of organic materials can be employed in PVD method. (iii) Compared to many other techniques, including electroplating, the method is ecologically favorable. One major drawbacks of PVD method coating is that it's high cost. Furthermore, coating PVD acts at a somewhat sluggish rate.

5.1.4.7 Sputtering Technique

Sputtering, also known as radio frequency (rf) sputtering (also known as radio frequency diode), magnetrons, and ion-beam sputtering, is the second type of active radiation-based PVD process [18]. It has a range of devices for thin-film deposition. This technique is also known as Plasma assisted technique as here the species are converted in to gaseous form and accelerated in to the plasma chamber where the plasma is created by the discharge of neutral gasses Helium or Argon. We see the ions are accelerating through potential gradient and are made to bombard with the target or cathode through momentum transfer. The substrate is located on the anode plate, and the source material, known as the target, is positioned in the vacuum chamber on the cathode plate. The chamber is then filled with sputtering gas, such as argon gas (Ar), an inert gas, at low pressure (4×10^{-2} torr). In this sort of deposition system, argon (Ar+) ions created during the radiance discharge are accelerated at a high rate toward the cathode (target) by an applied electric field, which causes the target to sputter, resulting in the deposition of thin layers on the substrates. The freedom to fundamentally deposit any amorphous and crystalline materials, superior control in maintaining stoichiometry, and regularity of film thickness compared to the evaporation method are the main advantages of the sputtering technology. Additionally, polycrystalline films' film grain structure, which frequently has a diversity of crystallographic orientations without a specific texture, is produced via sputter deposition. However, the

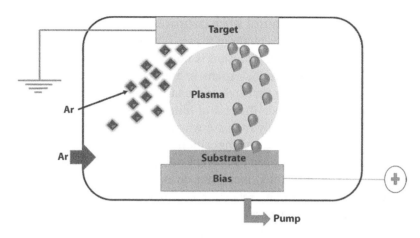

Figure 5.2 Schematic diagram of sputter deposition method for thin film formation.

sputtered atoms in the sputter deposition have greater kinetic energy than the atoms in the evaporative process. The surface of the deposited material in sputtering reveals additional defects and damage because argon gas (Ar) is present in the sputtering system. Additionally, concentration of impurity atoms in deposited films that were sputtered larger than films that are thermally deposited. This is the same technique as dry physical etching but here the energy is just sufficient to deposit the particles on to the film and the film grows at the surface via deposition. In ion sputtering, for sputter deposition the cathode is used as a target and anode is coated. The target at a high negative potential is bombarded with positive argon ions created in high-density plasma [19]. Condensed on to substrate placed at the anode. The schematic representation of sputtering technique for thin film fabrication is shown in the Figure 5.2.

5.1.4.8 Advantages and Disadvantages of Sputtering Technique

Elements, alloys, and compounds can all be sputtered or deposited. The sputtering target acts as a reliable, long-lasting source of vaporization. In some situations, the sputtering source can have a specific shape. For instance, a line, the surface of a rod, or a cylinder. In some setups, reactive deposition is easily accomplished using plasma-activated reactive gaseous species. There is hardly any radiant heat produced during the deposition process. The substrate and source may be in close proximity. The sputter deposition chamber could be compact.

Sputtering rates are minimal when compared to those of thermal evaporation. Moving fixturing is required to generate consistent thickness films since the deposition flux distribution is not uniform in many configurations. Sputtering targets can be expensive and have subpar material utilization. The bulk of energy that strikes the target is transformed into heat, which needs to be expelled. The risk of film contamination over vacuum evaporation can occasionally increase when gaseous pollutants are "activated" in the plasma. The gas composition in reactive sputter deposition needs to be carefully managed to prevent poisoning the sputtering target. The deposition of thin films metallization on semiconductor materials, architectural glass coatings, reflective coatings on polymers, magnetic films for storage media, transparent electrically conductive films on glass and flexible webs, dry-film lubricants, wear-resistant coatings on tools, and decorative coatings are all common applications for sputter deposition.

5.1.4.9 Chemical Vapor Deposition (CVD)

Two techniques for forming thin films that only involve chemical reactions in the gas or vapor phases are chemical vapor deposition and thermal oxidation. Chemical vapor deposition is a method of creating materials in which components of the vapor phase react chemically on or on the surface of a substrate to produce a solid result. Deposition technology is one of the most important methods for producing thin films and coatings of a range of materials required by advanced technology, particularly solid-state electronics, where some of the most strict purity and composition standards must be met. The main advantage of CVD is its ability to easily synthesize both simple and complex compounds at relatively low temperatures. Controlling the reaction chemistry and deposition conditions allows for tailoring of both chemical composition and physical structure [20].

Gas-phase reaction chemistry, thermodynamics, kinetics, transport processes, film growth phenomena, and reactor engineering are fundamental concepts in CVD. Fundamental CVD chemical reaction types include pyrolysis (heat decomposition), oxidation, reduction, hydrolysis, nitride and carbide production, synthesis processes, disproportionation, and chemical transport. In more complicated circumstances, a mix of various reaction types may be used to create a particular end result. The rate of deposition and the characteristics of the film deposit are influenced by temperature, pressure, input concentrations, gas flow rates, reactor geometry, and operating principle. In several industrial domains, CVD has become a crucial process technique. At low pressure and high temperatures, CVD has long been employed to coat substrates [21]. The schematic

SEMICONDUCTOR METAL OXIDE THIN FILMS 143

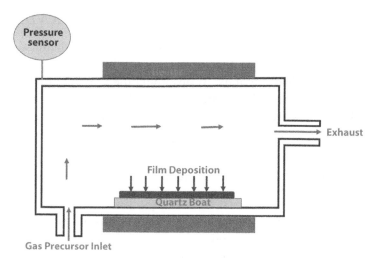

Figure 5.3 Schematic diagram of chemical vapor deposition (CVD) method for thin film deposition.

representation of CVD method for thin film fabrication is shown in the Figure 5.3.

5.1.4.10 Photo-Enhanced Chemical Vapor Deposition (PHCVD)

In photo-enhanced chemical vapor deposition (PHCVD), electromagnetic radiation, typically short-wave ultraviolet light, activates the reactants in the gas or vapor phase. Absorption of specific photonic energy by reactant molecules or Reactive free-radical species are produced by atoms to initiate the process then work together to produce the final cinematic result. Typically, mercury vapor is as a photosensitizer included in the mixture of reactant gases triggered by the powerful crystalline mercury radiation the resonance lamp (253.7 nm wavelength). Excited mercury atoms collide, and energy is transferred kinetically. In order to generate free radicals from the reactants [22].

Two benefits of this flexible and promising CVD method are the low temperature (usually 150°C) needed to create films like SiO2 and Si3N4 and the much reduced radiation damage (compared to PECVD). In order to achieve acceptable rates of film deposition, it is now necessary (in most situations) to photoactivate with mercury due to a lack of efficient production equipment.

5.1.4.11 Laser-Induced Chemical Vapor Deposition (LICVD)

A highly localized laser beam is used in laser-induced chemical vapor deposition(LICVD) to heat the substrate, resulting in layer deposition through CVD surface reactions. By absorbing the given photonic energy at a particular wavelength, gaseous reactant atoms or molecules can be activated using laser (or electron radiation). High purity film deposits are produced as a result of very specialized chemical gas phase processes [23].

5.1.4.12 Atmospheric Pressure Chemical Vapor Deposition (APCVD)

A CVD technique called atmospheric pressure chemical vapor deposition APCVD deposits both doped and undoped oxides at atmospheric pressure, which is normal pressure. The deposited oxide has a modest density and a moderate amount of coverage due to the low temperature. At rates ranging from 2000 to 3000/min, the surface reaction on the heated wafer, which was generally heated to 400°C, grew films. The process's low temperature, straightforward reactor construction, and rapid wafer throughput are all key benefits. Issues with APCVD include inadequate step coverage, rapid precursor flow, particle contamination, and incessant cleaning. With a greater understanding of the reaction processes and reactant injection, some of these step coverage difficulties could be reduced. These improvements have been incorporated into new reactor designs that are still in use today; regeneration is taking place [24].

5.1.4.13 Plasma Enhanced Chemical Vapor Deposition (PECVD)

Plasma Enhanced Chemical Vapor Deposition (PECVD) technique uses plasma and reactive gases in the deposition chamber to coat the substrate with the required solid surface. Any vapor in which a sizable fraction of the atoms or molecules are ionized, energetic, and settle on the surface is referred to as plasma. PECVD is performed at temperatures between 250 and 350 °C. The process gases cannot be thermally degraded at these low temperatures. Direct current (DC) or alternating current (AC) discharge between two electrodes, with reacting gases filling the space in between, produces the plasma. Si_3N_4 and SiO_2 are deposited on the upper of metal coatings using PECVD since metallization, like aluminum, cannot be exposed to high temperatures. It becomes possible to use PECVD, a low temperature, low pressure method that uses plasma as an energy source. Although the conformance (0.6 to 0.8) is not as good as in LPCVD, the rate of deposition is significantly higher (0.5 microns per minute). To create the appropriate quality films, process variables such as gas flow rates, power input, chamber pressure, reactor geometry, inter electrode spacing, substrate temperature, and others can be controlled. Low temperature, improved film density for higher dielectric and more compression, and ease of chamber cleaning are all advantages of PECVD [25].

5.1.4.14   Atomic Layer Deposition (ALD)

A modified CVD procedure called atomic layer deposition (ALD) is used to make thin films. Several gases are used in the procedure, and they are introduced into the chamber in alternating cycles. It is a technique for thin film deposition that gradually makes use of a chemical process in the gas phase. Each gas reacts in a manner that causes the reaction to stop once the current surface is saturated. This surface can respond to the alternative gas in a similar manner. The chamber is flushed out with an inert gas, such as nitrogen or argon, in between each of these gas reactions [26].

5.1.4.15   Electrolytic Anodization

Similar to heat oxidation, anodization creates an oxide layer from the substrate. The anode reacts with the electrolyte's negative ions in solution to produce a hydrated oxide layer on semiconductors and limited metals, while the cathode produces hydrogen gas. Nonporous oxides with good adhesion can form on silicon, niobium, tantalum, titanium, aluminum and zirconium. The most important applications are for aesthetic dyed coatings and corrosion-resistant films on aluminum and its alloys, as well as electrical insulation layers for tantalum and aluminum electrolyte capacitors [27].

5.1.4.16   Electroplating

The cathode of an electrolytic cell, which consists of an anode, a cathode, and an electrolyte solution (which contains metal ions through which electric current passes), is electrodeposited with a metallic coating. The quantitative parts of the process are governed by Faraday's laws. Important electroplating factors include pH, temperature, agitation, current density, current distribution, and solution composition. Aqueous solutions have been successfully used to electroplate a variety of metals and metal alloys. An industrial procedure known as electroplating can create deposits that range in thickness from very thin films to very thick coatings (electroforming) [28].

5.1.4.17 Chemical Reduction Plating

Chemical diminution a metal ion in solution is reduced during plating by a reducing agent that is applied just before usage. Because of the homogenous nature of the reaction, deposition happens everywhere across the solution rather than only on the substrate. Silver, copper, nickel, gold, and certain sulphide films are some examples of plated films. The oldest application of the procedure, which employs silver nitrate solutions and one of numerous reducing agents, such as hydrazine, is the silvering of glass and polymers for the creation of mirrors [29].

5.1.4.18 Electroless Plating

A reducing agent reduces metal ions in solution to a metallic coating during the selective deposition plating method known as autocatalytic plating, sometimes referred to as electroless plating. The term "autocatalysis" refers to the fact that plating only takes place on surfaces that are suited for catalysis, such as substrates made of the same metal as the metal being plated. Electroless (or electrodeless) plating, which is limited to a few metals and alloys, has a number of advantages over electroplating, including selective (patterned) deposition [30].

5.1.4.19 Electrophoretic Deposition

Colloidal particle dispersion is used to deposit a film from an electrophoretic coating onto a conductive substrate. The dispersion separates into positively charged ions (cations) and negatively charged colloidal particles in a conductive liquid, or vice versa. Colloidal particles migrate to the substrate, discharge, and form a film when an electric field between the positive substrate electrodes (anode) is applied [31].

5.1.4.20 Immersion Plating

The chemical displacement deposition of a metal film on a substrate from a dissolved salt of the coating metal without the need of external electrodes is known as displacement deposition or immersion plating. The electromotive force series states that any more noble metal is displaced from solution by a less noble (more electronegative) element. In actuality, the metal surface develops cathodic and anodic regions, with cathodic regions showing thicker coatings. There are just a few industrial uses for this method, mostly thin coatings on copper and copper alloys [32].

5.1.4.21 Advantages and Disadvantages of CVD Process

As stated in the introduction, there are both benefits and drawbacks to the CVD system. The ability of this system to produce and control crystal structure is its greatest benefit. Furthermore, uniform, high-density films constructed of pure materials. The CVD system is capable of producing films with high clone and repeatability at reasonable deposition rates, adhesion. This, however, is incorrect as the film stoichiometry cannot be controlled by more than one method because the properties of various materials differ in evaporation rates.

5.1.4.22 Sol-Gel Method

As part of the "sol-gel" process, the "sol" (or solution) progressively develops into a network of liquid and solid phases that resembles a gel. Typically, the hydrolysis and condensation of metal alkoxide precursors results in the creation of a sol. However, a sol may be described more broadly as a colloidal suspension, which applies to a larger variety of systems. A colloidal system is defined as a dispersion of one phase into another where "the molecules or poly-molecular particles distributed in a medium have at least in one direction a dimension generally between 1 nm and 1 m," according to the International Union of Pure and Applied Chemistry (IUPAC). Precursors may generally be either cast into the required shape of a suitable container or applied to a substrate to create a film. Due to the cheap processing costs, uniform and homogeneous film generation over broad regions, the possibility of a precise stoichiometry, and thickness control, this typical sol-gel chemistry is still one of the most widely used and investigated. Sol-gel is a straightforward, inexpensive technique for creating thin films with tiny grain sizes. But occasionally, the films exhibit poor compactness and many fractures. However, the produced thin films' quality might still be impacted by the volatile component and complexing agent breakdown during the annealing stage [33]. Therefore, it is crucial to know the solvent's evaporation temperature as well as the temperature of the complexing agents in the precursor.

5.1.5 Application of Semiconducting Metal Oxide Thin Films

Modern electronics rely heavily on semiconductor thin films. They are utilized in many different gadgets, including sensors, solar cells, and LED lights [34]. Due to industrialization and population expansion, emergency scenarios have recently raised demand for high-performance chemical gas sensors. Chemical gas sensors are now utilized mostly to enhance and save human life, and this is anticipated to continue for the foreseeable future. Chemical gas sensors are utilized in many different applications, including security, food and water safety, environmental protection, and medical diagnostics. Semiconducting metal-oxide conductometric sensors have demonstrated excellent sensing performances among the numerous chemical gas sensors that have been researched recently. Such devices' operation is dependent on the surface contact between the sensing material and the gas. The specific interaction that may be measured is an adsorption-desorption-driven surface reaction that modifies semiconducting metal-oxide's electrical conductance (or resistance) as a result of the transfer of charge carriers.

Nowadays, a wide range of electronic devices employ semiconductor thin films, primarily because they have a number of benefits over other types of semiconductor materials. For instance, they may be deposited utilizing affordable manufacturing techniques on a range of substrates. Additionally, compared to bulk semiconductor materials, semiconducting thin films frequently have greater electron mobilities, which increases their electrical current carrying capacity. Transistors, solar cells, light-emitting diodes (LEDs), and thin-film batteries may all be made using them [35]. Additionally, optical waveguides, photodetectors, and laser diodes are just a few of the optoelectronic devices that make use of semiconducting thin films. This outcome has led to the widespread usage of semiconducting thin films in high-speed electronic components like memory chips and digital switches.

5.1.5.1 Photovoltaic Cells

The development of photovoltaic cells has relied heavily on thin-film solar panels, which will only grow in popularity as society shifts to renewable energy sources. For people who prefer to live off the grid or bring their own electricity on camping vacations, thin-film solar cells are the ideal option. Since these flexible gadgets are even bendable, they are significantly safer than conventionally heavy, rigid materials like silicon and glass, which would be hard to store flat [36].

5.1.5.2 Thin-Film Transistors

Electronic gadgets like smartphones, tablets, and TVs frequently employ special sorts of transistors called semiconductor thin-film transistors. They are made up of thin semiconductor material films that are placed on a base material or substrate. Thin-film transistors' ability to be fitted into more compact and smaller electronic devices is one of its main advantages over conventional transistors. The greater performance of semiconducting thin-film transistors in comparison to other classes of electronic parts, such as resistors and capacitors, is another significant advantage. Thin-film transistors are perfect for applications that call for high power or quickly changing signals due to their superior conductivity and stability at high temperatures [37].

5.1.5.3 Computer Hardware

The creation and application of thin films in storage gear is an excellent example of cutting-edge technology. SSDs may now provide more memory storage space than ever thanks to developments in microprocessors and diodes. These developments have accelerated recently. With time, hard drive capabilities will continue to grow, especially in light of these recent developments in semiconducting thin-film technology [38].

5.1.5.4 LED and Optical Displays

In the case of LED and OLED displays, the different electrical and optical parts of the display are made using semiconducting thin films. For example, thin films can be used to create light-emitting diodes (LEDs), as well as the transparent conductive electrodes that are used to drive the LEDs. They can also be used to create optical filters, which are used to control the color of the light emitted by the display. Semiconducting thin films are often employed in a wide range of electrical devices and are a crucial component of LED and OLED technologies [39].

5.1.6 Limitations of Semiconductor Thin Films

The use of thin-film semiconductors in electronic devices is subject to several restrictions. It could be challenging to create these thin films with the required uniformity and accuracy. Furthermore, thin films may have greater performance variations while operating in hotter or colder environments because they are more sensitive to temperature changes than conventional silicon semiconductors. The lifespan of thin films may also be less than that of conventional electronics parts because they may be more susceptible to degradation over time. Thin-film semiconductors are a useful tool for electronics engineers and designers despite these drawbacks since they enable them to get past other limits and develop novel new devices.

5.2 Conclusion and Outlook

To fulfill the rising demand for even more sophisticated electronics devices, researchers are working nonstop to enhance thin-film semiconductor technology. These studies focus on figuring out how to make thin films that can more effectively support the operation of a variety of electronic devices and conduct electricity. To enable more accuracy in manufacturing and device design, new methods for controlled thin film deposition are currently being developed. The nanoscale structuring of thin films is one widely used technique. Researchers can more precisely regulate how they interact with electrical currents by modifying their composition and functionalities at this level. Another one is the use of polymer additives to improve organic thin films, which could pave the way for their incorporation into electrical devices. The study of metal oxide thin-film transistors for image display can also be included because it should be pertinent to electronics tools and gadgets. Overall, these initiatives continue to advance thin-film technology as a crucial element of next-generation electronics, assisting in the emergence of an innovative and prosperous new era.

Acknowledgement

KKS would like to thank all the co-authors for their collaboration and helpful discussions. All the authors would like to thank the editor of the book for giving such opportunity to publish the chapter in this reputed Wiley-Scrivener publishing group.

References

1. Kim, M.G., Kanatzidis, M.G., Facchetti, A., Marks, T.J., Low-temperature fabrication of high-performance metal oxide thin-film electronics via combustion processing. *Nat. Mater.*, 10, 5, 382–388, 2011.

2. Li, P., Jia, C., Guo, X., Molecule-based transistors: From macroscale to single molecule. *Chem. Rec.*, 21, 6, 1284–1299, 2021.
3. Zhang, S.R. and Zhou, Y., Introduction to semiconducting metal oxides, in: *Semiconducting Metal Oxide Thin-Film Transistors*, IOP Publishing, Bristol, UK, 2020.
4. Zhou, Y., *Semiconducting metal oxide thin-film transistors*, IOP Publishing, Bristol, UK, 2020.
5. Fortunato, E., Barquinha, P., Martins, R., Oxide semiconductor thin-film transistors: A review of recent advances. *Adv. Mater.*, 24, 22, 2945–2986, 2012.
6. Khan, A.I., Keshavarzi, A., Datta, S., The future of ferroelectric field-effect transistor technology. *Nat. Electron.*, 3, 10, 588–597, 2020.
7. Zargar, R.A., Arora, M., Alshahrani, T., Shkir, M., Screen printed novel ZnO/MWCNTs nanocomposite thick films. *Ceram. Int.*, 47, 6084–6093, 2020.
8. Jiao, S., Fu, X., Zhang, L., Zeng, Y.J., Huang, H., Point-defect-optimized electron distribution for enhanced electrocatalysis: Towards the perfection of the imperfections. *Nano Today*, 31, 100833, 2020.
9. Zargar, R.A., Arora, M., Bhat, R.A., Study of nanosized copper-doped ZnO dilute magnetic semiconductor thick films for spintronic device applications. *Appl. Phys. A*, 124, 1–9, 2018.
10. Zargar, R.A., Kumar, K., Mahmoud, Z.M.M., Shkir, M., AlFaify, S., Optical characteristics of ZnO films under different thickness: A MATLABbased computer calculation for photovoltaic applications. *Physica B*, 63, 414634, 2022.
11. Sancakoglu, O., Technological background and properties of thin film semiconductors, in: *21st Century Surface Science-a Handbook*, Intech Open, London, UK, 2020.
12. Rossnagel, S.M., Thin film deposition with physical vapor deposition and related technologies. *J. Vac. Sci. Technol. A: Vac. Surf. Films*, 21, 5, S74–S87, 2003.
13. Shahidi, S., Moazzenchi, B., Ghoranneviss, M., A review-application of physical vapor deposition (PVD) and related methods in the textile industry. *Eur. Phys. J. Appl. Phys.*, 71, 3, 31302, 2015.
14. Jilani, A., Abdel-Wahab, M.S., Hammad, A.H., Advance deposition techniques for thin film and coating, in: *Modern Technologies for Creating the Thin-film Systems and Coatings*, vol. 2, 3, pp. 137–149, 2017.
15. Cho, A.Y. and Arthur, J.R., Molecular beam epitaxy. *Prog. Solid State Chem.*, 10, 157–191, 1975.
16. Smidt, F.A., Use of ion beam assisted deposition to modify the microstructure and properties of thin films. *Int. Mater. Rev.*, 35, 1, 61–128, 1990.
17. Coelho, S.M., Archilla, J.F., Auret, F.D., Nel, J.M., The origin of defects induced in ultra-pure germanium by Electron Beam Deposition, in: *Quodons in Mica*, pp. 363–380, Springer, Cham, 2015.

18. Wasa, K., Kitabatake, M., Adachi, H., *Thin film materials technology: Sputtering of control compound materials*, Springer Science & Business Media, Germany, 2004.
19. Han, J.G., Recent progress in thin film processing by magnetron sputtering with plasma diagnostics. *J. Phys. D: Appl. Phys.*, 42, 4, 043001, 2009.
20. Morosanu, C.E., *Thin films by chemical vapour deposition*, vol. 7, Elsevier, Oxford, UK, 2016.
21. Kleijn, C.R., Dorsman, R., Kuijlaars, K.J., Okkerse, M., van Santen, H.V., Multi-scale modeling of chemical vapor deposition processes for thin film technology. *J. Cryst. Growth*, 303, 1, 362–380, 2007.
22. Lian, S., Fowler, B., Krishnan, S., Jung, L., Li, C., Manna, I., Banerjee, S., Photo-enhanced chemical vapor deposition: System design considerations. *J. Vac. Sci. Technol. A: Vac. Surf. Films*, 11, 6, 2914–2923, 1993.
23. Shah, A.H., Zargar, R.A., Arora, M., Sundar,P.B., Fabrication of pulsed laser-deposited Cr-doped zinc oxide thin films: Structural, morphological, and optical studies. *J. Mater. Sci.: Mater. Electron.*, 31, 21193–21202, 2020.
24. Alexandrov, S.E. and Hitchman, M.L., Chemical vapor deposition enhanced by atmospheric pressure non-thermal non-equilibrium plasmas. *Chem. Vap. Deposition*, 11, 11-12, 457–468, 2005.
25. Meyyappan, M., A review of plasma enhanced chemical vapour deposition of carbon nanotubes. *J. Phys. D: Appl. Phys.*, 42, 21, 213001, 2009.
26. Johnson, R.W., Hultqvist, A., Bent, S.F., A brief review of atomic layer deposition: From fundamentals to applications. *Mater. Today*, 17, 5, 236–246, 2014.
27. Gulati, K., Martinez, R.D.O., Czerwiński, M., Michalska-Domańska, M., Understanding the influence of electrolyte aging in electrochemical anodization of titanium. *Adv. Colloid Interface Sci.*, 02, 102615, 2022.
28. Giurlani, W., Zangari, G., Gambinossi, F., Passaponti, M., Salvietti, E., Di Benedetto, F., Innocenti, M., Electroplating for decorative applications: Recent trends in research and development. *Coatings*, 8, 8, 260, 2018.
29. Karam, A.F. and Stremsdoerfer, G., A novel dynamic process for chemical reduction plating. *Plat. Surf. Finish.*, 85, 1, 88–92, 1998.
30. Mallory, G.O. and Juan, B.H., *Electroless plating: Fundamentals and applications*, William Andrew, New York, USA, 1990.
31. Van der Biest, O.O. and Vandeperre, L.J., Electrophoretic deposition of materials. *Annu. Rev. Mater. Res.*, 29, 327, 1999.
32. Haddadi, I., Amor, S.B., Bousbih, R., El Whibi, S., Bardaoui, A., Dimassi, W., Ezzaouia, H., Metal deposition on porous silicon by immersion plating to improve photoluminescence properties. *J. Lumin.*, 173, 257–262, 2016.
33. Zargar, R.A., Chackrabarti, Malik, M.H., Sol-gel syringe spray coating: A novel approach for Rietveld, optical and electrical analysis of CdO film for optoelectronic applications. *Phys. Open*, 7, 10069, 2021.
34. Zargar, R.A., Fabrication and improved response of ZnO-CdO composite flms under diferent laser irradiation dose. *Sci. Rep.*, 12, 10096, 2022.

35. Rao, M.C. and Shekhawat, M.S., A brief survey on basic properties of thin films for device application. *Int. J. Mod. Phys.: Conf. Ser.*, World Scientific Publishing Company, 22, 576–582, 2013.
36. Zargar, R.A. et al., ZnCdO thick film films: Screen printed TiO2 film: A candidate for photovoltaic applications. *Mater. Res. Express*, 7, 065904, 2020.
37. Street, R.A., Thin-film transistors. *Adv. Mater.*, 21, 20, 2007–2022, 2009.
38. Seshan, K. and Schepis, D. (Eds.), *Handbook of thin film deposition*, William Andrew, Oxford, UK, 2018.
39. Zargar, R.A. and Kumar, K. et al., Structural, optical, photoluminescence, and EPR behaviour of novel $Zn_{0.80}Cd_{0.20}O$ thick films: An effect of different sintering temperatures. *J. Lumin.*, 245, 118769, 2022.

6
Thin Film Fabrication Techniques

Lankipalli Krishna Sai[1], Krishna Kumari Swain[2] and Sunil Kumar Pradhan[1]*

[1]School of Electronics Engineering, Vellore Institute of Technology, Chennai, India
[2]Department of Applied Mechanics, Indian Institute of Technology Madras, Chennai, India

Abstract

Thin film fabrication process involves many considerations in order to obtain desired devices in the electronics domain. The bottom-up techniques which are used to synthesize thin films have numerous advantages and applications as far as VLSI (very large-scale integration) technology and nano-electronics are considered. If we take into consideration of physical processes, then under evaporation, thermal, molecular beam epitaxy, e-beam, laser, ion-plating are considered. Similarly, under sputtering, DC, RF, Magnetron are taken into contemplation. Further with respect to chemical processes, which include thermal, MOCVD (metal-organic chemical vapor deposition), PECVD (plasma enhanced chemical vapor deposition); similarly, under plating, electroplating, electroless along with sol-gel method and ALE (Atomic layer epitaxy) are considered to be requisite thin film synthesis process. Modern thin film growth techniques like molecular beam epitaxy (MBE) have demonstrated the ability to produce samples with carrier mobilities that are higher than those of bulk crystals while having complete control over the growth rate. This technology is essential for establishing the correct stoichiometry and layer-by-layer growth. We may create a thin film hall bar device for the same sample holder using two physical masks: one for reactive ion etching (RIE), also known as an etching mask, and the other for depositing metal electrodes, also known as a metal mask. While the etching mask is used to specify the size of the thin film Hall bar, the metal mask is utilized to deposit the metal electrodes. After the sample is made from the MBE, the thin film Hall bar can be made using RIE and an etching mask placed over the sample on the sample holder. The etching can be done using the required gas for a predetermined period of time to produce the required Hall bar structure. The metal mask can also be utilized to

---

*Corresponding author: sunilpradha@gmail.com

Rayees Ahmad Zargar (ed.) Metal Oxide Nanocomposite Thin Films for Optoelectronic Device Applications, (155–178) © 2023 Scrivener Publishing LLC

construct metal ohmic connections by applying heat evaporation to a thin film Hall bar sample.

Keywords: Thin film, thermal evaporation, molecular beam epitaxy, chemical vapour deposition, physical vapour deposition, atomic layer deposition, electrolytic anodization, electroplating

6.1 Introduction

In most materials we have bulk materials with fixed properties but have limited applications. To alter their properties, we reduce their size with certain dimension called as nano technology. Non-self-aligned metal gate MOSFETs with gate lengths in the order of 10 m were typical in the late 1970s. With gate widths in the 20-nm range, current VLSI fabrication technology has already reached the physical scaling limit [1, 2, 4]. This indicates a nearly 1000x reduction in device size, as well as an even more amazing increased device density on a single VLSI chip. Future VLSI technology advancements must rely on novel material concepts and device designs that take into account quantum phenomena. Researchers are exploring new technology at a really exciting time, but we can also be it is guaranteed that "conventional" CMOS and BiCMOS (bipolar CMOS) fabrication technology for many more years, will remain the workhorse of the microelectronic industry [3].

Similarly in VLSI the bulk materials have different properties but they do exhibit different properties when deposited as thin films with much more applications which provide flexibility in device designing by these sizing properties in IC circuits as they occupy less space & material and are cost-effective [5, 9]. The films used in VLSI are expected to have good reproducibility, appropriate thickness, uniformity, good composition, integrity, minimum stresses and purity. They are expected to maintain a same layer and properties in multi-level wafers. In reality, optical coatings, electrical gadgets, aesthetic components, and instrument hard coatings have all been produced using thin films over the past 50 years [6, 7, 10]. Contrarily based on the numerous theories that underlie film deposition, the deposition process is currently divided into two forms, PVD (physical vapor deposition) and CVD (chemical vapor deposition). The first PVD technology is a combination of thermal evaporation and sputtering, while the second CVD technology comprises of PECVD (plasma enhanced chemical vapor deposition), LCVD (laser chemical vapor deposition), and MOCVD (metal organic chemical vapor deposition).

6.2 Thin Film – Types and Their Application

- **Polycrystalline Silicon**
    - These are used as gate electrode in MOS device.
    - Conducting material in multilevel metallization.
    - Contact material for device junctions.
- **Doped or Undoped Silicon Dioxide**
    - These are used as insulation between conducting layer in multilevel metallization isolation or proper insulation can be done using Silicon Dioxide layer.
    - Diffusion and ion implantation as a mask to protect other layers.
    - We can use as capping the doped films to prevent the dopant losses.
    - It can be used for gettering the impurities or surface passivation to protect device from impurities.
- **Stoichiometric or Plasma Deposited Silicon Nitride**
    - Used as barrier for sodium diffusion it is resistant to moisture and has low-oxidation rate and prevent etching.
    - Is also used for local oxidation to create a characteristic profile to carry out oxidation only in particular region.
    - Used as a dielectric for DRAM & MOS capacitors when combined with silicon dioxide.
    - Used as a preventive layer in surface etching and passivation.

6.3 Classification of Thin-Film Fabrication Techniques

When an individual atomic, molecular, or ionic species is physically or ultra-chemically manipulated to condense onto a solid substrate, the solid substance is interpreted to be thin film. There are numerous deposition processes available for material creation [8, 10]. Because the focus is on thin-film deposition methods for creating layers with thicknesses ranging from a few nanometers to around 10 micrometers, limiting the number of approaches to be considered simplifies the process of classifying the techniques. Procedures for forming a thin film are generally either entirely physical, like evaporative techniques, or entirely chemical, that happens to be in the gaseous or liquid phases. Physical-chemical procedures are any of the numerous processes based on glow discharges and reactive sputtering that combine physical and chemical reactions. Thin film materials, their deposition processing, and fabrication procedures are combined to generate the devices. These methods can be classified in two different ways.

- Physical Vapor Deposition (PVD)
- Chemical Vapor Deposition (CVD)

Physical methods include deposition techniques such as evaporation or sputtering that rely on the evaporation a material from a source, whereas chemical methods rely on physical properties and solid films are produced on the substrate as a consequence of a chemical reaction involving substances in the vapor phase which contain the necessary constituents [11, 13]. Hence the core aspects of any electronic devices and the foundation of thin film technologies are structure-property interactions. Techniques for manufacturing thin film components which are based on a specific chemical reaction support their performance and economy [12, 14]. As seen in vapor phase deposition and thermal growth, thermal effects can influence chemical reactions. To obtain the absolute film, a specific chemical reaction is required in all of these cases [15].

Physical Vapor Deposition (PVD)

A technique that allows the quick and simple synthesis of thin film materials on a substrate is physical vapor deposition, transfer of atoms from a solid or molten source to the substrate through evaporation, ion collisions, or sublimation on a target [16, 17, 19]. In general, four steps are involved in the PVD procedures, which are carried out in a vacuum. The target is first vaporized during this stage by high energy sources like resistive heating or electron beams. The second phase is transportation, in which the vaporized atoms travel from the target to the substrate's surface. The third phase is reaction, and in certain cases deposition [18]. If a gas like oxygen or nitrogen is present in the solution, the material's atoms will react with the gas in this step. However, this phase is not a component of the process if the coating exclusively consists of the target material and contains no gas at all [20]. Deposition is the last step, during which the substrate's surface is constructed. The two types of this technology are thermal evaporation and sputtering. The schematic of PVD method is shown in the Figure 6.1.

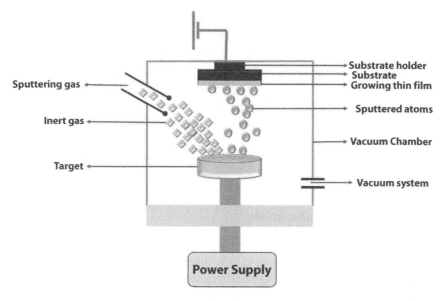

Figure 6.1 Schematic diagram of physical vapor deposition (PVD) method for thin film deposition.

6.4 Methodology

One of the earliest methods for creating thin films, thermal evaporation or vacuum evaporation is still widely employed in research labs and in industry to deposit metals and metal alloys. Ordered in terms of basic steps are: When a source material is boiled or sublimated, a vapor is formed, which is then (i) transferred to the substrate, (ii) transported to the surface of the substrate, and (iii) condensed to form a solid film. A wide range of chemical reactivity and vapor pressures are covered by evaporates [22, 24].

Due to this variability, a wide variety of source elements are produced, including arcs, exploding wires, lasers, and resistance heated filaments, electron beams, and crucibles heated by conduction, radiation, or rf-induction. Source-container interactions, the need for high vacuum, accurate substrate motion (to maintain homogeneity), and the requirement for process monitoring and control are now issues. During the evaporation process, heat is applied to the source material, in order to generate plume of vapor that goes directly to the substrate [21, 23].

Atoms, molecules, and molecular clusters condense from the vapor phase once they reach the substrate to form a solid layer [25, 26]. The substrate absorbs the condensation heat. The localized heating brought on by this mechanism can have significant effects on a small scale. When source-to-substrate distances and deposition rates are monitored, temperature sensitive substrates can be coated without melting [27].

The charge can be heated by induction, electric resistance, or electron beam heating, among other methods. The charge will evaporate as a result of this. Laser ablation and cathodic arc deposition of thin films might both be categorized as thermal processes in this sense [28]. The laser source is located outside the evaporation system, and after passing through a window, the laser beam is focused on the evaporated material, which is typically in the form of a fine powder.

6.4.1 Thermal Evaporation

The easiest method for producing thin films with few micrometres (μm) or less in thickness is heat evaporation, also known as vacuum deposition. In a vacuum, evaporating and condensing operations make up the thermal evaporation process. Firstly, the cylindrical chamber used in the PVD which has a furnace, designed cautiously to maintain a proper flow and have proper deposition, as this chamber is used for physical deposition of material in vapor form [29]. This technique is properly used for metals and the metals to be deposited are placed in inert crucible with a pressure up to 10^{-7} Torr. To evaporate the metal sources, the Tungsten or Electron Beam is used to heat the crucible and the evaporated material gets condensed on to the wafer to form a new layer over it [30]. This process of deposition will be moderate to fast with no damage to the substrate because it is a shear process of condensation. Resistive and electron beam sources are two different types of sources that can be used in this technique.

Diffusion pumps can be utilized as primary pumps to keep chamber pressure below 5×10^{-2} millibar and high diffusion pumps to maintain pressure below 5×10^{-5} millibar using diffusion lines. Through the roughing line, the chamber is evacuated from atmospheric pressure to a level where these pressures may be maintained using mechanical rotary pumps. In the electron beam evaporation, the vacuum coating system is made up of these elements [31].

The materials are heated using a resistively heated boat or filament, which is frequently constructed of refractory metals like tungsten, molybdenum, and tantalum with or without ceramic coating, in the resistive method to produce a thin layer on the substrate using the resistive process. First, a small amount of coating material is deposited in a boat inside a chamber, and the substrate is added to the boat. The chamber is then put into vacuum mode in order to reduce pressure and give free atoms and molecules an extremely long mean free path. The depositing material is then thermally vaporized by putting a high current (10–100 mA) through the boat, which is then heated resistively [32]. The huge vapor is then able to reach the substrate. Eventually, it returned to the solid state and formed thin film.

There are a few issues with thermal evaporation. The impurity layers that form as a result of beneficial substances reacting with the hot boat cause changes in optical characteristics. This approach subsequently has a significant impact on the material's stoichiometry, making the films produced by evaporation unsuitable for optoelectronic applications. Additionally, because the material dissolves at lower temperatures, resistive heating is unable to evaporate materials with high melting points, especially metal oxides like zirconia [33]. This substance will result in an impure layer on the surface if this technique is used. However, electronic beam (e-beam) heating eliminates these issues, making vacuum evaporation technology the preferred method for deposition films. Due to the low kinetic energy of the material's atoms during thermal evaporation, the deposition surface is protected against defect formation and damage.

6.4.2 Molecular Beam Epitaxy

A sophisticated, accurate technique for producing single-crystal epitaxial films in high vacuum is molecular beam epitaxy (MBE). The films are made on single-crystal substrates by slowly evaporating the molecules or elements from various Knudsen effusion source cells (deep crucibles in furnaces with cooled shrouds) onto substrates kept at a temperature suitable for chemical reaction, epitaxy, and excess reactant re-evaporation [34].

The Small-diameter atomic or molecule beams are produced by the furnaces and are focused on the heated substrate, usually gallium arsenide or silicon. Fast shutters are placed between the sources and the substrates. Super lattices with finely regulated uniformity, lattice match, composition, dopant concentrations, thickness, and interfaces down to the level of atomic layers can be created by varying these shutters. The schematic diagram of MBE is represented in Figure 6.2.

6.4.3 Electron Beam Evaporation

A stream of electrons is focused onto the surface of the material and accelerated via fields that are typically between 5 and 10 kV in electron beam evaporation (EBE). The substance melts and evaporates as a result of the electrons' quick loss of energy when they hit the surface. Unlike

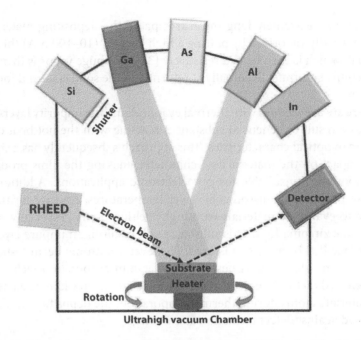

Figure 6.2 Schematic diagram of molecular beam epitaxy (MBE) method for thin film deposition.

conventional heating methods, impinging electrons immediately heat the surface. Materials from crucibles that have been cooled with water can be evaporated using direct heat. Such water-cooled crucibles are essential for evaporating reactive materials, in particular reactive refractory materials, in order to virtually minimize interactions with crucible walls. High purity films can be created when crucible materials or the by-products of their reaction are essentially excluded from the evaporation process. The two types of electron beam weapons are thermionic and plasma electrons. In the first kind, heated refractory metal filaments, rods, or discs are used to thermionically create electrons. The latter form involves the extraction of electron beams from plasma that is confined in a narrow area [35].

In the process of "electron beam evaporation" the high energy electron beam is accelerated using electric fields between 5 and 10 kV and directed at the surface of the substance to evaporate (EBE). The substance melts and evaporates as a result of the electrons' quick loss of energy when they hit the surface. Unlike traditional heating modalities, impinging electrons immediately heat the surface. Direct heating can evaporate materials from crucibles that have been cooled with water. In order to essentially minimize contacts with crucible walls when evaporating reactive materials, in particular reactive refractory materials, water-cooled crucibles are necessary. The two types of electron beam weapons are thermionic and plasma electrons. In the first kind, thermionically producing electrons is done using refractory metal. The other form involves the extraction of high energy electron beams that is limited in a narrow area. The difficulty in controlling inability to perform surface cleaning, the difficulty in improving step coverage, and the possibility of x-ray damage due to electron beam evaporation are the disadvantages.

6.4.4 Sputtering Technique

The second type of active radiation-based PVD process is sputtering, also known as radio frequency (RF) sputtering (also known as radio frequency diode), magnetrons, and ion-beam sputtering. It has a range of devices for thin-film deposition. This technique is also known as Plasma assisted technique as here the species are converted in to gaseous form and accelerated in to the plasma chamber where the plasma is created by the discharge of neutral gasses like Helium or Argon. We see the ions are accelerating through potential gradient and are made to bombard with the target or cathode through momentum transfer. The target, also known as the source material, is placed in the vacuum chamber on the cathode plate, while the substrate is situated on the anode plate. A low pressure (4×10^{-2} torr) sputtering gas, such as the inert gas argon (Ar), is then injected into the chamber. In this type of deposition system, the target (cathode) sputters as thin layers are deposited on the substrates as a result of the acceleration of argon (Ar+) ions produced during the radiance discharge toward the target by an applied electric field. The key benefits of sputtering technology over evaporation include the freedom to essentially deposit any amorphous and crystalline materials, improved control in maintaining stoichiometry, and regularity of film thickness. Additionally, polycrystalline film grain structure, which frequently has a diversity of crystallographic orientations without a specific texture, is produced via sputter deposition. However, the sputtered atoms in the sputter deposition have greater kinetic energy than the atoms in the evaporative process. The surface of the deposited material in sputtering reveals additional defects and damage because argon gas (Ar) is present in the sputtering system. Compared to thermal evaporation, nucleation. Additionally, concentration of impurity atoms in deposited films that were sputtered larger than films that are thermally deposited. Moreover, grain the sputtered films' dimensions are usually smaller than those of the coatings that evaporate and deposit. The sputtered films' dimensions are usually smaller than those of the coatings that evaporate and deposit. This is the

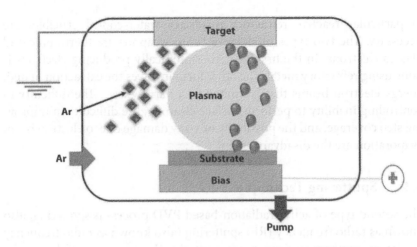

Figure 6.3 Schematic diagram of sputter deposition method for thin film deposition.

same technique as dry physical etching but here the energy is just sufficient to deposit the particles on to the film and the film grows at the surface via deposition. Ion sputtering uses a coated anode and a cathode as a target for sputter deposition. Positive argon ions produced in high-density plasma are blasted at the target, which is at a high negative potential. In the end, it condenses onto substrate that is positioned at the anode. The schematic of sputtering technique is illustrated in the Figure 6.3.

DC Sputtering

In this experiment, a direct current electric field is impressed over two water-cooled electrodes. To prevent charge accumulation on the surface during DC sputtering, positive ions impinging on the target surface must have the chance to recombine with electrons. A DC sputtering system's polarity can be reversed in order to etch the substrate before film deposition. Back sputtering is a very helpful technique for getting rid of thin surface layers like leftover oxides, which may have an impact on the electrical and mechanical characteristics of the film and substrate combination or hinder the adhesion of the film to the substrate.

RF Sputtering

The RF voltage used for radio frequency sputtering is typically between 1 and 3 kV peak to peak. The frequency of 13.36 MHz, which was designated specifically for equipment use, is commonly used for excitation.

DC sputtering cannot produce insulating films because the target surface becomes positively charged. This problem has been solved, and insulating material sputtering is now possible thanks to the use of an RF power source with an impedance matching network. The voltage applied at a frequency of 13.56 MHz can be used to charge and discharge the insulating target.

Sputtering by Diode

In a glow discharge, the cathode (or RF-powered) electrode is a plate that holds the material that needs to be deposited (target). After that, the substance can be transported from the target to a substrate and transformed into a film. Using noble gas discharges, pure metal or alloy layers can be applied on metal targets (usually Argon).

Reactive Sputtering

Reactive sputtering, which involves sputtering entities that can be used to generate compounds either directly from compound targets or indirectly from compound targets. Reactive sputtering can be used to deposit layers of substances like oxides or nitrides on pure metal targets. A reactive species, such as oxygen or nitrogen, is added to the sputtering gas (argon) in the growth chamber. The target atoms that are sputtered react with the gas to create the new substance.

Bias Sputtering

Bias sputtering, also known as ion plating, cleans the substrates by bombarding them with ions both during the deposition process and prior to the deposition of the layer. Ion bombardment may have one or more beneficial consequences during the film-deposition process, or changes to specific film properties. The source material could be an evaporation source, a reactive gas with condensable constituents, a mixture of reactive gases with condensable constituents and other gases that react with the condensed elements to form compounds, or a reactive gas with condensable constituents but no condensable constituents.

Magnetron Sputtering

A different kind of sputtering source uses magnetic fields that are perpendicular to the electric fields at the surfaces of the sputtering-targets. This type of procedure is referred to as magnetron sputtering. Transverse magnetic field sputtering modifies the fundamental processes in a number of ways. Due to their proximity to the target and their inability to bombard the substrate, secondary electrons produced by the target do not raise the temperature of the substrate or cause radiation damage. This makes it possible to use substrates that are surface- and temperature-sensitive (such metal-oxide semiconductor devices) and plastic materials that are sensitive to both. Additionally, compared to conventional sources, this kind of sputtering source yields higher deposition rates, making it appropriate for industrial applications requiring a broad surface area. For diverse purposes, each type of magnetron source—cylindrical, conical, and planar—has certain advantages and disadvantages.

Similar to other types of sputtering, reactive sputtering can make use of magnetron sources. Magnetron sources can be used as high-rate sources instead of low temperature and low radiation-damage based on the working in a bias-sputtering mode.

Electrons spiral in the direction of the magnetic field lines when a magnetic field is applied to plasma. Magnetron sputtering uses magnetic fields to boost deposition rates while keeping operating pressures and temperatures low. When a planar magnetron is used, electrons in the glow discharge take a helical path, increasing the rate of collisions and ionization. As a result, at lower operating pressures, magnetron sputtering can produce high-quality films.

Sputtering by Ion-Beam

Sputtering research is crucial because ion beams generated and recovered from glow discharges in a differentially pumped system are proving to be good film deposition technologies for specialized materials on relatively small substrate surfaces.

The deposition of ions using an ion beam has several advantages. Rather of a high-pressure glow discharge, the material which is going to be sputtered (target) and substrate are located in a well-controlled vacuum condition. Films with higher purity are frequently produced, and glow discharge artifacts are avoided.

Advantages and Disadvantages of Sputtering Technique

Sputtered or deposited materials can include elements, alloys, and compounds. The sputtering target serves as a steady, long-lasting vaporization source.

The sputtering source may take on a particular shape in various circumstances. Consider a line, a rod's surface, or a cylinder as examples. Reactive gaseous species that have been plasma activated can sometimes be used to readily achieve reactive deposition. Almost no radiant heat is generated during the deposition process. Both the substrate and the source might be close by. There may be a small sputter deposition chamber.

Sputtering rates are minimal when compared to those of thermal evaporation. Moving fixturing is required to generate consistent thickness films since the deposition flux distribution is not uniform in many configurations. Sputtering targets can be expensive and have subpar material utilization. The bulk of energy that strikes the target is transformed into heat, which needs to be expelled. The risk of film contamination over vacuum evaporation can occasionally increase when gaseous pollutants are "activated" in the plasma. To avoid poisoning the sputtering target, reactive sputter deposition's gas composition needs to be carefully controlled.

Sputter deposition is frequently used to deposit thin films on materials like semiconductors, architectural glass coatings, reflective coatings on polymers, magnetic films for storage media, transparent electrically conductive films on glass and flexible webs, wear-resistant coatings on tools, decorative coatings, and dry-film lubricants.

6.4.5 Chemical Vapor Deposition (CVD)

Two techniques for producing films that simply require chemical reactions in the gas or vapor phases are thermal oxidation and chemical vapor deposition.

CVD is a process for making materials in which elements of the vapor interact chemically with a substrate or the surface of it to create solid materials. Deposition technology is one of the most important methods for fabricating thin films and coatings of a range of materials required by advanced technology, particularly solid-state electronics, where some of the toughest purity and composition standards must be met. The main advantage of CVD is how easily both simple and complex molecules can be created at comparatively low temperatures. By controlling the reaction chemistry and deposition conditions, it is possible to customize the chemical composition and physical structure.

Gas-phase reaction chemistry, thermodynamics, kinetics, transport processes, film growth phenomena, and reactor engineering are key concepts in CVD. Examples of fundamental CVD chemical reaction types include pyrolysis (heat breakdown), oxidation, reduction, hydrolysis, production of nitride and carbide, synthesis processes, disproportionation,

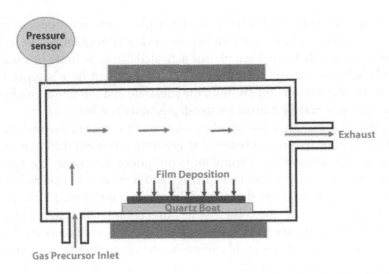

Figure 6.4 Schematic diagram of chemical vapor deposition (CVD) method for thin film deposition.

and chemical transport. In more complicated circumstances, a combination of many response types may be used to achieve a particular result. The rate of deposition and the characteristics of the film deposit are both influenced by temperature, pressure, input concentrations, gas flow rates, reactor geometry, and operating principle. As a crucial process technique in a number of industrial industries, CVD has become increasingly prevalent. For a very long time, CVD has been utilized to coat substrates at high temperatures and low pressure. Schematic diagram of chemical vapor deposition (CVD) method for thin film deposition is shown in the Figure 6.4.

Vapor-Phase Epitaxy

For the production of optoelectronic devices, compound semiconductors are deposited in epitaxial layers using metal-organic chemical vapor deposition (MOCVD) and vapor-phase epitaxy (VPE). It is necessary to develop composite layers with carefully controlled thickness and dopant profile in order to construct the structures with the best design for manufacturing devices.

Photo-Enhanced Chemical Vapor Deposition

In photo-enhanced chemical vapor deposition, electromagnetic radiation, primarily short-wave ultraviolet light, activates the reactants in the gas or vapor phase. Reactant molecules or reactive free-radical species are formed by atoms to start the process, which is ultimately combined to produce the final cinematic outcome. Typically, mercury vapor is added as a photo-sensitizer to the mixture of reactant gases sparked by the resonance lamp's potent crystalline mercury radiation (254 nm wavelength). The atoms of Mercury clash when they are excited, and energy is transferred kinetically to cause the reactants to produce free radicals.

Two benefits of this flexible and promising CVD method are the low temperature (about 150°C) required to create films like SiO_2 and Si_3N_4 and the greatly reduced radiation damage (compared to PECVD). Due to a lack of effective production equipment, it is currently required to photoactivate with mercury in order to obtain adequate rate of thin film deposition.

Laser-Induced Chemical Vapor Deposition (LICVD)

A highly localized laser beam is used in Laser-Induced Chemical Vapor Deposition (LICVD) to heat the substrate, resulting in layer deposition through CVD surface reactions.

By absorbing the given photonic energy at a particular wavelength, gaseous reactant atoms or molecules can be activated using laser (or electron radiation). High purity film deposits are produced as a result of very specialized chemical gas phase processes.

Atmospheric Pressure Chemical Vapor Deposition

A chemical vapor deposition technique called, Atmospheric Pressure Chemical Vapor Deposition (APCVD) deposits both doped and undoped oxides at atmospheric pressure, which is normal pressure. Due to the low temperature, the deposited oxide has a moderate density and coverage. At rates ranging from 2000 to 3000Å/min, the surface reaction is there on the heated wafer, which is generally heated to 400°C, to grow films. The process's low temperature, straightforward reactor construction, and rapid wafer throughput are all key benefits. Problems with APCVD include insufficient step coverage, rapid precursor flow, particle contamination, and ongoing cleaning. With a greater understanding of the reaction processes and reactant injection, some of these step coverage issues can be reduced. New reactor designs that are still in use today incorporate these advances

Low Pressure Chemical Vapor Deposition (LPCVD)

In Low pressure chemical vapor deposition (LPCVD), a vacuum is employed. Processes like LPCVD produce practically perfect conformance. This is because non-uniform particle movement results from the air pressure, which ranges from 10 to 100 Pa. The particles spread out and cover both vertical and horizontal surfaces as a result of collisions. Up to 900°C of high temperature can support the compliance. The stability and density are astronomically great when compared to APCVD. The process of LPCVD is used to create thin films on semiconductors with thicknesses ranging from a few nanometers to many micrometers. A wide variety of film compositions with great conformal step coverage are deposited using LPCVD.

Plasma Enhanced Chemical Vapor Deposition (PECVD)

This CVD technique uses plasma and reactive gases in the deposition chamber to coat the substrate with the required solid surface. Any vapor in which a sizable fraction of the atoms or molecules are ionized, energetic, and settle on the surface is referred to as plasma. PECVD is performed at temperatures between 250 and 350°C. The process gases cannot be thermally degraded at these low temperatures. Direct current (DC) or alternating current (AC) discharge between two electrodes, with reacting gases filling the space in between, produces the plasma. Si_3N_4 and SiO_2 are deposited on the upper of metal coatings using PECVD since metallization, like aluminum.

The PECVD process, which uses plasma as an energy source, can be used at low temperatures and low pressures. Despite the conformance being less than ideal compared to LPCVD, the rate of deposition is significantly higher. To create films of the proper quality, process variables including gas flow rates, power input, chamber pressure, reactor geometry, inter electrode spacing, substrate temperature, and others can be controlled. Low temperature, improved film density for better dielectric and more compression, and ease of chamber cleaning are all advantages of PECVD.

6.4.6 Atomic Layer Deposition (ALD)

Making thin films involves a modified CVD process known as ALD (Atomic Layer Deposition). The technique employs a number of gases, which are cycled in and out of the chamber. It is a method for thin film deposition that uses a chemical reaction in the gas phase over time. Each gas reacts in a manner that causes the reaction to stop once the current surface is saturated. This surface can respond to the alternative gas in a similar manner. Between each of these gas reactions, the chamber is flushed out with an inert gas, such as nitrogen or argon.

An easy ALD procedure might resemble this:

1. Surface reaction with initial gas that is self-limiting
2. Purging with an inert gas
3. Surface reaction with a self-limiting second gas
4. For purging inert gas to be used

Atomic-scale control is possible because to the self-limiting and surface reaction characteristics of ALD film production. ALD and CVD have the same chemistry; the difference is that ALD divides the CVD reaction into two half-reactions, retaining the separation of the precursor materials throughout the reaction. The coating method allows for atomic layer control of film production down to 0.1 (10 pm) each cycle by maintaining the separation of the precursors.

After each precursor pulse, a purge gas (usually nitrogen or argon) is pulsed to clear the process chamber of excess precursor and stop "parasitic" CVD deposition on the substrate.

6.4.7 Liquid Phase Chemical Formation Technique

By using electrochemical techniques and chemical deposition techniques, inorganic thin films are primarily produced from liquid phases. There are numerous in-depth evaluations of various filmmaking techniques that cover both theory and practice.

Another approach to liquid-phase film creation relies on mechanically deposited chemically reactive films. And finally, a variety of single crystal semiconductors are still grown via liquid phase epitaxy.

6.4.8 Electrolytic Anodization

An oxide layer is produced from the substrate by anodization, much like it is with heat oxidation. A hydrated oxide layer on semiconductors and a limited metal are produced by the anode's reaction with the electrolyte's negative ions in solution, while H_2 gas is produced by the cathode.

Oxides which are nonporous with good adhesion can form on silicon, niobium, tantalum, titanium, aluminum and zirconium. The most crucial uses are for tantalum and aluminum electrolyte capacitor electrical insulation layers, cosmetic-colored coatings, and corrosion-resistant films on aluminum and related alloys.

6.4.9 Electroplating

The coating of metal is electrodeposited on the cathode of an electrolytic cell, which consists of an anode, a cathode, and an electrolyte solution (which includes metal ions through which electric current travels). Faraday's laws govern the process' quantitative components. pH, temperature, agitation, current density, current distribution, and solution composition are all significant electroplating variables. Aqueous solutions have been successfully used to electroplate a variety of metals and metal alloys. The industrial process of electroplating can result in deposits with thicknesses ranging from very thin sheets to very thick coatings.

Plating by Chemical Reduction

Chemical diminution A metal ion in solution is reduced during plating by a reducing agent that is applied just before usage. Because of the homogenous nature of the reaction, deposition happens everywhere across the solution rather than only on the substrate.

Some examples of plated films include films made of silver, copper, nickel, gold, and certain sulphides. Silvering glass and polymers for the creation of mirrors, which uses silver nitrate solutions and one of many reducing agents, including hydrazine, is the process's earliest application.

Electroless Plating

A reducing agent reduces metal ions in solution to a metallic coating during the selective deposition plating method known as autocatalytic plating, sometimes referred to as electroless plating. The term "autocatalysis" refers to the fact that plating only takes place on surfaces that are suited for catalysis, such as substrates made of the same metal as the metal being plated. Selective (patterned) deposition is one of the benefits of electroless (or electrodeless) plating, which is only applicable to selective materials of interest.

Electrophoretic Deposition

Colloidal particle dispersion is used to deposit a film from an electrophoretic coating onto a conductive substrate. The dispersion separates into positively charged ions (cations) and negatively charged colloidal particles in a conductive liquid, or vice versa.

Colloidal particles travel to the substrate and discharge when an electric field is provided between the positive substrate electrodes, causing them to then form a film.

Immersion Plating

Displacement deposition, also known as immersion plating, is the chemical displacement deposition of a metal layer on a substrate from a metal coating's dissolved salt without the use of external electrodes. The electromotive force series states that any more noble metal is displaced from solution by a less noble (more electronegative) element. In actuality, the metal surface develops cathodic and anodic regions, with cathodic regions showing thicker coatings. There are just a few industrial uses for this method, mostly thin coatings on copper and copper alloys.

6.5 Advantages of CVD Process

As stated in the introduction, there are both benefits and drawbacks to the CVD system. The ability of this system to produce and control crystal structure is its greatest benefit. Furthermore, uniform, high-density films constructed of pure materials [36]. The CVD system is capable of producing films with high clone and repeatability at reasonable deposition rates, adhesion. This, however, is incorrect as film stoichiometry cannot be controlled by more than one method because the properties of various materials differ in evaporation rates.

6.6 Comparison Between PVD and CVD

Physical vapor deposition and chemical vapor deposition systems share a variety of similarities and distinctions. PVD and CVD vacuum systems are comparable to beginning with. They are fitted with a vacuum chamber, a high-vacuum diffusion pump, and a low-vacuum rotary pump. The PVD process, on the other hand, typically operates under high vacuum, whereas the CVD process operates under low vacuum [37].

The fact that both PVD and CVD processes typically deposition at temperatures greater than 150 °C is the second most notable commonality between PVD and CVD techniques in terms of temperature. While CVD coating is applied at temperatures between about 450°C and roughly 1050°C, PVD coating is applied at temperatures between 250°C and 450°C.

Last but not least, whereas in PVD the material is put onto the substrate in solid form, in CVD it is introduced into the substrate in gaseous form. However, in PVD, atoms travel and deposit on the substrate, whereas in CVD, gaseous molecules react with the substrate. Overall, it seems that the temperature and vacuum requirements for PVD and CVD systems are very different [38].

6.7 Conclusion

The technology utilized to deposit thin film materials have advanced significantly over the last 195 years. The two primary methods for creating thin films are physical vapor deposition and chemical vapor deposition. In this project, several important aspects and methods of the thin film deposition process have been presented. These include a brief explanation of the PVD process, several PVD systems, including thermal evaporation and sputtering, a brief explanation of the CVD process, a crucial CVD technique known as PECVD, and the benefits and drawbacks of both systems. Finally, PVD and CVD have been compared and contrasted. Processes for thin film deposition using PVD and CVD are described in depth. It can be said that the thermal evaporation method using an electron beam works well.

Source is the best method for depositing thin films on both high- and low-melting-point materials. Sputtering, on the other hand, is useful for depositing thin polycrystalline films. Whereas the PECVD technology performs better for dielectric films than other CVD methods. Numerous studies have shown that the PVD method is superior to the CDV method for producing thin films. The vacuum system and deposition temperature were found to be the two aspects of the PVD and CVD processes that are most similar to one another.

But each system has a very distinct temperature and vacuum level. On the other hand, the way that materials are introduced is where there is the most obvious difference. For instance, the substance is delivered onto the substrate in solid and gaseous forms, respectively, in PVD and CVD. It seems that PVD and CVD are somewhat comparable.

References

1. Fu, L., Kane, C.L., Mele, E.J., Topological insulators in three dimensions. *Phys. Rev. Lett.*, 98, 106803, 2007.
2. Zhang, H., Liu, C.-X., Qi, X.-L., Dai, X., Fang, Z., Zhang, S.-C., Topological insulators in Bi2Se3, Bi2Te3and Sb2Te3 with a single Dirac cone on the surface. *Nat. Phys.*, 5, 438, 2009.
3. Moore, J.E. and Balents, L., Topological invariants of time-reversal-invariant band structures. *Phys. Rev. B*, 75, 121306, 2007.
4. Xia, Y., Qian, D., Hsieh, D., Wray, L., Pal, A., Lin, H., Bansil, A., Grauer, D., Hor, Y.S., Cava, R.J., Hasan, M.Z., Observation of a large-gap topological-insulator class with a single Dirac cone on a surface. *Nat. Phys.*, 5, 398, 2009.
5. Hasan, M.Z. and Kane, C.L., Colloquium: Topological insulators. *Rev. Mod. Phys.*, 82, 3045, 2010.
6. Akhmerov, A.R., Nilsson, J., Beenakker, C.W.J., Electrically detected interferometry of majorana fermions in a topological insulator. *Phys. Rev. Lett.*, 102, 216404, 2009.
7. Yazyev, O.V., Moore, J.E., Louie, S.G., Spin polarization and transport of surface states in the topological insulators Bi2Se3 and Bi2Te3 from first principles. *Phys. Rev. Lett.*, 105, 266806, 2010.
8. Checkelsky, J.G., Hor, Y.S., Liu, M.-H., Qu, D.-X., Cava, R.J., Ong, N.P., Quantum interference in macroscopic crystals of non-metallic Bi2Se3. *Phys. Rev. Lett.*, 103, 246601, 2009.
9. Checkelsky, J.G., Hor, Y.S., Cava, R.J., Ong, N.P., Bulk band gap and surface state conduction observed in voltage-tuned crystals of the topological insulator Bi_2Se_3. *Phys. Rev. Lett.*, 106, 196801, 2011.
10. Chen, J., He, X.Y., Wu, K.H., Ji, Z.Q., Lu, L., Shi, J.R., Smet, J.H., Li, Y.Q., Tunable surface conductivity in Bi_2Se_3 revealed in diffusive electron transport. *Phys. Rev. B*, 83, 241304, 2011.
11. Onose, Y., Yoshimi, R., Tsukazaki, A., Yuan, H., Hidaka, T., Iwasa, Y., Kawasaki, M., Tokura, Y., Pulsed laser deposition and ionic liquid gate control of epitaxial Bi_2Se_3 thin films. *Appl. Phys. Express*, 4, 083001, 2011.

12. Qu, F., Yang, F., Chen, J., Shen, J., Ding, Y., Lu, J., Song, Y., Yang, H., Liu, G., Fan, J., Li, Y., Ji, Z., Yang, C., Lu, L., Strong superconducting proximity effect in Pb-Bi_2Te_3 hybrid structures. *Phys. Rev. Lett.*, 107, 016802, 2011.
13. Husmann, A., Betts, J., Boebinger, G., Megagausssensor. *Nature*, 417, 421, 2002.
14. Qu, D.X., Hor, Y.S., Xiong, J., Cava, R.J., Ong, N.P., Quantum oscillations and hall anomaly of surface states in the topological insulator Bi_2Te_3. *Science*, 329, 821, 2010.
15. Wang, X.L., Du, Y., Du, S.X., Zhang., C., Room temperature giant and linear magneto resistance in topological insulator Bi_2Te_3 Nanosheets. *Phys. Rev. Lett.*, 108, 266806, 2012.
16. He, H., Li, B., Liu, H., Guo, X., Wang, Z., Xie, M., Wang, J., High-field linear magneto-resistance in topological insulator Bi_2Se_3 thin films. *Appl. Phys. Lett.*, 100, 032105, 2012.
17. Tang, H., Liang, D., Qiu, R.L.J., Gao, X.P.A., Two-dimensional transport-induced linear magneto-resistance in topological insulator Bi_2Se_3 nanoribbons. *ACS Nano*, 5, 7510, 2011.
18. Zhang, G.H., Qin, H.J., Teng, J., Guo, J.D., Guo, Q., Dai, X., Fang, Z., Wu, K.H., Topological insulator Bi_2Se_3 thin films grown on double-layer graphene by molecular beam epitaxy. *Appl. Phys. Lett.*, 95, 053114, 2009.
19. He, H., Li, B., Liu, H., Guo, X., Wang, Z., Xie, M., Wang, J., High-field linear magneto-resistance in topological insulator Bi2Se3 thin films. *Appl. Phys. Lett.*, 100, 032105, 2012.
20. Kim, D., Syers, P., Butch, N.P., Paglione, J., Fuhrer, M.S., Coherent topological transport on the surface of Bi_2Se_3. *Nat. Commun.*, 4, 2040, 2013.
21. Yan, Y., Wang, L.X., Yu, D., Liao, Z., Large magnetoresistance in high mobility topological insulator Bi_2Se_3. *Appl. Phys. Lett.*, 103, 033106, 2013.
22. Zhang, G. and Qin, H., Growth of topological insulator Bi_2Se_3 thin films on $SrTiO_3$ with large tunability in chemical potential. *Adv. Funct. Mater.*, 21, 2351, 2011.
23. Garate, I. and Glazman, L., Weak localization and antilocalization in topological insulator thin films with coherent bulk-surface coupling. *Phys. Rev. B*, 86, 035422, 2012.
24. Bergmann, G., Physical interpretation of weak localization: A time-of-flight experiment with conduction electrons. *Phys. Rev. B*, 28, 2914, 1983.
25. Steinberg, H., Laloe, J.B., Fatemi, V., Moodera, J.S., Jarillo-Herrero, P., Electrically tunable surface-to-bulk coherent coupling in topological insulator thin films. *Phys. Rev. B*, 84, 233101, 2011.
26. Novoselov, K.S., Coherent topological transport on the surface of Bi_2Se_3. *Nature*, 438, 197, 2005.
27. Qi, X.L. and Zhang, S.C., Topological insulators and superconductors. *Rev. Mod. Phys.*, 83, 1057, 2011.
28. Xia, Y., Observation of a large-gap topological-insulator class with a single Dirac cone on the surface. *Nat. Phys.*, 5, 6, 398–402Y, 2009.

29. Hikami, S., Larkin, A.I., Nagaoka, Y., Spin-orbit interaction and magnetoresistance in the two-dimensional random system. *Prog. Theor. Phys.*, 63, 707, 1980.
30. Wenhong, W. and Yin, D., Large linear magnetoresistance and shubnikov-de hass oscillations in single crystals of YPdBi heusler topological insulators. *Sci. Rep.*, 3, 2181, 2013.
31. Abrikosov, A.A., Quantum magnetoresistance. *Phys. Rev. B*, 58, 2788, 1998.
32. Abrikosov, A.A., Quantum linear magnetoresistance. *Europhys. Lett.*, 49, 789, 2000.
33. Hu, J. and Rosenbaum, T.F., Classical and quantum routes to linear magnetoresistance. *Nat. Mater.*, 7, 697, 2008.
34. Lee, M., Rosenbaum, T.F., Saboungi, M.L., Schnyders, H.S., Band-gap tuning and linear magnetoresistance in the silver chalcogenides. *Phys. Rev. Lett.*, 88, 066602, 2002.
35. Zhang, Y., He, K., Chang, C.-Z., Song, C.-L., Wang, L.-L., Chen, X., Jia, J.-F., Fang, Z., Dai, X., Shan, W.-Y., Shen, S.Q., Niu, Q., Qi, X.L., Zhang, S.C., Ma, X.C., Xue, Q.-K., Crossover of the three-dimensional topological insulator Bi_2Se_3 to the two-dimensional limit. *Nat. Phys.*, 6, 584, 2010.
36. Analytis, J.G., McDonald, R.D., Riggs, S.C., Chu, J.H., Boebinger, G.S., Fisher, I.R., Two-dimensional Dirac fermions in a topological insulator: Transport in the quantumlimit. *Nat. Phys.*, 6, 705, 2010.
37. He, L., Surface-dominated conduction in a 6 nm thick Bi_2Se_3 thin film. *Nano Lett.*, 12, 3, 1486–1490, 2012.
38. Pradhan, S.K. and Barik, R., Observation of the magneto-transport property in a millimeter-long topological insulator Bi_2Te_3 thin-film hall bar device. *Appl. Mater. Today*, 7, 55–59, 2017.

7
Printable Photovoltaic Solar Cells

Tuiba Mearaj[1], Faisal Bashir[2], Rayees Ahmad Zargar[3*], Santosh Chacrabarti[1] and Aurangzeb Khurrem Hafiz[1]

[1]Centre for Nanoscience and Nanotechnology, JMI, New Delhi, India
[2]Department of Electronics and Intstrumentation Technology, Kashmir University, Srinagar (J&K), India
[3]Department of Physics, BGSB University Rajouri (J&K), India

Abstract

Printable photovoltaic modules and other electronic components like light-emitting diodes, thin-film transistors, capacitors, coils, resistors, and so on are less expensive than deposited devices. The printable solar cell (PSC) has potential in many application domains due to its ease of fabrication and ability to use large-area flexible substrates. PSC's light-absorbing layer is usually thinner than that of commonly used conventional Si solar cells in several orders of magnitude. This allows less consumption of material for PCS manufacturing and producing and there is significantly less wastage. Large area flexible substrates enable smart clothing and solar-powered window coverings, among other cutting-edge applications. In this chapter, the metal oxide electron and hole transport materials used in PSCs are reviewed and preparation of these materials is summarized. Finally, the challenges and future research direction for metal oxide-based charge transport materials are described.

Keywords: Metal oxide, solar cells, thin film, screen printing

7.1 Introduction

Regardless of their shape or thermal stability, laminated photovoltaics and printed electronics in general allow for the efficient deposit of components over a large area on a variety of substrates. Electrically functioning

---

Corresponding author: rayeesphy12@gmail.com

Rayees Ahmad Zargar (ed.) Metal Oxide Nanocomposite Thin Films for Optoelectronic Device Applications, (179–202) © 2023 Scrivener Publishing LLC

electronic or photonic inks either conductive or semiconducting are deposited on the substrate to build active or passive devices like thin film transistors, capacitors, coils, and resistors. Organic materials are frequently used in printed electronics (both polymers and tiny molecules). Perovskite, nanocarbon, and other novel paste or hybrid materials have been created. These inorganic materials have the ability to be imprinted (like silicon, silver, TiO2, and many more) [1].

It is possible to create thin, flexible, eco-friendly goods well as printed electrical components that can be used in existing systems using printed electronics. Organic light-emitting diode (OLED) displays, smart clothes, solar cells, flexible displays, printed batteries, supercapacitors, memory, or label-protection tags are a few examples of possible applications based on natural electronics systems. If these novel materials are dissolved in an appropriate organic liquid, they can be imprinted in massive quantities. This is an important consideration for the use of printed electronic devices in a variety of applications because they can be quite cheaper than inorganic devices manufactured using traditional methods [2–5].

7.2 Working Principle of Printable Solar Cells

The ability to wet process, or use procedures for solution processing, is a requirement for printed solar cells. A solar cell that meets this requirement can be divided into three types based on its application potential.

The first type of solar cell is copper indium gallium selenide (CIGS), also known as a photovoltaic (PV) cell. Amorphous silicon and cadmium telluride are two other thin-film PV technologies. There are already large quantities of solar cells produced in this form, including those deposited onto flexible substrates for solar panels. In the traditional manufacturing process, primers are deposited by vacuum, which increases the cost of the final product. In addition, all previous record efficiencies were achieved on glass support.

The second type is the perovskite solar cell [6], which has emerged as a "superstar" in photovoltaic technology and has drawn the attention of the scientific community in particular for its distinctive qualities. Record efficiencies of perovskite solar cells have been achieved using solution processing, in contrast to CIGS solar cells. This form of photovoltaic has several absorber layers, which are commercially inexpensive, and has excellent efficiencies. However, there are still some stability [8] and toxicity [7] problems that need to be fixed.

Solar cells with bulk heterojunctions are the third type of solar cells. These solar cells key benefits include cheap manufacturing costs, adaptability, nontoxicity, and color tunability. The average power conversion efficiency for organic solar cells has remained constant at 11% among the three types of printed photovoltaics [9, 10]. However, organic photovoltaic cells continue to be feasible in the worldwide market for their nontoxicity and unrivalled low weight [11].

7.3 Wide Band Gap Semiconductors

Solar cells based on wide-band-gap semiconductors work better under water than the narrow-band-gap ones used in conventional silicon photovoltaic devices. The various semiconductor solar cells are given below.

7.3.1 Cadmium Telluride Solar Cells (CIGS)

Cadmium telluride (CdTe), amorphous thin-film silicon, and copper indium gallium selenide (CIGS) are the three main components of thin-film photovoltaic cells [12]. The most common method for making these solar cells is vacuum deposition, but it may also be effectively produced from precursors [13, 14].

Actively bridging the efficiency gap between silicon wafer cells and other photovoltaic systems is one of the solar technologies with the most potential. Due to its wide spectral range, which absorbs the majority of sunlight [15], and it's extremely high extinction coefficient, a CIGS requires a significantly thinner coating than other inorganic semiconductor materials. Manufacturers have reported power conversion efficiencies of above 20% at glass substrates; Flexible CIGS solar cells on plastic, however, have a power conversion efficiency of 18.7%, according to experts from the Swiss Federal Laboratories for Materials Science and Technology. All solar-powered cells operate primarily on the basis of the photoactive semiconducting material (CIGS absorber) absorbing sunlight. When photons are absorbed, electron-hole pairs or excitons are created, which awaken electrons from their ground-state molecular/atomic energy orbitals. In order to discharge its energy, an excited electron has two options: either it will release it as heat and return its initial orbital, or it will go through the cell until it reaches the electrode. To attain the material's potential equilibrium, current flows through it, and this electricity is collected.

7.3.2 Perovskite Solar Cells

Perovskite solar cells with lead halide offer many advantages over the more common Si solar cells. The processing of perovskite photovoltaics is quite straightforward, which is one of the key benefits. Perovskite solar cells, as opposed to silicon cells, don't need pricey, multi-step procedures. Simultaneously, composite organic-inorganic halide perovskite solar cells may be created using solution process technology, including ink printing in a conventional laboratory [16]. Additionally, thorough balancing analysis [17] reveals that perovskite solar cells have an efficiency limit of roughly 31%, which is close to the Shockley-Queisser limit (33%) with the aim of sustainable energy, the aforementioned characteristics of perovskite photovoltaics make them a financially appealing choice.

Due to the remarkably rapid increase in power conversion efficiencies from less than 4% in 2010 to a record efficiency of 22% in 2016, perovskite solar cells have become one of the most divisive photovoltaic technologies. For effective light absorption, the lead halide perovskite active layer may be very thin, just a few hundred nanometers thick, comparable to the other technologies mentioned in this chapter.

7.3.3 Solar Cells Based on Additive Free Materials

A photovoltaic device known as an organic solar cell employs semiconducting organic materials (polymers or tiny molecules) to absorb light and transfer charges [18]. The bulk-heterojunction solar cell made of polymer and fullerene is one of the most popular types of organic photovoltaics. In comparison to inorganic solar collector cells like the often used Si, CdTe, or GaAs solar cells, this form of solar cell is simple to produce and possibly affordable, but it still has certain drawbacks, including poor efficiency [19–21] and long-term stability [22]. On the contrary side, the bulk of the components used in organic solar cells is soluble and may, as a result, be deposited using a variety of printing and coating techniques.

Since organic molecules have a high optical absorption coefficient, majority of the solar energy may be soaked up by the surface that is just a few hundred nanometers thick, on average. As a result, they may be made in large quantities at a low cost using a wide range of substrates; for example, organic photovoltaics can be placed on paper [24] and cloth [23]. Additionally, as demonstrated in Solar Park [25], the installation of organic solar cells that are bendable is very quick and simple. Additionally, organic solar cells have an unmatched weight-to-power ratio, which might be crucial in industries with such high demands as aeronautical engineering. Frederic C. Krebs from the Technical University of Denmark is one of the world's top researchers in the field of mass manufacturing of additive free solar cells. Krebs has published a number of articles on printing and coating methods [26–29]. The capacity of organic conjugated materials to be modified via molecular engineering is another distinctive quality (e.g., changing the length and functional group of polymers). In this way, modifying a molecule's functional groups can affect the bandgap, enabling electrical and optical tunability. Because solar cells may be manufactured in any hue, tunable organic materials are beneficial not just in terms of their electrical qualities but also in terms of their aesthetic appeal. For instance, organic semitransparent solar cells of any aesthetically pleasing hue can be used to create photovoltaic stained glass windows, whereas inorganic or hybrid solar cells are typically black, brown, or grey. A transparent anode constructed of one or two layers of organic semiconductor, or a mixture of similar materials, and a metallic conductor, such as aluminum cathode, makes up a regular organic solar cell. They perform an objective and are based on photoelectric phenomena. Charge carriers are formed after light has been absorbed, to put it as simply as possible. Current moves through the circuit as charge carriers are delivered to the electrodes. In the case of organic solar cells, the created particle is referred to as an exciton [30], which is an electron-hole pair bonded together with energy of around 0.4 eV. Excitons can dissociate in an electric field [31] at around 10^6 V cm^{-1}. Oftentimes, light absorption does not result in the creation of charge carriers [32]. The exciton needs to be broken apart in order for the electrodes to carry the charge. The diffusion length of an exciton generally varies by around 10 nm in organic materials such as polymers and pigments [33]. The active layer has to be roughly 100 nm thick to provide adequate absorption of light in the visible spectrum, which results in more excitons that can combine. Schottky diodes, which had an anode, a cathode, and one organic layer, were the original solar cells. As the photogeneration could only be accomplished in a tiny area at the organic layer and metal electrode contact, the devices were somewhat useless. The heterojunction, which is the interface between two organic semiconductors, provides the foundation for more efficient solar cells. Despite the fact that the heterojunction in the case of a bilayer is tiny and difficult for charge carriers to access, For this reason, bulk heterojunction solar cells are more appropriate . There are several connectors for organic solar cells, it must be highlighted. There are discrete heterojunctions, bulk heterojunctions, continuous junctions, discrete heterojunctions, bilayers, and single layers. We shall concentrate on the bulk heterojunction kind of junction, though. A solution method is mostly used to handle this sort of junction. This type of connector is the most promising when mass production is considered because of how easy it is to manufacture [34].

7.3.4 Charge-Carrier Selective Layers That Can Be Printed

The three main forms of printed solar cells addressed in this chapter all require an effective charge carrier separation at the anode and cathode, therefore high electron/hole selectivity of both contacts is required. Contact selectivity is greatly influenced by the band structure (or HOMO/LUMO levels for organic materials) of the interfacial materials, as well as the high mobilities of one kind of charge carrier and effective blocking of the other, i.e., unipolarity is required.

Metal oxides, such as MoO_x, WO_3, NiO_x, Al_2O_3, etc., provide the basis for charge carrier selective connections in the majority of widely used materials, whereas ZnO, TiO_2, and LiO are employed for electron injection and transport. Typically, these materials may be deposited via vacuum evaporation (for example, MoO_x, WO_3) or the sol-gel method (for example, ZnO, TiO_2), However these processes are ineffective if completely printable devices are to be taken into account.

Contrary to the notion of low processing temperatures for printable electronics, sol-gel is a solution-processable technology that frequently requires a high-temperature post-deposition treatment (up to 300°C and more) and isn't suitable for many flexible polymeric or textile substrates. However, there are already available metal oxide nanoparticle dispersion inks that do not require high-temperature processing. As an alternative, one of the most promising options for a printed charge transfer layer is carbon-based materials such as graphene [35], graphene oxide [36], or carbon nanotubes [37, 38] aqueous dispersions.

7.4 Metal Oxide-Based Printable Solar Cell

Si-based photovoltaic technology has been extensively studied to lower production costs and increase device efficiency. Due to its exceptional qualities in electrical and optoelectronic applications, titanium dioxide (TiO2) has received a lot of attention. Due to its wide band gap energy of 3.35 eV, TiO2 is only able to absorb UV radiation from the sun [39, 40]. As a result, by doping TiO2 with the appropriate metal oxides, such as NiO, Fe2O3, Al2O3, CdO, and others, the absorption range of TiO2 must be increased. With a straight band gap of 2.2 eV and great transparency, cadmium oxide (CdO) is a desired II-VI semiconductor that produces better results in optoelectronic applications by absorbing most of the sun spectrum.

By enabling more light to flow through, metal oxides having a wide bandgap are useful for increasing the amount of light that the perovskite layer can absorb. The performance of the device is also impacted by the morphology and crystallinity of the transport layers. Doping or surface modification can further improve the electrical and optical properties. The band alignment of the representative metal oxides, which are frequently utilized as the charge transport layers in PSCs, is summarized in Figure 7.1. Due to its broad bandgap, excellent chemical stability, and acceptable energy levels compatible with light-absorbing layers including dyes, quantum dots, etc. The best materials that fulfill these needs are metal oxides. The metal oxides used in PSCs as the transport layers must meet a number of requirements in order to produce highly efficient PSCs. To successfully move electrons or holes from the perovskite layer to the transport layer and enhance the built-in potential, they must first have a compatible band alignment with the perovskitelayer [41]. In order to transfer electrons and holes and minimize charge recombination inside the transport layer, a good charge mobility is necessary [42].

Solar cells are already often constructed using titanium dioxide. TiO2 is not only a material that is beneficial to the environment and non-toxic, but

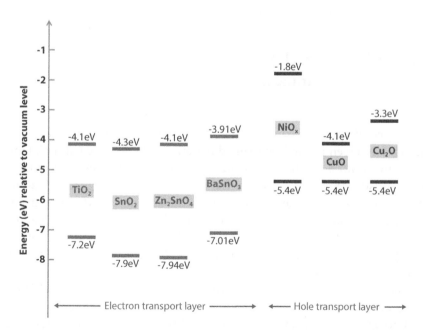

Figure 7.1 Band alignment of the representative metal oxide electron transport layer (ETL) and hole transport layer (HTL).

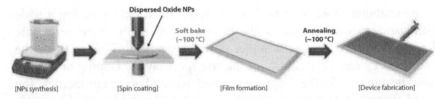

Figure 7.2 Schematic illustration of the low-temperature process through colloidal NPs.

it also has great electrical qualities, according to the university. Another semiconductor with high optical transparency is Cadmium oxide. CdO is a fantastic material to create eco-friendly cells because cadmium is one of the most common elements on Earth and its oxide is easily produced at low industrial temperatures.

Utilizing well-dispersed TiO2 NPs and spin-coating on the substrate are the simplest methods for creating planar type PSCs, as shown in Figure 7.2. Because of the interconnectedness, low-temperature produced TiO2 NPs films for planar PSCs exhibit worse electron transport properties (the necking of NPs) inadequate compared to mp-TiO2 based PSCs between individual NPs. Wang *et al.* looked at nanocomposites of hydrothermally produced TiO2 NPs and high grade graphene flakes as low-temperature processed ETLs to solve this issue [43].

7.5 What is Thick Film, Its Technology with Advantages

Thick film is a 'printed-and-fired' film deposition technology that makes use of conductive, resistive, and insulating pastes containing glass frit, which are screen printed on defined patterns and then fused at high temperatures onto a ceramic substrate. Thick films with resistivity ranging from 10/square to 10 MΩ/square have thicknesses ranging from 5–20 m. Multi-layer structures are frequently made with them. Microelectronic devices used thick films in the 1950s. It has been discovered to be a viable alternative to printed circuit board technology for producing reliable integrated circuits. Since the 1960s, the era of silicon technology has seen a boom in this technology.

The great Chinese dynasties of around a thousand years ago used thick films deposition by screen printing (stenciling) and graphic reproduction technologies. There is evidence that primitive stenciling techniques were also used to create Palaeolithic cave paintings from around 15000 BC.

Thick film technology drew attention in the 1980s when it came to the development of surface mount devices (SMD), which are circuits that are built without the use of through-hole components. This technology has grown rapidly in recent years, and it is now the most widely used technique for representing microelectronics on a small scale.

The concept of a screen is used in modern thick-film printing equipment, with the mesh material typically being fine stainless steel or polyester strands. Because of its low cost, ease of handling, large surface area, and ability to operate in a wide range of temperatures, thick film technology is more widely used than thin film technology [44, 45]. The easy fabrication of materials such as piezoelectrics and magnetostrictives, which can be combined with MEMS to produce novel solutions to a variety of sensor, actuator, and other components that serve a wide range of optoelectronic applications [46], is a fundamental aspect of thick film technology.

7.5.1 Thick Film Materials Substrates

Ceramics, such as alumina with particle sizes in the 3–5 m range, are the most commonly used substrate materials for thick film deposition. The can be used as a glass substrate or as fired, polished ceramics. The insulation resistance of the glass is critical, and glasses with low free-alkali percentages are required if high-temperature film sintering is desired. Many new substrate materials, such as 'porcelainized steel' (vitreous enameled steel), organic materials like epoxies, flexible substrates, and even synthetic diamond, have been proposed and used in various applications.

7.5.2 Thick Film Inks

Thick film technology is essentially an additive process in which various components are formed on the substrate by layering 'inks' (or 'pastes') to create the desired conductor, resistor, or dielectrics designs.

Depending on the device fabrication requirement, different types of paste containing a binder: glass frit, a carrier: organic solvents, plasticizers, and the material: pure metals, alloys, and metal oxides whose film to be deposited have been used.

This paste is applied in a pattern to the substrate, then sintered in a furnace to achieve the desired stochiometry and structure.

The surface of the films is generally not uniform or homogeneous at the micro level after the firing process, which causes wirebonding issues.

Inks are being developed to provide optimal low viscosity for screen printing for uniform spreading in order to solve this problem. Generally, the paste manufacturer will recommend a mesh type for his paste, and this is always a good place to start: The most common stainless steel mesh sizes are 200 mesh and 325 mesh. For screen printing, there are four useful "rules of thumb": There are a number of factors that contribute to high-quality printing. The topography of the surface plays an important role in screen printing. For thin film applications, the substrate must pass the following characteristics: an excessively smooth surface results in poor adhesion, while a very rough surface results in poor thickness reproducibility. A smooth, uniform surface that allows the fired thin film layers to adhere properly. The substrate should be free of distortion or bowing, and it should be able to withstand normal firing temperatures, which are usually between 500 and 1000°C.

7.6 To Select Suitable Technology for Film Deposition by Considering the Economy, Flexibility, Reliability, and Performance Aspects

Films that are thick in this chapter, screen printing technology has been chosen.

This basic screen printing technique was first used to make hybrid circuits with thick film layers, semiconductor devices, monolithic integrated circuits, and other discrete devices about 46 years ago. The substrate is a flat piece of alumina that is typically one to six inches square and 0.025 to 0.040 inches thick. Other substrate materials are used as well, but they are not significantly different from a screen printing standpoint. All of them are abrasive, brittle, and easily marked inadvertently. These characteristics have an impact on jig design and handling practices: there are few completely automated lines, especially in the high-reliability area. Screens are made of stainless steel or polyester mesh that is highly tensioned. Generally, the paste manufacturer will recommend a mesh type for his paste, and this is always a good place to start: The most common stainless steel mesh sizes are 200 mesh and 325 mesh. For screen printing, there are four useful "rules of thumb": There are a number of factors that contribute to high-quality printing. The topography of the surface plays an important role in screen printing. For thin film applications, the substrate must pass the following characteristics: an excessively smooth surface results in poor adhesion, while a very rough surface results in poor thickness reproducibility. A smooth, uniform surface that allows the fired thin film layers to adhere properly. The substrate should be free of distortion or bowing, and it should be able to withstand normal firing temperatures, which are usually between 500 and 1000°C.

7.6.1 Experimental Procedure for Preparation of Thick Films by Screen Printing Process

In the screen printing process, the subtraction is placed on a carriage that is then lowered below the screen, causing the subtraction to be out of register with the pattern on the screen. The open mesh area in the screen is represented by the pattern on the screen. The substrate is placed at a short distance below the screen in the printing position, which corresponds to the configuration to be printed. Snap off distance is the distance between the screen and the substrate. A small amount of paste or ink is applied to the upper surface of the screen, and a very flexible wiper known as squeeze is then moved across the screen surface, forcing it vertically into contact with the substrate, allowing the paste to pass through the mesh area. The printing pattern on the substrate is left behind by the natural tension mesh. After that, the substrate is separated from the screen, a new substrate is installed, and the process is repeated (Figure 7.3) [47].

The components of a screen printer are as follows:

i. The screen and its mountain frame work.
ii. The substrate carriage and associated mechanical feed system, both of which can be controlled manually or automatically.
iii. The mechanism that allows the screen to be precisely positioned in relation to the substrate.

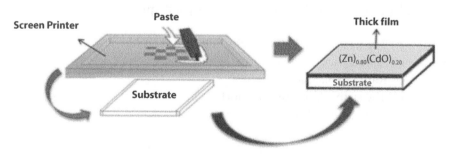

Figure 7.3 Screen printing method.

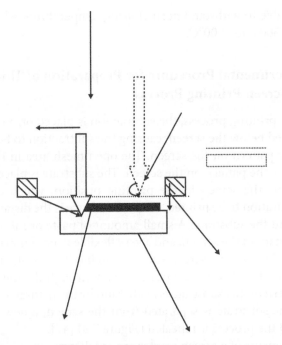

Figure 7.4 Schematic representation of the screen printing process.

The squeeze pressure should be just enough to remove the paste from the print pattern areas while leaving a thick film on the screen's blocked areas (Figure 7.4) [48].

7.6.2 Quality of Printing

Past characteristics such as viscosity, surface tension, and thixotropy had a significant impact on print quality. These variables influence the extent to which the paste welts both the substrate and the screen. The following considerations should be made for a high-quality print:

- The paste has a low viscosity.
- Squeeze with a strong grip.
- Rapid stroke rate.
- The snap-off distance is short.
- Squeeze bleed with a small angle of attack.

End-to-end and side-to-side non-uniformity are caused by:

i. Inconsistent stroke speed.
ii. There isn't enough parallelism between the screen and the substrate.
iii. There isn't enough substance on the screen.

7.6.3 The Following Factors Contribute to Incomplete Filling

i. Paste viscosity.
ii. Squeeze blade (hard or warm).
iii. Low attack angle.
iv. Printing at an excessive rate.

Smearing occurs when:

i. the paste's viscosity is too low.
ii. The snap-off distance is excessively short.
iii. The print is excessively slow.
iv. There is an excessive amount of squeeze pressure.

7.7 Procedures for Firing

There are four parts to the firing procedure:

1. Coalescence of paste.
2. The drying process.
3. Get rid of the organic binder.
4. A high sintering temperature is required.

The pattern immediately after screen printing consists of a series of discrete paste spots, each corresponding to a mesh opening in the screen. After that, the substrates are left at room temperature for a few minutes to allow the paste to coalesce and form a coherent level film. The amount of time it takes for coalescence to occur is largely determined by the paste's composition. The drying of the paste film above is the next step. Temperatures of 700°C to 1500°C are commonly used for periods of 1 to 2 hours. For good results, close control of the drying step is required during this process because more volatile components of the paste are removed. Improper drying can result in flaws such as blisters, cracks, and crazing.

After drying, the organic material in the composite material is removed by carbonizing and oxidizing at a relatively low temperature, i.e. 400°C. As a result, an oxidizing environment is required. The blinder, such as cadmium chloride or zinc chloride, is usually removed as the first step in the firing process. The print is taken to the maximum firing temperature, which can be up to 100°C, during the final firing or high sintering process. The chemical reactions that take place in the furnace's high temperature zone determine the composite's important properties. The time temperature must be precisely controlled for good results. For dry operation, a low temperature tunnel kiln with resistance heating is used.

7.7.1 Thick Film Technology has Four Distinct Advantages

• Economic Situation (Cost-Effectiveness) Performance

Thick film technology has the advantage of reduced parasitic. Organic material in the composite material is generally removed after drying at a relatively low temperature, 400°C, by capacitive coupling between components, and lead resistance and inductance are minimized. This improves the performance of high-speed and high-frequency systems.

• Flexibility

1. Existing breadboard or PCB designs can be converted with minimal design changes on a direct one-on-one basis
2. Quick turnaround times and design changes are possible to be completed in the shortest amount of time and with the least amount of effort
3. A large number of active and passive components parameters under strict control, both packaged and unpackaged.

• Reliability

1. The reduction in the number of physical interconnections improves reliability.
2. Wiring errors, shock, vibration, and acceleration damage are reduced because solder connections are replaced by chemically bonded material interfaces.
3. Resistive elements and hot spots in resistors are closely bound.

- Economy

    1. Initial equipment and personal training costs, as well as development costs, are minimal.
    2. The thick film process is well suited to mass production, and prototype and evaluation modules can be put together at a low cost.
    3. Circuit changes are simple to make, and final product assembly time is cut in half.

7.8 Deposition of Thin Film Layers via Solution-Based Process

It is important to note the following in order to give a thorough overview of coating and printing methods for the commercial manufacturing of printed solar cells:

- Thin films can be made by depositing an active material on a flexible substrate (such as polymer foil, paper, or textile) or a non-flexible substrate (such as glass).
- Thin films made from the deposition of an active material on a flexible substrate (such as polymer foil, paper, or textile) can be produced with or without contact.

There are a wide range of techniques that may be used to create thin films from a solution, but some of these methods have practical limits that make them less viable. First a lot of techniques create a lot of waste and demand a lot of resources. Second, achieving strong repeatability, which is a difficult undertaking, is crucial for large-scale production. Thirdly, many of these approaches cannot be used to large-scale manufacturing and are best suited for small-scale laboratory applications. Spin coating is an excellent illustration of this issue, cannot be applied to manufacturing because the largest sample size is just a few square centimeters. For the preparation of large-volume devices, many printing methods have been optimized, and they will be briefly discussed in this chapter. As was already noted, there are fundamentally two types of wet processing thin film deposition methods: coating procedures and printing processes. While printing requires the formation of a highly well-defined film, coating may be defined as the act of generating a film on a substrate without the creation of a sophisticated design. One may argue that the only characteristic that matters is the thickness of a layer created by coating when the films are assumed to be uniform [47]. Indirect or direct contact between the substrate and the deposition unit can be used to classify techniques. In these situations, contact procedures are the topic; contactless methods are the alternative.

7.8.1 Approaches for Coating

As previously established, coated films lack patterning in contrast to printed films. The thin film can be prepared by casting, spreading, spraying, or painting it on. Spin coating, doctor blade coating, spray coating, slot-die coating, curtain coating, and knife-over-edge coating are a few coating techniques [48].

7.8.2 Casting

The easiest method for layer preparation is undoubtedly casting. The only steps in this process are the transportation of ink onto a substrate and solvent drying. The fact that no specialized equipment is required is the sole true benefit of this procedure. Though it is possible to produce both thick and high-quality films, the inability to control film thickness, picture framing at the edges, and occasionally even an ink downpour during drying are significant drawbacks of employing casting. Another challenge might be managing the interfacial tension between the fluid and the substrate. The drying process is uneven when the interfacial tension of an ink exceeds the interfacial tension of a medium.

7.8.3 Spin Coating

Spin coating is without a doubt the most widely used method for creating thin layers from solutions. When making transistors, it is utilized in microelectronics to apply polymeric photoresist on silicon wafers. Additionally, it participates in a few phases in the production of CDs and DVDs. Numerous research have examined this method, such as Norrman *et al.* [49]. Substrate is often placed on a vacuum pad during a spin-coating operation, and solution is then applied on top. The main variable in this process is the rotational speed of the pad holding the substrate, which can range from a few hundred to several thousand spins per minute. The solvent soon evaporates, causing a layer to develop. The abundant material is ejected from the substrate as soon as rotation starts due to the fast angular speed, leaving just the desired thin layer. The features of the layers (thickness, homogeneity, topology, etc.) produced while using the same solvent and concentration of solid active substance show exceptional reproducibility. It should be emphasized that a variety of thickness values can be produced for the layer rather than a single precise number. Most commercially available spin coaters do not provide this option when deposition at a high temperature is required.

The amount of material deposited and the length of the spinning process have less of an impact on the thickness and shape of the emerging layers than do factors such as rotational speed, the ink's viscoelastic characteristics, volatility, molecular weight, and others. The empirical formula may be used to roughly determine the layer's thickness:

$$d = k_{spin} \cdot \omega^a \qquad (7.1)$$

where k_{spin} and a are empirical constants that take into account the physical characteristics of the solvent, solute, and substrate and is rotational speed (rev.min^{-1}) Constant a usually reaches a value of about -0.5 [50]. Additionally, the rotation might begin while the solution is being applied to the spinning substrate. When the solution is based on a volatile solvent that would start to evaporate too fast to generate a homogeneous layer, the second method is preferred. With very good layer homogeneity, a big area might be covered all at once (substrates up to 550 mm in diameter). Since the morphology, topography, and creation of thin films with comparatively set thicknesses can be easily replicated, spin coating has found particular applicability in the field of photovoltaics. The preparation of P3HT-PCBM layer using 1,2-dichlorobenzene is a superb illustration of spin coating in polymer solar cells [51, 52]. Although spin coating has a wide range of lab-scale applications, it is yet unknown if the technological challenges of mass manufacturing a variety of goods via this method will be resolved.

7.8.4 Blade Coating

Blade coating involves placing a sharp blade at a precise location (10–500 nm) over the substrate to be coated, this is followed by the placing the coating solution in front of the blade. The blade is then moved across the substrate at a linear speed (typically several cm/s1) to create the wet thin film layer of a certain height. Theoretically, the wet layer's thickness should be roughly equivalent to the blade's substrate's width [53]. The surface energy of the substrate, the surface tension of the solution, and its viscosity are just a few examples of the physical processes that can cause this value to vary significantly. The development of the meniscus between the blade and the solution is another key factor under consideration.

An empirical formula may be used to determine the thickness of a dry layer:

$$d = \frac{1}{2}\left(g\frac{c}{p}\right) \qquad (7.2)$$

where g is the gap distance, c is the solid material's concentration in solution (measured in g.cm^{-3}), and is the solution's density (also measured in g.cm^{-3}). All of the aforementioned factors can be valid explanations for why spin-coating technology is more often used than blade coating.

In comparison to other coating processes (like spin coating), blade coating is typically described as being highly frugal and causing essentially little useful material loss. However, a lot of material must be lost in order to identify the coated solution's ideal qualities. Additionally, using this approach to create prototype devices is not recommended. In the end, it is simple to compare the quantity of material lost by this procedure to that of spin coating or other approaches. The advantage of this strategy over spin coating is how quickly blade coating can be transformed into the roll-to-roll technique, in which the flexible substrate is put on a cylinder. Over the past 20 years, there has been a moderate growth in interest in the development of lab equipment using blade coating [54–59].

The layers produced by this technique have higher crystallinity than spin-coated layers [60]. Currently, only fully generated ZnO layers on ITO are printed using roll-to-roll (R2R) technology.

Slot-Die Coating

Slot-die coating is a contactless high area deposition method for the creation of thin layers with outstanding uniformity. Since it is a one-dimensional method, it may be used to create electrodes for printed electronics [61], secondary batteries, and multilayer polymer solar cells. This approach combines three crucial characteristics in a single way:

- **Premetered:** The coating thickness is predetermined by the feeding pump, coating speed, and mask width; additional head modifications are not required.
- **Contactless:** Only the coating liquid comes into touch with the substrate, which has the advantage of reducing substrate scratching.

- **Impervious:** Due to the coating system's total closure, solvent evaporation is prevented and the ink concentration is kept constant.

A rather straightforward process is used to get the ink into the printing head in a very precisely specified shape. The true challenge is in the customarily complex design of the printing head.

The internal mask is a vital element that affects the resolution and dimensions of the slots (shim). The wedge and the complete printing head are made of stainless steel, and the printing mask's typical width ranges from 10 to 100 μm. Masks with a diameter of more than 60 μm can be difficult to handle and may require a more viscous ink for efficient deposition [62]. When employing ink with a relatively low viscosity (<20 mPa.s), the ideal width of the mask is between 25 and 50 μm, although theoretically, a very wide range of viscosities from less than 1 mPa.s to several thousand Pa.s may be used. A second supporting internal mask can be applied when employing the low-viscosity ink to help avoid the merging of the menisci of the two neighboring stripes. The R2R type of slot-die coating for high viscosity inks at high deposition rates (>11 mm/min) has been the subject of several investigations [58]. The thickness of a wet film can be roughly estimated by the empirical formula:

$$d = \frac{f}{v.w} \cdot \frac{c}{\rho} \qquad (7.3)$$

where f is the flow rate (in $cm^3.min^{-1}$), v is the coating (or web speed) speed (in $cm.min^{-1}$), and w is the coating width (in cm) and c is the concentration of the particles in the ink (in $g.cm^{-3}$) and ρ is the density of the active dry material in the final film (in $g.cm^{-3}$).

In Krebs' fascinating 2009 investigation [63], a polymer solar cell based on the active component P3HT was produced almost totally from the solution via slot-die coating. On a polyethyleneterephthalate (PEN) substrate, the bottom electrode was deposited. It was based on nanoparticle silver.

The other layers—an active layer of P3HT:PCBM, an electron transporting layer of nanoparticle ZnO, an intermediate layer of active material, and a hole transporting layer of PEDOT:PSS—were all coated by slot-die coating, leaving just the top electrode to be screen printed.

Although the lack of an ITO layer had significant economic benefits, overall performance was a dismal 0.3%. All of the layers were deposited in air throughout the manufacturing process since there was no vacuum, which had a negative effect on performance. In a similar work, Krebs introduced ITO using screen printing, while the other layers were deposited by slot-die coating, knife-over-edge deposition, screen printing, or knife-over-edge deposition with the ITO. The top silver electrode was covered with a slot-die coating. The width and length of each individual stripe were 0.9 cm and 25 cm, respectively. The active area of each stripe is 15 cm^2. Performance was 10 times better than devices with fully screen-printed surfaces. Processing speed for each layer ranged from 40 to 50 m.h.$^{-1}$. Again, no further vacuum steps were introduced. Additionally, slot-die coating may be used to manufacture polymer-based light-emitting diodes (OLEDs) [64] or more process-tolerant light-emitting cells (LECs) [65].

Conclusion

In conclusion, printable circuitry and photovoltaic cells can both be employed in the future. In the long run, typical organic photovoltaic solar cells with a performance of 12 to 15% are anticipated, and these cells will be fully integrated into buildings. It is obvious that printing and coating processes are good ways to produce different devices in large quantities and quickly using this new approach to electronics. It is noteworthy that the creation of entirely printed electrical devices requires a great deal of effort, and the outcomes so far have not been particularly convincing.

References

1. Zargar, R.A., Micro size developed TiO2-CdO composite film: Exhibits diode characteristics. *Mater. Lett.*, 335, 15, 133813, 2022.
2. Sharma, G.D., Agrawal, A., Pradhan, R., Keshtov, M.L., Singhal, R., Liu, W., Zhu, X., Mishra, A., Fullerene-free all-small-molecule ternary organic solar cells with two compatible fullerene-free acceptors and a coumarin donor enabling a power conversion efficiency of 14.5%. *ACS Appl. Energy Mater.*, 4, 10, 11537–11544, 2021.
3. Wen, X., Nowak-Król, A., Nagler, O., Kraus, F., Zhu, N., Zheng, N., Müller, M., Schmidt, D., Xie, Z., Würthner, F., Tetrahydroxy-perylenebisimide embedded in a zinc oxide thin film as an electron-transporting layer for high-performance non-fullerene organic solar cells. *Angew. Chem.*, 131, 37, 13185–13189, 2019.
4. Wang, G., Melkonyan, F.S., Facchetti, A., Marks, T.J., All-polymer solar cells: Recent progress, challenges, and prospects. *Angew. Chem. Int. Ed.*, 58, 13, 4129–4142, 2019.

5. Zargar, R.A., Arora, M., Alshahrani, T., Shkir, M., Screen printed novel ZnO/MWCNTs nanocomposite thick films. *Ceram. Int.*, 47, 5, 6084–6093, 2021.
6. Niu, G., Guo, X., Wang, L., Review of recent progress in chemical stability of perovskite solar cells. *J. Mater. Chem. A*, 3, 17, 8970–8980, 2015.
7. Babayigit, A., Ethirajan, A., Muller, M., Conings, B., Toxicity of organometal halide perovskite solar cells. *Nat. Mater.*, 15, 3, 247–251, 2016.
8. Berhe, T.A., Su, W.N., Chen, C.H., Pan, C.J., Cheng, J.H., Chen, H.M., Tsai, M.C., Chen, L.Y., Dubale, A.A., Hwang, B.J., Organometal halide perovskite solar cells: Degradation and stability. *Energy Environ. Sci.*, 9, 2, 323–356, 2016.
9. Nian, L., Gao, K., Liu, F., Kan, Y., Jiang, X., Liu, L., Xie, Z., Peng, X., Russell, T.P., Ma, Y., 11% efficient ternary organic solar cells with high composition tolerance via integrated near-IR sensitization and interface engineering. *Adv. Mater.*, 28, 37, 8184–8190, 2016.
10. Zhang, S., Ye, L., Hou, J., Breaking the 10% efficiency barrier in organic photovoltaics: Morphology and device optimization of well-known PBDTTT polymers. *Adv. Energy Mater.*, 6, 11, 1502529, 2016.
11. Kaltenbrunner, M., White, M.S., Głowacki, E.D., Sekitani, T., Someya, T., Sariciftci, N.S., Bauer, S., Ultrathin and lightweight organic solar cells with high flexibility. *Nat. Commun.*, 3, 1, 1–7, 2012.
12. Singh, U.P. and Patra, S.P., Progress in polycrystalline thin-film Cu (In, Ga) solar cells. *Int. J. Photoenergy*, 2010, 468147, 19pp., 2010.
13. Lin, X., Klenk, R., Wang, L., Köhler, T., Albert, J., Fiechter, S., Ennaoui, A., Lux-Steiner, M.C., 11.3% efficiency Cu (In, Ga)(S, Se) 2 thin film solar cells via drop-on-demand inkjet printing. *Energy Environ. Sci.*, 9, 6, 2037–2043, 2016.
14. Singh, M., Jiu, J., Sugahara, T., Suganuma, K., Thin-film copper indium gallium selenide solar cell based on low-temperature all-printing process. *ACS Appl. Mater. Interfaces*, 6, 18, 16297–16303, 2014.
15. Matur, U.C., Akyol, S., Baydoğan, N., Cimenoglu, H., The optical properties of CIGS thin films derived by sol-gel dip coating process at different withdrawal speed. *Procedia Soc. Behav. Sci.*, 195, 1762–1767, 2015.
16. Li, S.G., Jiang, K.J., Su, M.J., Cui, X.P., Huang, J.H., Zhang, Q.Q., Zhou, X.Q., Yang, L.M., Song, Y.L., Inkjet printing of CH3NH3PbI3 on a mesoscopic TiO2 film for highly efficient perovskite solar cells. *J. Mater. Chem. A*, 3, 17, 9092–9097, 2015.
17. Sha, W.E., Ren, X., Chen, L., Choy, W.C., The efficiency limit of CH3NH3PbI3 perovskite solar cells. *Appl. Phys. Lett.*, 106, 22, 221104, 2015.
18. Lu, L., Zheng, T., Wu, Q., Schneider, A.M., Zhao, D., Yu, L., Recent advances in bulk heterojunction polymer solar cells. *Chem. Rev.*, 115, 23, 12666–12731, 2015.
19. Scharber, M.C. and Sariciftci, N.S., Efficiency of bulk-heterojunction organic solar cells. *Prog. Polym. Sci.*, 38, 12, 1929–1940, 2013.
20. Kirchartz, T., Taretto, K., Rau, U., Efficiency limits of organic bulk heterojunction solar cells. *J. Phys. Chem. C*, 113, 41, 17958–17966, 2009.

21. Janssen, R.A. and Nelson, J., Factors limiting device efficiency in organic photovoltaics. *Adv. Mater.*, 25, 13, 1847–1858, 2013.
22. Cheng, P. and Zhan, X., Stability of organic solar cells: Challenges and strategies. *Chem. Soc. Rev.*, 45, 9, 2544–2582, 2016.
23. Arumugam, S., Li, Y., Senthilarasu, S., Torah, R., Kanibolotsky, A.L., Inigo, A.R., Skabara, P.J., Beeby, S.P., Fully spray-coated organic solar cells on woven polyester cotton fabrics for wearable energy harvesting applications. *J. Mater. Chem. A*, 4, 15, 5561–5568, 2016.
24. Leonat, L., White, M.S., Głowacki, E.D., Scharber, M.C., Zillger, T., Rühling, J., Hübler, A., Sariciftci, N.S., 4% efficient polymer solar cells on paper substrates. *J. Phys. Chem. C*, 118, 30, 16813–16817, 2014.
25. Kovalenko, A. and Hrabal, M., Printable solar cells, in: *Printable Solar Cells*, pp. 163–202, 2017.
26. Krebs, F.C., Fabrication and processing of polymer solar cells: A review of printing and coating techniques. *Sol. Energy Mater. Sol. Cells*, 93, 4, 394–412, 2009.
27. Krebs, F.C., Polymer solar cell modules prepared using roll-to-roll methods: Knife-over-edge coating, slot-die coating and screen printing. *Sol. Energy Mater. Sol. Cells*, 93, 4, 465–475, 2009.
28. Krebs, F.C., Jørgensen, M., Norrman, K., Hagemann, O., Alstrup, J., Nielsen, T.D., Fyenbo, J., Larsen, K., Kristensen, J., A complete process for production of flexible large area polymer solar cells entirely using screen printing—First public demonstration. *Sol. Energy Mater. Sol. Cells*, 93, 4, 422–441, 2009.
29. Krebs, F.C., Alstrup, J., Spanggaard, H., Larsen, K., Kold, E., Production of large-area polymer solar cells by industrial silk screen printing, lifetime considerations and lamination with polyethyleneterephthalate. *Sol. Energy Mater. Sol. Cells*, 83, 2-3, 293–300, 2004.
30. Tipnis, R., Bernkopf, J., Jia, S., Krieg, J., Li, S., Storch, M., Laird, D., Large-area organic photovoltaic module—Fabrication and performance. *Sol. Energy Mater. Sol. Cells*, 93, 4, 442–446, 2009.
31. Mazhari, B., An improved solar cell circuit model for organic solar cells. *Sol. Energy Mater. Sol. Cells*, 90, 7-8, 1021–1033, 2006.
32. Nelson, J., Organic photovoltaic films. *Curr. Opin. Solid State Mater. Sci.*, 6, 1, 87–95, 2002.
33. Saunders, B.R. and Turner, M.L., Nanoparticle–polymer photovoltaic cells. *Adv. Colloid Interface Sci.*, 138, 1, 1–23, 2008.
34. Photovoltaics, O., *Mechanisms, materials, and devices*, CRC, Boca Raton, Edition 1, 666, 2005.
35. Arapov, K., Abbel, R., Friedrich, H., Inkjet printing of graphene. *Faraday Discuss.*, 173, 323–336, 2014.
36. Kim, H., Jang, J.I., Kim, H.H., Lee, G.W., Lim, J.A., Han, J.T., Cho, K., Sheet size-induced evaporation behaviors of inkjet-printed graphene oxide for printed electronics. *ACS Appl. Mater. Interfaces*, 8, 5, 3193–3199, 2016.
37. Ujjain, S.K., Bhatia, R., Ahuja, P., Attri, P., Highly conductive aromatic functionalized multi-walled carbon nanotube for inkjet printable high performance supercapacitor electrodes. *PLoS One*, 10, 7, e0131475, 2015.

38. Takagi, Y., Nobusa, Y., Gocho, S., Kudou, H., Yanagi, K., Kataura, H., Takenobu, T., Inkjet printing of aligned single-walled carbon-nanotube thin films. *Appl. Phys. Lett.*, *102*, 14, 143107, 2013.
39. Shin, S.S., Lee, S.J., Seok, S.I., Exploring wide bandgap metal oxides for perovskite solar cells. *APL Mater.*, *7*, 2, 022401, 2019.
40. Liu, D. and Kelly, T.L., Perovskite solar cells with a planar heterojunction structure prepared using room-temperature solution processing techniques. *Nat. Photonics*, *8*, 2, 133–138, 2014.
41. Zargar, R.A., Arora, M., Bhat, R.A., Study of nanosized copper-doped ZnO dilute magnetic semiconductor thick films for spintronic device applications. *Appl. Phys. A*, *124*, 1, 1–9, 2018.
42. Zargar, R.A., ZnCdO thick film: A material for energy conversion devices. *Mater. Res. Express*, *6*, 9, 095909, 2019.
43. Shah, A.H., Zargar, R.A., Arora, M., Sundar, P.B., Fabrication of pulsed laser-deposited Cr-doped zinc oxide thin films: Structural, morphological, and optical studies. *J. Mater. Sci.: Mater. Electron.*, *31*, 23, 21193–21202, 2020.
44. Zargar, R.A., Fabrication and improved response of ZnO-CdO composite films under different laser irradiation dose. *Sci. Rep.*, *12*, 1, 1–10, 2022.
45. Zargar, R.A., Kumar, K., Mahmoud, Z.M., Shkir, M., AlFaify, S., Optical characteristics of ZnO films under different thickness: A MATLAB-based computer calculation for photovoltaic applications. *Phys. B: Condens. Matter*, *631*, 413614, 2022.
46. Zargar, R.A., Kumar, K., Arora, M., Shkir, M., Somaily, H.H., Algarni, H., AlFaify, S., Structural, optical, photoluminescence, and EPR behaviour of novel ZnO· 80Cd0· 20O thick films: An effect of different sintering temperatures. *J. Lumin.*, *245*, 118769, 2022.
47. a. Zargar, R.A., Chackrabarti, S., Malik, M.H., Hafiz, A.K., Sol-gel syringe spray coating: A novel approach for Rietveld, optical and electrical analysis of CdO film for optoelectronic applications. *Phys. Open*, *7*, 100069, 2021.
    b. Swerdlow, R.M., Step-by-step guide to screen process, Printing, Prentice Hall Direct, Hoboken, New Jersey, U.S. 2012.
48. a. Zargar, R.A., Boora, N., Hassan, M.M., Khan, A., Hafiz, A.K., Screen printed TiO2 film: A candidate for photovoltaic applications. *Mater. Res. Express*, *7*, 6, 065904, 2020.
    b. Zargar R.A., Ahmad P.A., Gogre M.A.S. Sceen printing coating of (ZnO)0.8(CdO)0.2 material for optoelectronic applications. *Opt. Quant. Electron*, *52*, 401, 2020.
49. Shin, S.S., Yang, W.S., Noh, J.H., Suk, J.H., Jeon, N.J., Park, J.H., Kim, J.S., Seong, W.M., Seok, S.I., High-performance flexible perovskite solar cells exploiting Zn2SnO4 prepared in solution below 100° C. *Nat. Commun.*, *6*, 1, 1–8, 2015.
50. Lupo, D., Clemens, W., Breitung, S., Hecker, K., OE-A roadmap for organic and printed electronics, in: *Applications of Organic and Printed Electronics*, pp. 1–26, Springer, Boston, MA, 2013.

51. Krebs, F.C., Fabrication and processing of polymer solar cells: A review of printing and coating techniques. *Sol. Energy Mater. Sol. Cells*, 93, 4, 394–412, 2009.
52. Norrman, K., Ghanbari-Siahkali, A., Larsen, N.B., 6 Studies of spin-coated polymer films. *Annu. Rep. Sect." C"(Phys. Chem.)*, 101, 174–201, 2005.
53. Li, G., Shrotriya, V., Huang, J., Yao, Y., Moriarty, T., Emery, K., Yang, Y., High-efficiency solution processable polymer photovoltaic cells by self-organization of polymer blends, in: *Materials For Sustainable Energy: A Collection of Peer-Reviewed Research and Review Articles from Nature Publishing Group*, pp. 80–84, 2011.
54. Ma, W., Yang, C., Gong, X., Lee, K., Heeger, A.J., Thermally stable, efficient polymer solar cells with nanoscale control of the interpenetrating network morphology. *Adv. Funct. Mater.*, 15, 10, 1617–1622, 2005.
55. Søndergaard, R.R., Hösel, M., Krebs, F.C., Roll-to-roll fabrication of large area functional organic materials. *J. Polym. Sci. Part B: Polym. Phys.*, 51, 1, 16–34, 2013.
56. Mens, R., Adriaensens, P., Lutsen, L., Swinnen, A., Bertho, S., Ruttens, B., D'Haen, J., Manca, J., Cleij, T., Vanderzande, D., Gelan, J., NMR study of the nanomorphology in thin films of polymer blends used in organic PV devices: MDMO-PPV/PCBM. *J. Polym. Sci. Part A: Polym. Chem.*, 46, 1, 138–145, 2008.
57. Han, G.H., Lee, S.H., Ahn, W.G., Nam, J., Jung, H.W., Effect of shim configuration on flow dynamics and operability windows in stripe slot coating process. *J. Coat. Technol. Res.*, 11, 1, 19–29, 2014.
58. Fara, L. (Ed.), *Advanced solar cell materials, technology, modeling, and simulation*, IGI Global, USA, Edition 1, 354, 2012.
59. Krebs, F.C., All solution roll-to-roll processed polymer solar cells free from indium-tin-oxide and vacuum coating steps. *Org. Electron.*, 10, 5, 761–768, 2009.
60. Shin, S., Yang, M., Guo, L.J., Youn, H., Roll-to-roll cohesive, coated, flexible, high-efficiency polymer light-emitting diodes utilizing ITO-free polymer anodes. *Small*, 9, 23, 4036–4044, 2013.
61. Sandström, A., Dam, H.F., Krebs, F.C., Edman, L., Ambient fabrication of flexible and large-area organic light-emitting devices using slot-die coating. *Nat. Commun.*, 3, 1, 1–5, 2012.
62. Riemer, D.E., The theoretical fundamentals of the screen printing process. *Microelectron. Int.*, 6, 1, 8–17, 1989.
63. Shaheen, S.E., Radspinner, R., Peyghambarian, N., Jabbour, G.E., Fabrication of bulk heterojunction plastic solar cells by screen printing. *Appl. Phys. Lett.*, 79, 18, 2996–2998, 2001.
64. Aernouts, T., Vanlaeke, P., Poortmans, J., Heremans, P., Polymer solar cells: Screen-printing as a novel deposition technique. *MRS Online Proc. Lib. (OPL)*, 836, L3-9, 2004.
65. Bendoni, R., Sangiorgi, N., Sangiorgi, A., Sanson, A., Role of water in TiO2 screen-printing inks for dye-sensitized solar cells. *Sol. Energy*, 122, 497–507, 2015.

8

Response of Metal Oxide Thin Films Under Laser Irradiation

Rayees Ahmad Zargar

Department of Physics, BGSBU University, Rajouri, India

Abstract

New directions in material processing have been made possible by recent technological advancements in the creation of a wide variety of lasers. Metal oxides are attracting a lot of attention due to their potential applications in microelectronics and medicine. Numerous studies have focused on the synthesis, characterization, and patterning of metal oxide films and nanostructures. When materials are processed by lasers, conditions of electronic and thermodynamic non-equilibrium result from exposure to rapid, localized energy. Excellent control over the manipulation of materials is possible due to the laser's ability to localize heat both in space and time. This chapter reports the recent work on laser irradiating metal oxide films and nanostructures for various applications and provides a brief overview of the underlying laser-material principles.

Keywords: Laser irradiation, metal oxide, screen printing, optical and electrical parameters

8.1 Introduction

In-depth study of low-cost fabrication and laser irradiation has recently come to light as an effective technique for changing the material's structural, optical, and electrical properties. The laser irradiation technique has a number of benefits over the conventional furnace annealing method, including quick crystallization at room temperature, substrates with low melting points, less thermal exposure of the sample, and an increase in

---

Email: rayeesphy12@gmail.com

Rayees Ahmad Zargar (ed.) Metal Oxide Nanocomposite Thin Films for Optoelectronic Device Applications, (203–220) © 2023 Scrivener Publishing LLC

charge carriers [1–3]. One or more of the four well-known processes reflection, absorption, scattering, and transmission can occur simultaneously when a laser beam interacts with a solid material. However, only the photons that are absorbed during a laser-matter interaction can change the chemical and physical characteristics of any material [4]. Ternary metal oxide (TMO) alloys have drawn a lot of attention from the research and development community, particularly because of their applications in energy storage, chemical sensing, and photonic devices [5–7].

The unique physical and chemical characteristics that metal oxides (MOs) are known to display are of great interest for scientific and technological advancements. Different crystal structures can be used to synthesize MOs, which can have electronic properties ranging from insulator to semiconductor to conductor and contain at least one metal cation and an oxygen anion [8, 9]. MOs' thin films and nanostructures have developed into crucial design elements for practical devices. These oxide films and nanostructures are typically created using deposition techniques with precursors that are liquid, gaseous, ionized vapor, or plasma [10, 11].

The advantages of liquid-precursor-based synthesis methods over the others include their speed, low temperature, vacuum-free nature, cost-effectiveness, and scaling [12]. MO thin films are typically grown using wet-chemical techniques like metal-organic deposition (MOD) and the sol-gel method. For the creation of nanostructured MOs, coprecipitation and hydrothermal methods are more common. To obtain the desired behavior in all of these methods, the solution-derived precursor layers or nanostructures must typically be thermally treated or annealed [13, 14]. The following enhancements are possible with thermal processing methods: The hydrous skeleton is eliminated by dissolving the remaining solvents and organic binders; phase purity and chemical homogeneity are improved; stoichiometry, composition (doping), and concentration/type of defects are controlled; and phase structure (amorphous/crystalline) and microstructure (imperfections in crystals and lattices) are modified. For the aforementioned reason, thermal treatment is also necessary for oxide films that are deposited using other synthesis techniques that use gaseous and ionized vapor or plasma precursors. Traditionally, MO films and nanostructures are thermally treated in a furnace or oven at temperatures between 300 and 1000 °C, depending on the material system, thermodynamics, and the desired outcome [15–17]. Due to the fact that the sample dimensions are much smaller than the heated volume and almost all of the energy is used to raise and lower the temperature of the substrate and the furnace, this process has a high thermal budget (thermal power), lengthy processing times, and a high degree of energy loss. Additionally, the high temperatures involved in this process can result in micro structural changes and thermal-expansion mismatch, which can cause mechanical failures, on thermally delicate substrates (such as Si, glass, amorphous alloys, and polymers), as well as being incompatible with these substrates. Additionally, thermal treatment procedures based on conventional heating techniques are unable to produce spatially resolved the physical separation of the films or nanostructures from the electronic circuitry is required due to these effects. The direct incorporation of MO films or nanostructures into the CMOS fabrication process is constrained by these problems. The above problems may be resolved by using laser irradiation as an alternative to conventional thermal treatment, which also enables the highly compatible on-chip integration of MO films and nanostructures. This method's primary foundation is the photothermal effect caused by a focused laser, which produces a confined temperature field at a desired location with excellent control over distance [8]. To produce the desired thermal effect, a number of laser process parameters, including laser intensity, pulse width, and scanning rate, can be changed. The material properties can be precisely controlled because of the process's notably quick and localized thermal effects. Rapid material fabrication with little energy loss is made possible by the laser's heating and cooling rates, which are orders of magnitude higher than those of conventional annealing and rapid thermal annealing (RTA). It is possible to selectively anneal the films and nanostructures without thermal interference with the underlying substrates and nearby structures because laser-induced heat can be focused to a particular region in both the in-plane and thickness directions. Additionally, using lasers that are sufficiently efficient (defined as the ratio of optical output power to electrical input power) can significantly lower the amount of energy needed for thermal processing [18, 19].

8.2 Interaction of Laser with Material

Theoretically, stimulated radiation leads to laser. According to Figure 8.2a, an incident photon will specifically cause the decay of an excited electron in a material to its ground state if they have the same energy and are followed by the emission of a second photon with the same frequency and phase as the incident one [20]. After that, these two photons will spread through constructive interference, causing another excited electron with a similar energy to decay. This process can be repeated to create laser light, which has a clearly defined frequency and a significantly higher intensity. To date, a wide variety of active materials have been used to create lasers, allowing for highly tunable wavelengths from ultraviolet to infrared (Figure 8.2b) [21]. High coherence, monochromaticity, and directionality are all characteristics of the resulting laser [22]. The monochromaticity allows for selective processing, while the directionality and coherence enable localized superheating [23]. These characteristics make laser stand out as a cutting-edge technology that is making remarkable strides in a variety of applications, such as the synthesis, modification, and processing of materials.

8.3 Radiation Causes Modification

Radiation-induced processes in solids have received a lot of attention due to their potential to modify or improve material properties as well as their role in the degradation of material qualities. It is generally known that radiation damage can cause defects to form on surfaces and in the bulk as well as cause sputtering and collision cascades that are started by knock-on processes. The number of primary defects and sputtered atoms can really be predicted using theoretical calculations and computer simulations [24, 25]. Experimental irradiation of a wide range of materials reveals that, depending on the material under investigation, the mass of irradiating ions, the irradiation temperature for a specific material, and other factors, the saturation of the damage occurs at different levels [26]. This behavior shows that the irradiation-induced changes in a material's response depend on both the material's physico-chemical characteristics and the radiation circumstances [27]. Except when Auger recombination predominates and electron-hole concentrations are exceedingly high, electron-hole pairs have relatively long lives. An electron-hole pair's band gap energy, whose wave functions are delocalized, should be used by any atomic process that results from electronic excitation. Localizing electronic excitation energy is crucial for converting it to the energy required for bond scission because the energy required to break a bond is of the same order of magnitude as the band gap energy. Either self-trapping or defect-induced trapping causes the localization of electronic excitation energy [24]. The creation and subsequent trapping of the holes in oxide sheets results in significant performance changes for the device. It has been widely demonstrated that optical absorption analysis is a crucial and effective technique for examining and analyzing the many phenomena of electronic structure and processes in radiation-exposed materials [28–30]. The extensive theoretical research on the optical behavior of thin films focuses mainly on the optical properties of reflection, transmission, and adsorption and how they relate to the optical constants of films [28]. This technique's capacity to provide details on the fundamental gap, electronic transition, trapping levels, and localized states highlights the significance of investigating a material's optical properties. Films are typically amorphous and, at their most, polycrystalline in form. Understanding the issue of how disorder in amorphous materials affects the band structure and consequently the electrical and optical properties of the material has advanced over the past few decades. The acute structure in the valence and conduction bands and the abrupt terminations at the valence band maximum and conduction band minimum are the fundamental characteristics of the energy distribution of electronic states density of crystalline solids for semiconductors. The sharp edges in the density of states curves produce a well-defined forbidden energy gap. Amorphous solids are substances that lack structural order. The amorphous state is partially unstable or metastable and often exhibits a gradual or rapid transition to an ordered crystalline state. Despite the high density of disorder present in amorphous materials, they only deviate slightly from the ideal crystalline structure [31]. In other words, the rigidity of the chemical bonds allows us to assume short-range order, and the principle of crystalline ribbon structure still applies to amorphous solids [10]. A nominally amorphous film can have different electrical and other properties, depending on how it is manufactured. In particular, it is known that the deposition rate of vapor-deposited films has a large effect on the dielectric constant and conductivity [32].

8.4 Application Laser Irradiated Films

Applications ranging from communications to material processing have made use of laser-based processing. Lasers have historically been used extensively in the manufacturing sector for tasks like cutting, welding, patterning, drilling, etc. In recent years, the use of additive manufacturing and micro/nanofabrication technologies has opened up new opportunities for laser processing. These brand-new industries for laser processing have only recently begun to grow. A wide range of MO films and films have been laser irradiated for use in dielectric, ferroelectric, piezoelectric, and multiferroic magnetoelectric applications. Currently, the production of low-temperature devices is one of many commercial, large-area device applications for excimer laser systems. Transparent conductors, field-effect transistors, solar cells, thermistors, photocatalysis, gas sensors, and other technologies [33] are examples of technological advancements. Figure 8.1 represents the laser induced coated films and their applications.

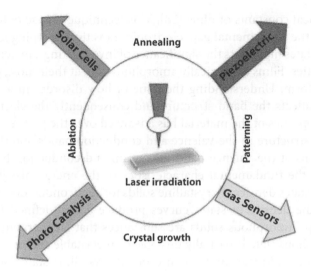

Figure 8.1 Laser-induced phenomena in metal oxide films and their applications in various functional devices.

8.5 Wavelength Range of Radiation

Lasers have wavelengths that range from the far infrared to the deep ultraviolet (Figure 8.2). Laser radiation differs from other light sources due to its monochromaticity, coherence, and collimation [34]. The laser's spot size, spatial/temporal distribution, and power output can be precisely controlled to range from very low (mW) to extremely high (1-100 kW) [35]. There are various types of lasers, such as solid-state lasers, liquid-dye lasers, gas lasers, semiconductor lasers, fiber lasers, and free-electron lasers, depending on the lasing medium. The properties of the optical resonator and the laser medium determine the laser beam's wavelength and power. Information processing, medical diagnosis, surgery, and therapy, metrology, holography, spectroscopy, and materials processing are just a few of the many uses for lasers [36]. Due to its adaptability for roll-to-roll manufacturing and compatibility with a wide range of materials, laser-based material processing is a very alluring fabrication method. The laser process parameters can be digitalized, making it easier to incorporate the procedure with computer-aided design and a flexible manufacturing system [37]. Laser processing of materials primarily relies on the material's ability to absorb laser radiation, which causes a variety of effects like heating, melting, vaporization, plasma formation, and ablation. The attenuation and spatial redistribution (diffusion) of the beam energy caused by the absorption phenomenon are

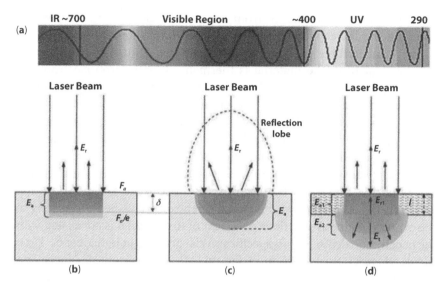

Figure 8.2 (a) Wavelengths used for lasers in the electromagnetic spectrum of light, (b) absorbing material, (c) diffusing material, and (d) material with adjacent absorbing and diffusing layers [38].

typically linked to the scattering of laser radiation. Figure 8.2b–d illustrates the redistribution of laser energy during laser–material interaction for various absorption scenarios [38].

Energy redistribution during laser-material interaction represented: b) an absorbing substance, c) a diffusing substance, and d) a substance with adjacent diffusing and absorbing layers. Er: energy reflected, Et: energy transmitted, Ea: b-d energy absorbed [38].

8.6 Laser Irradiation Mechanism

The electronic structure of a material affects the method of photon-energy (hν) absorption [39–42]. In this case, the photon frequency is, and h is the Planck constant. The absorption in insulators and semiconductors with bandgap (E_g) is caused by resonant excitation, which primarily involves interband electronic transitions. Only weak phonon excitation and absorption are feasible if hν < E_g. However, sequential multiphoton excitation via defect states or coherent multiphoton excitation may cause interband transitions at higher laser intensities. Additionally, sub-bandgap excitation is frequently permitted by defect or impurity states. If hν ≥ E_g, strong interband excitation of electrons will occur, creating electron–hole pairs.

The phonons receive energy from the excited electronic states, which amplifies the lattice vibrations. The availability of more free electrons at higher temperatures can lead to intraband transitions in semiconductors or insulators. The material is thermalized by electron-electron and electron-phonon interactions brought about by the laser-induced interband or intraband electronic transition. phonon-electron interactions. The process of electron-hole recombination restores the equilibrium state over a period of time that depends on the properties of the material [43]. The overall result is the transformation of electronic energy from the incident laser beam's beam into heat. Based on the corresponding phenomena' respective timescales, the electron and lattice excitation and relaxation processes in a semiconductor material can be distinguished. According to the expression, the absorption coefficient of the material is correlated with the penetration depth (δ) of a specific wavelength of light radiation: $\delta = 1/\alpha = \lambda/4\pi k$, where k is the extinction coefficient [44]. Furthermore, α is dependent on the thickness (t) and optical properties of the materials as per the relationship, $\alpha = -1/t \ln(T/1-R)$, where T and R stand for the material's transmittance and reflectance, respectively [45]. The extent of laser penetration is quite constrained in metals due to their high values and nontransparency to almost all laser wavelengths. On the other hand, while being transmitted through semiconductors and insulators, laser energy may be absorbed over a sizable depth [46]. Since these oxides typically exhibit much higher UV absorbance than in the vis and IR ranges, high-energy, short-wavelength UV lasers have been used in the majority of studies on MO thin-film fabrication. Due to its shallow penetration depth, UV laser radiation can only be absorbed by films between 200 and 300 nm thick (Figure 8.3) [47].

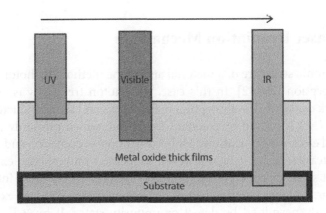

Figure 8.3 Comparison of optical penetration depths of laser radiation of different wavelengths in a metal oxide film (≈ 1–10 μm thick).

8.7 Experimental Procedure

8.7.1 Thin Film Technologies

There are several different methods for depositing thin films. The processes include molecular beam epitaxy (MBE), chemical vapor deposition (CVD), magnetron sputtering, electron-beam evaporation, thermal evaporation (also known as vacuum vapor deposition), and electron-beam evaporation [48]. In a highly evacuated room, several different types of materials can be deposited using the vacuum evaporation technique. It entails heating a solid substance to a high enough temperature to cause it to vaporize, then re-condensing it onto a cooler substrate to create a thin film. The material is heated when a significant current flows through a filament container with finite electrical resistance (often shaped as a basket, boat, or crucible). The temperature of evaporation and its resistance to alloying and chemical reaction with the evaporant dictate the choice of this filament material. Once the substance has evaporated, the vapor inside the evaporation chamber collides with the surrounding gas molecules. As a result, while being transferred through the ambient gas, a portion is spread over a specific area. At pressures of 10^{-4} torr and 10^{-6} torr, the mean free path for air at 298 K is roughly 45 cm and 4500 cm, respectively. In order to guarantee a straight-line path for the majority of the evaporated species and for a substrate-to-source distance of around 10 cm to 50 cm in a vacuum chamber, pressures lower than 10^{-5} torr are required. The substrates are positioned above and a distance away from the material that is being evaporated, each with the proper mask. The vacuum is let go after the procedure is finished, and the masks are then taken off the substrates. On every area of the substrates exposed by the open areas of the mask, this procedure leaves a thin, uniform layer of the deposition material. The deposition of materials that are challenging to evaporate in air is best suited for the vacuum evaporation approach. The process is clean and promotes better contact between the material layer and the surface it has been applied on. Evaporation beams can also make extremely exact patterns since they move in straight lines. Thermal vacuum deposition typically results in films with structural flaws including lattice faults or grain boundaries [49]. Dislocation loops, stacking-fault tetrahedral, and small triangular defects are among the so-called minor defects that are frequently seen in deposited films; they are all typically attributed to vacancy collapse [50]. The severity of such defects can be changed by adjusting the deposition parameters, including pressure, deposition rate, substrate temperature, and surface make-up. Depending on the type of material being deposited and desired film qualities, such as thickness and conductivity, also the evaporation and procedure's settings change.

8.7.2 What is Thick Film, Its Technology with Advantages

Thick film is a 'printed-and-fired' film deposition technology that makes use of conductive, resistive, and insulating pastes containing glass frit, which are screen printed on defined patterns and then fused at high temperatures onto a ceramic substrate. Thick films with resistivities ranging from 10/square to 10 MΩ/square have thicknesses ranging from 5–20 m. Multi-layer structures are frequently made with them. Microelectronic devices used thick films in the 1950s. It has been discovered to be a viable alternative to printed circuit board technology for producing reliable integrated circuits. Since the 1960s, the era of silicon technology has seen a boom in this technology [51].

8.7.3 Experimental Detail of Screen Printing and Preparation of $Zn_{0.80}Cd_{0.20}O$ Paste for Coated Film

Here the molecular weight of ZnO = 81.408 and molecular weight of CdO = 128.4104.

Hence the calculated amounts of $Zn_{0.80}Cd_{0.20}O$ compositions are
Weight of ZnO = (81.408) × (0.80 = 65.1264gm &
weight of CdO = (128.4104) × (0.20 = 25.68208gm.
And weight of $ZnCl_2$ = (10/100) × (90.80848 = 9.080848gm,

Due to the size of the aforementioned calculation, we reduce all weights by the same preposition. All the above three were mix properly and a paste was prepared with ethylene Glycol, The prepared paste was used for screen printing on pre cleaned glass substrates. The details of film casting procedure were described in [52]. The films were irradiated by using IR laser and setup schematic for laser striking is clearly depicted in Figure 8.4.

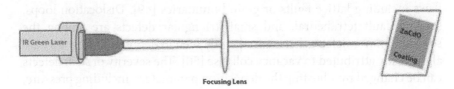

Figure 8.4 Laser set up for irradiation on ZnCdO coated films.

The synthesized samples were further heated at higher temperature and were kept in a muffle furnace in air at 550°C for 10 min for proper adherence and stability and for the decomposition of organic material.

8.7.4 Variation of Optical Properties

The absorbance spectra of coated films were measured to investigate their optical properties are shown in Figure 8.5a, in the wavelength range 300–700 nm. From the absorbance data we can calculate transmittance (T) as,

$$T = Log\frac{1}{A} \quad (8.1)$$

So, if T is the Transmittance and A is the absorbance of the film then, we can calculate reflectance of the film by using following equations [53]. Figure 8.5b shows the wavelength dependence of reflectance of the coated films obtained from above relations.

$$A + T + R = 1, \text{ or } R = 1 - (A + T) \quad (8.2)$$

As seen in Figure (8.8a and 8.8b), the absorbance and reflectance spectrum of the ZnCdO coated film shows a sharp absorption edge in the wavelength range 380–400 nm. Further observation shows the increase in

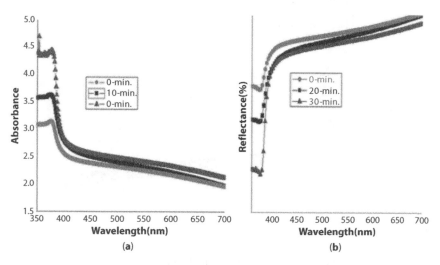

Figure 8.5 (a) Absorbance and (b) reflectance spectra of ZnCdO films under laser irradiation.

absorbance and decrease in reflectance with a red shift tend may be probably ascribed due to decrease in crystalline size and phase transforms leads to the decrease of band gap [54].

The absorption coefficient α is related to the absorbance A in the strong absorption region by

Beer-Lambert's law:

$$\log \frac{(I_0)}{(I)} = 2.303 \, A = \alpha \, t \tag{8.3}$$

where Io and I are the intensity of incident and transmitted light respectively, and t the film thickness.

$$\alpha = 2.303 \frac{A}{t} \tag{8.4}$$

The absorption curve can be separated into three regions: a strong region for wavelengths below 390 nm, an absorption tail region near the band edge for wavelengths between 390 and 570 nm, and a weak region

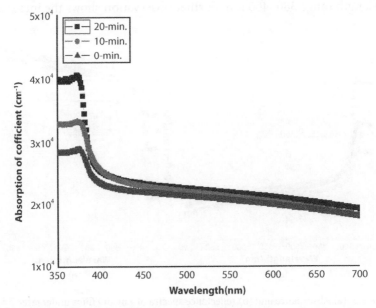

Figure 8.6 Absorption coefficient vs. wavelength of coated films.

for wavelengths above 570 nm. An exciton is created when an electron in the valence band absorbs some photon energy in the strong absorption region and moves from there to the conduction band. Figure 8.6 illustrates the absorption coefficient α for ZnCdO coated films in the spectral region from 350 nm to 700 nm. The peak value of the absorption coefficient is about $4.0 \times 10^4 cm^{-1}$ this indicate that, the material is becoming more absorbing upon laser irradiation [55]. The result shows all the films exhibit a low absorption in the visible range but exhibits high absorption in the ultra violet (UV) rang upon laser irradiation and these characteristics may be due to wide band gap property [56].

We computed the first derivative of optical reflectance in order to identify the absorption band edge of the films, and this result is shown in Figure 8.7. The plots of $dR/d\lambda$ versus wavelength give a peak corresponding to the absorption band edge. Longer wavelengths ,shifting of curves peak position changes. This suggests that the absorption band edge is consistent with reported reference [57] and shifts from 3.10 to 2.96 eV with increasing irradiation time.

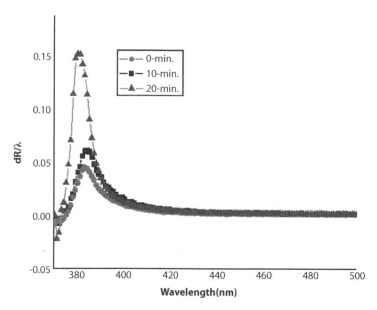

Figure 8.7 First derivative ($dR/d\lambda$) plot of the reflectance spectra.

8.7.5 Electrical Conduction Mechanism

To determine the type of electrical conductivity of the deposited material, an electrical study was conducted on all of the films. In the temperature range of 300-400K, the ZnCdO coated films' electrical conductivity was measured. All of the films that were deposited in this work were discovered to have electrical conductivity of this type. Figure 8.8 shows the well-known Arhenius plot of ZnCdO films and it is observed that the conductivity increases with increase in temperature along with laser irradiation. The thermally activated conduction mechanism can be expressed as [58];

$$\sigma = \sigma_0 exp(\frac{-\Delta E}{KT}) \quad (8.4)$$

Where σ is conductivity, σ_0 is the pre-exponential factor and E is the thermal activation energy for generation process is Boltzmann's constant and T is the temperature (in Kelvin). Trap levels below the conduction band are represented by the activation energy. Most likely, an increase in conductivity is the result of several parallel processes. The annihilation of defects is primarily responsible for a sharp rise in sheet conductivity and

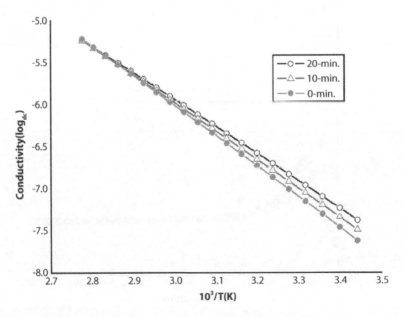

Figure 8.8 Arhenius plot of DC conductivity of ZnCdO coated films.

a sharp fall in activation energy with increasing irradiation time. This is in line with the PL study [59] regarding the zinc vacancies defects and the noticed annealing-induced reduction in grain size. The values of activation energies have been calculated from the slope of graph and come out 0.33eV, 0.29eV and 0.26eV such values are suitable value for hetero-junction in the solar cell [60].

8.7.6 Conclusion and Prospects

In addition to its conventional application, laser irradiation has discovered expanded use in the controlled manipulation of electrode materials for electrochemical energy storage and conversion. This is primarily made possible by the laser-driven rapid, selective, and programmable materials processing at low thermal budgets. This Review summarizes the recent developments in laser-mediated engineering of electrode materials, placing particular emphasis on its capacity to precisely fabricate heterostructures, integrate electrode architectures, and introduce controlled structural defects—all of which are highly desired for many electrochemical processes but challenging to precisely synthesize via conventional technologies. Following a brief explanation of the basic workings of laser processing, its practical application for the structural control of electrode materials is covered in detail. Additionally, the compatibility of laser-mediated structural regulation with numerous industrial processes gives this technology promising potential for quick application in the near future.

References

1. Overschelde, O.V., Guisbiers, G., Hamadi, F., Hemberg, A., Snyders, R., Wautelet, M., Alternative to classic annealing treatments for fractally patterned TiO2 thin films. *J. Appl. Phys.*, 104, 103106, 2008.
2. Nakajima, T., Shinoda, K., Tsuchiya, T., UV-assisted nucleation and growth of oxide films from chemical solutions. *Chem. Soc. Rev.*, 43, 2027, 2014.
3. Hwang, D., Ryu, S.G., Misra, N., Jeon, H., Grigoropoulos, C.P., Nano scale laser processing and diagnostics. *Appl. Phys. A*, 96, 289, 2009.
4. Pauleau, Y., Material surface processing by directed energy techniques, in: *Handbook of Advanced Ceramics*, p. 744, Elsevier, Amsterdam, Netherlands, 2005.
5. Zargar, R.A. *et al.*, Development and characterization of (ZnO) 0.90 (CNT) 0.10 thick film for photovoltaic application. *Optik*, 148, 167975, 2021.
6. Lee, Y.C., Hu, S.Y., Feng, Z.C., Yang, C.S., Huang, C.C., Temperature-dependent excitonic luminescence in ZnO thin film grown by metal organic chemical vapor deposition. *Jpn. J. Appl. Phys.*, 48, 112302, 2009.

7. Li, L. et al., Laser irradiation of metal oxide films and nanostructures: Applications and advances. *Adv. Funct. Mater.*, 28, 1705389, 2018.
8. Ogale, S.B., Venkatesan, T.V., Blamire, M., *Functional metal oxides: New science and novel applications*, Wiley-VCH, Weinheim, Germany, 2013.
9. Srivastava, A.K., *Oxide nanostructures: Growth, microstructures, and properties*, Pan Stanford Publishing, CRC Press, Taylor & Francis Group, Boca Raton, FL, USA, 2014.
10. Zargar, R.A. et al., Alcohol vapor sensing by cadmium-doped zinc oxide thick films based chemical sensor. *Mod. Phys. Lett. B*, 30, 12, 1650244, 2016.
11. Pacchioni, G. and Valeri, S., *Oxide ultrathin films: Science and technology*, Wiley-VCH, Weinheim, Germany, 2012.
12. Thin film metal-oxides: Fundamentals and applications in electronics and energy, S. Ramanathan (Ed.), p. 354, Springer, Berlin, 2010.
13. Hwang, D., Ryu, S.-G., Misra, N., Jeon, H., Grigoropoulos, C.P., Nanoscale laser processing and diagnostics. *Appl. Phys. A*, 96, 289, 2009.
14. Joe, D.J., Kim, S., Park, J.H., Park, D.Y., Lee, H.E., Im, T.H., Choi, I., Ruoff, R.S., Lee, K.J., Laser–material interactions for flexible applications. *Adv. Mater.*, 29, 1606586, 2017.
15. Cheng, J.-G., Wang, J., Dechakupt, T., Trolier-McKinstry, S., Low temperature crystallized pyrochlore bismuth zinc niobate thin films by excimer laser annealing. *Appl. Phys. Lett.*, 87, 232905, 2005.
16. Wang, H., Pyatenko, A., Kawaguchi, K., Li, X., Swiatkowska Warkocka, Z., Koshizaki, N., Selective pulsed heating for the synthesis of semiconductor and metal submicrometer spheres. *Angew. Chem. Int. Ed.*, 49, 6361, 2010.
17. Queraltó, A., Pérez del Pino, A., de la Mata, M., Arbiol, J., Tristany, M., Obradors, X., Puig, T., Ultrafast epitaxial growth kinetics in functional oxide thin films grown by pulsed laser annealing of chemical solutions. *Chem. Mater.*, 28, 6136, 2016.
18. Hong, S., Lee, H., Yeo, J., Ko, S.H., Digital selective laser methods for nanomaterials: From synthesis to processing. *Nano Today*, 11, 547, 2016.
19. Mincuzzi, G., Palma, A.L., Di Carlo, A., Brown, T.M., Laser processing in the manufacture of dye-sensitized and perovskite solar cell technologies. *ChemElectroChem*, 3, 9, 2016.
20. Thomas, S., Grohens, Y., Pottathara, Y.B. (Eds.), Industrial applications of nanomaterials, pp. 181–203, Elsevier, Amsterdam, Netherlands, 2019.
21. Li, G., Direct laser writing of graphene electrodes. *J. Appl. Phys.*, 127, 010901, 2020.
22. Rung, S., Christiansen, A., Hellmann, R., Influence of film thickness on laser ablation threshold of transparent conducting oxide thin-films. *Appl. Surf. Sci.*, 305, 347–351, 2014.
23. Bäuerle, D., *Laser processing and chemistry*, 4th edn, pp. 3–12, Springer-Verlag, Berlin, 2011.
24. Kittel, C., *Introduction to solid state physics*, pp. 329–330, John Wiley & Sons, New York, 1968.

25. Eckstein, W., *Computer simulation of ion-solid interactions*, Springer, Berlin, 1991.
26. Khan, M.I. and Ali, A., Effect of laser irradiation on the structural, morphological and electrical properties of polycrystalline TiO2 thin films. *Results Phys.*, 7, 3455–3458, 2017.
27. Thompson, D.A., High density cascade effects. *Radiat. Eff.*, 56, 105, 1981.
28. Chopra, K.L., *Thin film phenomena*, Krieger Publishing Company, Florida, 1979.
29. Mott, N.F., Conduction in non-crystalline systems. *Philos. Mag.*, 24, 190, 911–934, 1971.
31. Mott, N.F., *Conduction in non-crystalline materials*, Clarendon Press, Oxford, 1997.
32. Jonscher, A.K., Electronic properties of amorphous dielectric films. *Thin Solid Films*, 1, 213, 1967.
33. Palneedi, H. et al., Laser irradiation of metal oxide films and nanostructures: Applications and advances. *Adv. Mater.*, 28, 1705148, 2018.
34. Bogdan, A.I., and Kaufman, J., *Basics in dermatological laser applications*, A.I. Bogdan and D.J. Goldberg (Eds.), pp. 7–23, Karger, Basel, Switzerland, 2011.
35. Majumdar, J.D. and Manna, I., *Laser-assisted fabrication materials*, pp. 1–67, Springer-Verlag, Heidelberg, 2013.
36. Majumdar, J.D. and Manna, I. (eds.), Laser processing of materials. *Sadhana*, 28, 495, 2003.
37. Lee, D., Pan, H., Ko, S.II., Park, H.K., Kim, E., Grigoropoulos, C.P., Non-vacuum, single-step conductive transparent ZnO patterning by ultra-short pulsed laser annealing of solution-deposited nanoparticles. *Appl. Phys. A*, 107, 161, 2012.
38. Siano, S., *Handbook on the use of lasers in conservation and conservation science*, M. Schreiner, M. Strlic, R. Salimbeni (Eds.), pp. 1–26, COST Office, Brussels, Belgium, 2008.
39. Brown, M.S. and Arnold, C.B., *Laser precision microfabrication*, K. Sugioka, M. Meunier, A. Piqué (Eds.), pp. 91–120, Springer, Heidelberg, Germany, 2010.
40. Bäuerle, D., *Laser processing chemistry*, pp. 13–38, Springer, Heidelberg, Germany, 2011.
41. Ganeev, R.A., *Laser–surface interactions*, pp. 1–21, Springer Netherlands, Dordrecht, The Netherlands, 2014.
42. Cullis, A.G., Transient annealing of semiconductors by laser, electron beam and radiant heating techniques. *Rep. Prog. Phys.*, 48, 1155, 1985.
43. Carpene, E., Hoche, D., Schaaf, P., Laser processing of materials, P. Schaaf (Ed.), pp. 21–48, Springer-Verlag, Heidelberg, Germany, 2010.
44. Sundaram, S.K. and Mazur, E., Inducing and probing non-thermal transitions in semiconductors using femtosecond laser pulses. *Nat. Mater.*, 1, 4, 217, 2002.

45. Kim, H., Gilmore, C.M., Piqué, A., Horwitz, J.S., Mattoussi, H., Murata, H., Kafafi, Z.H., Chrisey, D.B., Electrical, optical, and structural properties of indium–tin–oxide thin films for organic light-emitting devices. *J. Appl. Phys.*, 86, 6451, 1999.
46. Zhang, W. and Yao, Y.L., *Manufacturing engineering handbook*, 2nd ed, P. Hwaiyu Geng (Ed.), pp. 1–32, McGraw Hill Professional, Access Engineering, New York, 2016.
47. Bogdan, A.I., and Kaufman, J., *Basics in dermatological laser applications*, A.I. Bogdan and D.J. Goldberg (Eds.), pp. 7–23, Karger, Basel, Switzerland, 2011.
48. Maissel, L.I. and Glang, R., *Handbook of thin film technology*, McGraw-Hill Book Company, New York, 1983.
49. Francis, M., Debeda, H., Lucat, C., Screen-printed thick-films: From materials to functional devices. *J. Eur. Ceram. Soc.*, 25, 2105–2113, 2005.
50. Jabir Saad, A.A., *Thick film electronic ceramic sensors for civil structures health monitoring*, Edinburgh Napier University (Thesis), Munich, Germany, 2011.
51. Zargar, R.A. et al., Growth of TiO2-CdO coated films: A brief study for optoelectronic applications. J. Phys. Chem. Solids, 179, 111390, 2023.
52. Zargar, R.A., Arora, M., Bhat, R.A., Study of nanosized copper-doped ZnO dilute magnetic semiconductor thick films for spintronic device applications. *J. Appl. Phys. -A*, 124, 36, 2018.
53. Zargar, R.A., ZnCdO thick film: A material for energy conversion devices. *Mater. Res. Express*, 6, 9, 095909, 2019.
54. Mamat, M.H., Sahdan, M.Z., Amizam, A. et al., Optical and electrical properties of aluminum doped zinc oxide thin films at various doping concentrations. *J. Ceram. Soc. Jpn.*, 117, 1263–1267, 2009.
55. Zargar, R.A. et al., Optical characteristics of ZnO films under different thickness: A MATLAB-based computer calculation for photovoltaic applications. *Phys. B: Condens. Matter.*, 631, 413614, 2022.
56. Mott, N.F. and Davis, E., Electronics processes in non-crystalline materials, p. 428, Clarendon, Oxford, 1979.
57. Gupta, R.K., Cavas, M., Yakuphanoglu, F., Strucural and optical properties nanostructure CdZnO films. *Spectrochim. Acta Part A: Mol. Biomol. Spectrosc.*, 95, 107, 2012.
58. Zargar, R.A., Fabrication and improved response of ZnO–CdO composite flms under diferent laser irradiation dose. *Sci. Rep.*, 12, 1, 10096, 2022.
59. Zargar, R.A., Arora, M., Hafiz, A.K., Investigation of physical properties of screen printed nanosized ZnO films for optoelectronic applications. *Eur. Phys. J. Appl. Phys.*, 70, 10403, 2015.
60. Zaragr, R.A., Micro size developed TiO2-CdO composite film: Exhibits diode characteristics. *Mater. Lett.*, 335, 15, 133813, 2023.

Part III
PHOTOVOLTAIC AND STORAGE DEVICES

9

Basic Physics and Design of Photovoltaic Devices

Rayees Ahmad Zargar[1]*, Muzaffar Iqbal Khan[1†], Yasar Arfat[2], Vipin Kumar[3] and Joginder Singh[4]

[1]Department of Physics, BGSB University, Rajouri (J&K), India
[2]Department of Electrical Engineering, SOET, BGSB University, Rajouri (J&K), India
[3]Department of Physics, KIET Group of Institutions, Gaziabad (U.P.), India
[4]Department of Physics, Government Degree College Nowshera, Rajouri (JK), India

Abstract

Photovoltaic (PV) technology is becoming more and more important as a result of the oncoming energy crisis, due to the depletion of natural occurring fossil fuel resources, and the associated greenhouse effect caused by the use of carbon. Semiconductors have ability to absorb photons and deliver its energy to the carriers in this away recombination of charge carrier's takes place. But up until now, the development of thin film solar cells especially metal oxide thin films solar cells and lack understanding of the materials and devices is still infancy. This chapter discusses the basic physics of semiconductors and the findings that assist in resolving the issue of energy crises. In order to attain higher energy conversion efficiency at lower prices, researchers around the world are searching for less expensive materials, less expensive production techniques, and smaller or more effective device topologies. In order to increase the performance/price ratio and make it more competitive with conventional energy, new materials and structures are therefore sought by using proper physics knowledge.

Keywords: Solar cell, physics equations, p-n junction, electrical parameters

---

*Corresponding author: rayeesphy12@gmail.com
†Corresponding author: muzaffariqbalkhan786@gmail.com

9.1 Introduction: Solar Cell

The basic design of semiconductor solar cells is relatively straightforward. The electrons and holes that carry current in the electrical system can get part of the photons' energy from semiconductors after they have absorbed it. An electrical current created by a semiconductor diode is conducted preferentially in one direction while isolating and gathering the carriers. A solar cell is thus just a semiconductor diode that has been expertly modified and created to effectively capture and transform solar light energy into electrical energy. Figure 9.1 depicts the basic design of a common solar cell; on the top of the solar cell's front, sunlight incident. The metallic grid that makes up one of the diode's electrical contacts allows light to pass through the grid and land on the semiconductor, where it is absorbed and converted into electrical energy [1]. The amount of light that reaches the semiconductor is increased by an antireflective layer positioned in between the grid lines. When an n-type semiconductor and a p-type semiconductor are connected to create a metallurgical junction, a semiconductor diode is created. This is frequently done by distributing, inserting, or depositing particular contaminants. The additional electrical photons generated by the diode contain the tiny, energetic photons that make up all electromagnetic radiation, including sunlight. Each photon's energy content is influenced by the spectral behaviors of its source [2]. Additionally, photons have wave-like properties, with their energy and wavelength related by

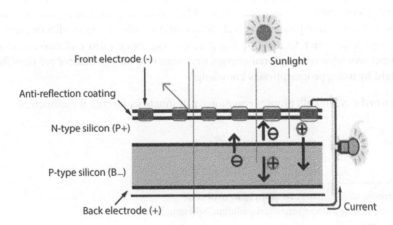

Figure 9.1 A simple typical solar cell diagram. This diagram shows how electron-hole pairs, e⁻ and h⁺, are created.

$$E = h\frac{c}{\lambda} \quad (9.1)$$

where quantity h is Plank's constant and the quantity c is the speed of light. Only photons with energies larger than the semiconductor band gap or high enough to produce an electron-hole pair can perform the energy conversion process. Therefore, the sun's spectrum is necessary for the development of effective solar cells. Since the sun releases electromagnetic radiation at a surface temperature of 5762 K, which can generate a huge number of electron-hole pairs.

9.2 Semiconductor Physics

Solid-state physics fundamentals must be understood in order to comprehend how semiconductor solar cells work. Here, an introduction to the key ideas required to study the physics of solar cells is given. Various sources provide more thorough and rigorous treatments. Many semiconductor materials can be used to make solar cells, but silicon (Si) crystalline, polycrystalline, and amorphous are the most used types. In addition to CdTe, other materials used to make solar cells include GaAs, GaInP, ZnO/CdS, and GaAs etc. Materials for solar cells are selected in large part based on how effectively they fabricate. Due to the fact that above materials especially, metal oxides-based materials absorption characteristics fit the sun spectrum reasonably well and that is why silicon fabrication technology is widely used in semiconductors, silicon has been a popular choice [3]. The time-independent Schrodinger equation can be used to derive the electron wave function (ψ), which can be used to deduce the electron's dynamic behavior.

$$\nabla^2 \psi + \frac{2m}{\hbar}[E - U(\vec{r})]\psi = 0 \quad (9.2)$$

where E denotes the energy of an electron, U(r) denotes the periodic potential energy within a semiconductor, m denotes the mass of an electron, and P denotes the diminished Planck constant. Surprisingly, the answer demonstrates that, if the electron's mass, m*, is replaced, the quantum mechanically predicted motion of the electron in the crystal is, to a good approximation, similar to that of an electron in free space. The response also describes the band structure of the semiconductor, along with the permitted electron energies and the relationship between electron velocity and energy. This quantum mechanical equation is not attempted to be solved in this paper.

$$F = \{m * a\} \tag{9.3}$$

where F denotes the electron's acceleration as well as the applied force. Figure 9.2 displays compacted energy band architecture. Using the equation p = hk, where k is the wave vector (here simplified to a scalar) corresponding to the solutions of the wave function of the Schrodinger equation, the permitted electron energies vs. crystal momentum are displayed. It is assumed that the energy bands below the valence band are completely filled with electrons and that the energy bands above the conduction band are empty. We only display the relevant energy bands at this time. Consequently, the definition of the electron effective mass is

$$m^* = \left[\frac{d^2 E}{dp^2}\right]^{-1}$$

$$= \left[\frac{1}{(h/2\pi)^2} \frac{d^2 E}{dk^2}\right]^{-1} \tag{9.4}$$

The effective mass varies between each band, which should be noted. Additionally, as the valence band near its maximum, the effective mass is really negative. The states at the top of the valence band are vacant because some electrons are thermally driven into the conduction band, where they are filled by electrons from bottom to top. These empty states can be simply described as positively charged holes with positive effective mass. Due to the fact that these few positive effective mass holes behave like regular positively charged particles, dealing with them theoretically is considerably simpler. Because of the nearly parabolic geometry between the top of the valence band and the bottom of the conduction band, the electron effective mass (mn) curves like a parabolic curve [3, 4].

The hole effective mass (MP) is constant at the top of the valence band and the bottom of the conduction band. The analysis of semiconductors is significantly simplified by this logical premise. When they do, as they do in Figure 9.2, the valence band maximum and conduction band minimum occur at the same crystal momentum value, indicating a direct band

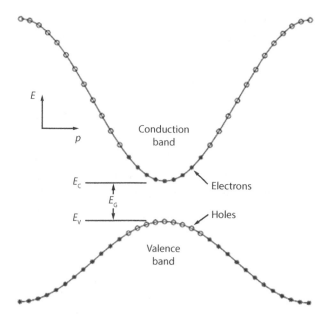

Figure 9.2 A condensed energy band diagram for a semiconductor with a direct band gap (E_G) at T > 0 K [4].

gap semiconductor. The semiconductor is referred to as an indirect band gap semiconductor when they are not aligned. This is crucial because a subsequent section of this chapter will discuss how well a semiconductor can absorb light. Even amorphous materials have a band structure comparable to this. The atoms are grouped in a periodic fashion over small distance. Since the arrangement of the atoms is periodic over relatively short distances, it is possible to define the electron wave function. It is possible to establish a mobility gap by combining the wave functions from these tiny locations, with holes below the mobility gap represented the valence band and electrons above it defining the conduction band. The analysis of devices made from these materials is made more difficult by the presence of a significant figure of localized energy states within mobility gap, which contrasts with crystalline materials.

9.3 Carrier Concentrations in Equilibrium

The ratio of filled states to accessible states at each energy is determined by the Fermi function, which is generated when the semiconductor is in thermal equilibrium, which is defined as remaining at a constant temperature without any external injection or creation of carriers.

$$F = \left[\frac{1}{1 + \frac{e^{\{E - E_F\}}}{kT}} \right] \qquad (9.5)$$

where the Boltzmann constant is k, the Fermi energy is EF, and the Kelvin temperature is T. The Fermi function exhibits a significant temperature dependence, as shown in Figure 9.3. It is a step function at absolute zero, where all of the states are completely empty above EF and all of the levels below EF are occupied with electrons [5]. Thermal excitation causes some of the states below EF to become vacant as the temperature rises, while simultaneously filling the respectively number of states above EF along excited electrons. By adding particular impurities, or dopants, referred to as donors and acceptors, one can regulate the quantity of electrons and holes in each band such that subsequently the conductivity of the semiconductor [6].

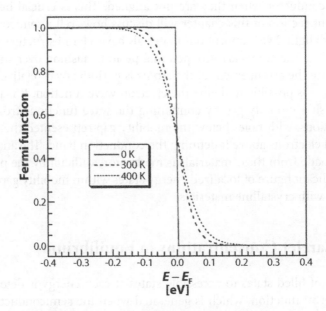

Figure 9.3 Shows how the Fermi function changes with temperature [7].

9.4 p-n Junction Formation

When two materials attain equilibrium and form a junction, the electrons and holes in any two of them, including n- and p-type semiconductors, metals, and insulators, will diffuse and drift apart from one another to create an energy-steady state across the junction. In their static condition, the two materials have identical Fermi energy levels, or EFs. Here, EF stands for the material energy level where, by the Fermi-Dirac Distribution Function equation, the probability of an electron occupying a given space is exactly 1/2:

$$F(E) = \left[1 + e^{\frac{(E-E_F)}{kT}}\right]^{-1} \qquad (9.6)$$

where E, k, and T, respectively, stand for the state's energy, the Boltzmann constant and thermal (measured in Kelvin). The likelihood that an electron will be in the given energy state is called F(E). When an n-type and a p-type semiconductor are mixed, carrier diffusion and drift cause a space charge zone to form. This homo-breadth, or W for the junction region, can be written as [8].

$$W = \sqrt{\frac{2\varepsilon_s}{q}\left(\frac{N_A + N_B}{N_A N_B}\right) V_{bi}} \qquad (9.7)$$

The letters ε_s, q, V_{bi}, N_A, and N_D, respectively, stand in for a semiconductor's relative constant, electron charge, inherent electric applied field, concentration acceptor in a p-type semiconductor, and concentration donor in an n-type semiconductor.

9.5 Process of Carrier Production and Recombination

The charge development rate, G, for the development process is represented as

$$G = G_L + G_{th} \qquad (9.8)$$

where G_{th} is the heat generation development rate and G_L is the small-induced mobile generation rate. The recombination rate (R) is total to the development rate under equilibrium conditions [9].

$$R = R_D + R_{ID} = G \tag{9.9}$$

Where the symbols R_D and R_{ID} called the corresponding direct and the indirect recombination rates.

9.6 Equations for Poisson's and Continuity Equation

Poisson's equation and the continuity equations are the two fundamental equations that guide the charge conveying process. The simple Poisson's differential equation defined by

$$-\frac{d^2\psi}{dx^2} = \frac{dE}{dx} = \frac{\rho_s(x)}{\varepsilon_s} \tag{9.10}$$

Where the electrostatic potential, an external electric field, and concentration of space charges at location (x) in developed device, are denoted by, E, and $\rho_s(x)$. Equation (9.10) for solar cells can also be written as,

$$-\frac{d^2\psi}{dx^2} = \frac{dE}{dx} = \frac{q}{\varepsilon_s}[p - n + N_D^+ - N_D^- + p_t - n_t] \tag{9.11}$$

The concentrations of free hole, free electron, ionized donor-like doping, ionized acceptor-like doping, trapped hole, and trapped electron at location x are denoted by the letters p, n, N_D^+, N_A^-, p_t, & n_t. For electrons and holes, the fundamental one-dimensional continuity equations are as follows:

$$\frac{\partial n}{\partial t} = \frac{1}{q}\frac{\partial J_n}{\partial x} + (G_n - R_n) \tag{9.12}$$

$$\frac{\partial p}{\partial t} = -\frac{1}{q}\frac{\partial J_p}{\partial x} + (G_p - R_p) \tag{9.13}$$

where the quantities J_n and J_p represent the electron & corresponding hole currents [10]. The letters G_n, Rn, G_p, and R_p, respectively, stand for the rates of electron development generation, electron reproduction, hole development generation, and hole reproduction. The total current for the p-n junction also called solar cell device is total sum of the electron current and the hole current. Due to continuity of the current, the addition of J_n and J_p at every position x indicates entire current flowing through the diode. Below are presented the relevant equations [11].

$$J_p(x_n) = \frac{qD_p p_{n0}}{L_p}\left(e^{\frac{qV}{kT}} - 1\right) \quad (9.14)$$

$$J_n(-x_p) = \frac{qD_n n_{n0}}{L_n}\left(e^{\frac{qV}{kT}} - 1\right) \quad (9.15)$$

$$J_{total} = J_p(x_n) + J_n(-x_p) = J_s\left(e^{\frac{qV}{kT}} - 1\right) \quad (9.16)$$

$$J_s = \frac{qD_n n_{n0}}{L_n} + \frac{qD_p p_{n0}}{L_p} \quad (9.17)$$

where quantities such as $J_p(x_n)$, $J_n(-x_p)$ are represents the hole current and electron current, respectively, at the n- and p-side edges of the depletion area. D_n and D_p are the electron & the hole diffusivities. In the p-side and n-side, respectively, diffusion lengths of electrons and holes (minor carriers) are L_n and L_p. J_s stands for the saturation current density.

9.7 Photovoltaic (Solar Power) Systems

The photovoltaic aspect has been well-known since 1839, when French physicist Edmond Becquerel discovered how to generate electricity by shining light on a metal electrode dipped in a power less electrolyte substance. Adam and Day developed the first selenium solar cell in 1876 [12] with an efficiency of 1–2% in order to investigate the photovoltaic effect in solids. Albert Einstein's photon theory helped to explain the photovoltaic phenomenon in 1904. Polish scientist Jan Czochralski made a fundamental contribution to modern electronics in 1916 when he discovered a way to produce pure crystalline silicon. Compared to modern solar cells, which have an efficiency of between 14% and 20%, first generation silicon cells had a relatively low 6% efficiency. Due to the high device prices, early attempts to make photovoltaic cells a practical way to generate electricity for terrestrial applications failed. The "energy crisis" of the 1970s sparked a fresh wave of attempts in numerous nations to lower the cost of solar systems, particularly for off-grid uses. Recent years have seen a sharp decline in photovoltaic cell prices, which has rekindled interest in the technology. For instance, since 2000, annual growth in the manufacturing of PV systems has increased by more than 40%, and around 22 GW of installed capacity is currently available globally [12].

9.8 Types of Photovoltaic Installations and Technology

The four channel PV installation types are grid-tied distributed (small installations on the roof or ground), grid-tied centralized (big power plants) and off-grid commercial. Each installation has its own specific requirements for system balance. For instance, battery banks or other forms of alternate electrical storage are typically needed for off-grid independent applications [12]. Based on the development of solar cells, photovoltaic systems can be further differentiated (Figure 9.4). The most advanced technologies are those based on silicon (Si), which can be break into two categories: amorphous or thin film silicon coexists with crystalline silicon. Mono-crystalline, multi-crystalline, and ribbon cast multi-crystalline silicon, among other crystalline forms, can be used to create crystalline silicon cells [12]. The instantaneous direct conversion of solar energy into electricity without the need for sophisticated mechanical components or integration is a significant characteristic of photovoltaic systems [12].

Currently, the majority of solar cells are used with huge power development generation, either core power plants or as "building integrated photovoltaics" (BIPV). BIPV is getting a lot of attention since employing photovoltaic cells in this fashion reduces the amount of land needed and makes up for the high manufacturing costs by utilizing the cells as construction materials. Crystalline silicon made solar cells are most widely used class of solar cell, but throughout the majority of the second half of the 20th century, other cell kinds that finish in terms of lower manufacturing costs or higher efficiencies were developed.

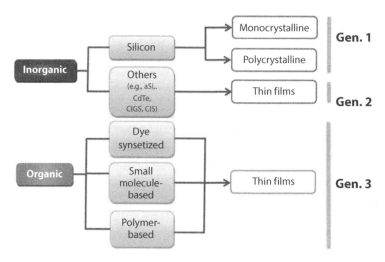

Figure 9.4 A broad classification of photovoltaic technologies.

9.9 Electrical Characteristics Parameters

Under the process of illumination, the photovoltaic parameters are testified for solar cell performance similarly J_{sc}, V_{oc}, FF, and PCE. Photovoltaic parameters are used to validate the J-V behavior of photovoltaic cells.

9.9.1 Open-Circuit Voltage

Solar cells are subjected to a bias voltage (also known as an open-circuit voltage, or Voc) in order to destroy the current produced during radiance. The applied current (J = 0) does not flow at the Voc when it is illuminated. The Voc in DSSC is dependent on the photo anode and electrolyte's work function of Fermi energy [13]. The experimentally determined V_{oc} is a little low due to charge carrier recombination. It is possible to reach the theoretical limit if all recombination systems are minimized. Under open-source circuit conditions, all charge carriers will pair inside the photoactive structure. The thermodynamic type of surface drainage in the depletion zone, which serves as the basis for the creation of dark current and is a perfect diode current, has been recognized as making it challenging to avoid recombination events [14]. When a delivered bias load is applied, the potential difference between the solar cell's terminals is created. The total current is minimized from its short circuit value as a result of the potential difference, resulting in an opposite to the photocurrent current. Dark current is the slang term for the reverse current, I dark (V), that moves through

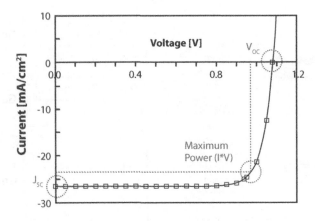

Figure 9.5 I-V plot of a TiO2/CuO/Cu2O solar cell develop by AFORS-HET software [16].

a device when a voltage is applied while it is dark. Solar cells function similarly to diodes in the dark, accepting more of the equilibrium between the whole recombination of photogeneration development and charge carriers [15]. The conversion power over illumination power determines a solar cell's efficiency. Experimentally, it is often determined by measuring the current-voltage under specific solar illumination conditions, as shown in Figure 9.5.

9.9.2 Density of Short-Circuit Current (Jsc)

The current density produced in a solar cell at applied zero bias is referred to as J_{sc}. The generation of the built-in potential is mediated by exciton dissociation and charge transport [16]. Incident light and excitons produced by the solar cell have an impact on its Jsc. Second, the photoactive layer's broad absorption range enables the solar cell's Jsc to be increased by capturing more excitons from the spectrum of the terrestrial sun. The Jsc of a solar cell is impacted by charge carrier mobility of active layer [17].

9.9.3 Fill Factor Percentage (FF%)

The solar cell's J-V properties, A solar cell's main job is to convert solar energy into electrical energy. The maximum deliverable power, or Pmax, can be calculated using the J-V curve. The following calculation is used to get the solar cell's fill factor. The behaviors of a solar cell diode can be measured using the fill factor. The fill factor needs to be unity for the diode to be ideal. The diode will be more ideal as FF increases. High fill factor FF is caused solar cells are characterized by high shunt resistance and low series resistance.

$$FF\% = \frac{P_{max}}{J_{sc} + V_{oc}} = \frac{P_{max} + V_{max}}{J_{sc} + V_{oc}} \quad (9.18)$$

The FF can be used to measure the solar cell diode characteristics. The FF needs to be unity for the diode to be ideal. The diode will be largely ideal as FF increases. Due to recombination and transport losses.

9.9.4 Power Conversion Efficiency (η)

The difference among putted power (P_{max}) tithe incident light power is what is known as a solar cell efficiency (P_{in}). The following expression is used to calculate the solar cell's photo conversion efficiency;

$$n = \frac{P_{max}}{P_{in}} = \frac{J_{sc} + V_{oc}}{P_{in}} \times FF \quad (9.19)$$

The equation provides information on the cell output power as well as information on photovoltaic parameters and how light is converted into electricity.

9.9.5 Dark Current

When it is dark, the current allowing over the diode referred to known dark current. Dark currents are generated in the depletion region by sort of surface drainage or by the decompensation of charge carriers [14]. The potential difference between the solar cell's terminals is produced when a forward bias load is adopted. The potential difference causes the net current to decrease from its short circuit value, producing an opposite to the photocurrent current. When a voltage is applied to a component in the dark during low light, as is the case with solar cells, the phrase "dark current" refers to the reverse current I dark (V) that crosses the component.

9.9.6 Standard Test Conditions

Since the photovoltaic parameters are affected by temperature, irradiance, humidity, and light intensity, test conditions were created to achieve results that were similar and meaningful. The solar emission spectrum and spectral distribution on a sunny day with a radiant intensity of 100 mW/cm² and an angle of incidence of 48.2 are the basis for these device test parameters. The corresponding atmosphere spectrum known as "Air Mass 1.5 Global" for this Sun additionally includes specific concentrations of moisture, carbon dioxide, and aerosols.

9.10 Basic p-n Junction Diode Parameters

With the aid of the following equations [18], the Schottky barrier height as well as ideality factor derived adopting I-V characteristics.

$$J = J_0 \exp(\frac{qV}{nk_BT} - 1) \qquad (9.20)$$

$$J_0 = R^*T^2 e^{\left\{\frac{-q\varphi_b}{k_BT}\right\}} \qquad (9.21)$$

Here, the Schottky barrier height represented by s, Boltzmann constant is s, and actual Richardson constant is R*. The ideality factor is n. The charge of an electron is q. T is used to represent absolute temperature. The Schottky barrier is s in height. S is the Boltzmann constant. The barrier height can be calculated using the formula below:

$$\varphi_b = \left[\frac{k_BT}{q}\right] ln\{R^*T^2/J_s\} \qquad (9.22)$$

The junction's ideality factor is provided by

$$n = (\frac{e}{kT})[\frac{\partial V}{\partial (\ln J)}] \qquad (9.23)$$

The values of the Schottky junction barrier height (φ_b) & ideality factor (n) obtained from these formulae.

9.11 Conclusion

This chapter's main objective is to give the reader a fundamental understanding of physics. This has been accomplished through a consideration of the fundamental physical properties of solar cell materials that allow for the conversion of light into energy. These characteristics include the ability of semiconductors to conduct electrical current and their capability to absorb photons by converting their energy to electrical current carriers. The chapter includes theoretical calculations in this area as well as general physics. The essential operating principles of the solar cell, a well-constructed pn-junction diode, were derived using the equations describing the dynamics of holes and electrons in semiconductors. This led to the definition of the solar cell's open-circuit voltage (V_{OC}), short-circuit current (I_{SC}), fill factor (FF), and cell efficiency (η). The creation of electron-hole pairs and recombination were the two key factors that have been identified and considered as affecting solar cells' efficiency. Examples were given to emphasize the necessity of minimizing all potential recombination causes in solar cells.

References

1. Bruce, C., The photovoltaic generation of electricity. *Sci. Am.*, 235, 4, 34–43, 1976.
2. Lewis, G.N., The conservation of photons. *Nat.*, 118, 2981, 874–875, 1926.
3. Szlufcik, J. et al., Low-cost industrial technologies of crystalline silicon solar cells. *Proc. IEEE*, 85, 5, 711–730, 1997.
4. Rice, T.M., The electron-hole liquid in semiconductors: Theoretical aspects. *Solid State Phys.*, 32, 1–86, 4978.
5. Kerkhove, D., *Silicon production technology*, Technische Universiteit Delft, Delft, Netherlands, 1994.
6. Forwald, K., Dissertation NTNU, Norway, MI-47, 1997.
7. Dosaj, V., *Kirk-Othmer encyclopedia of chemical technology*, 4th Edition, vol. 21, pp. 1104–1122, Wiley, New York, 1997.
8. Gray, J., *Two-dimensional modeling of silicon solar cells*, Ph.D. thesis, Purdue University, West Lafayette, IN, 1982.
9. Gray, J. and Schwartz, R., *Proc. 18th IEEE Photovoltaic Specialist Conf.*, pp. 568–572, 1985.
10. Sze, S.M., *Semiconductor devices: Physics and technology*, p.523 Wiley, USA, 1985.
11. Kano, K., *Semiconductor devices*, p.480, Pearson Education, New Delhi, 1998.

12. Kumar, R. and Rosen, M.A., A critical review of photovoltaic-thermal solar collectors for air heating. *Appl. Energy*, 88, 11, 3603–3614, New Delhi, 2011.
13. Khan, M. I., and Upadhyay, T. C., Ferroelectric phase transitions in PbHPO4-type crystals, *J. Mater. Sci. Mater. Electron.*, 32, 11, 14569–14583, 2021.
14. Rhee, J.H., Chung, C.C., Diau, E.W.G., A perspective of mesoscopic solar cells based on metal chalcogenide quantum dots and organometal-halide perovskites. *NPG Asia Mater.*, 5, 68, 2013.
15. Dyakonov, V., Mechanisms controlling the efficiency of polymer solar cells. *Appl Phys. A.*, 79, 21–25, 2004.
16. Fonash S.J. *et al.*, A manual for AMPS-1D for Windows 95/NT: A one-dimensional device simulation program for the analysis of microelectronic and photonic structures, Pennsylvania State University, 1997.See also http://www.psu/edu/dep/AMPS.
17. Shin, S. *et al.*, *Energy Environ.* Energy-level engineering of the electron transporting layer for improving open-circuit voltage in dye and perovskite-based solar cells. *Energy Environ. Sci.*, 12, 958–964, 2019.
18. Zaragr, R.A., Micro size developed TiO_2-CdO composite film: Exhibits diode characteristics. *Mater. Lett.*, 335, 15, 133813, 2023.

10

Measurement and Characterization of Photovoltaic Devices

Saleem Khan[1]*, Vaishali Misra[2], Ayesha Bhandri[3] and Suresh Kumar[4]

[1]*R&D Sensor Division, Multi Nano Sense Technologies, Maharashtra, India*
[2]*Dept. of Nanosciences and Materials, Central University of Jammu, J&K, India*
[3]*Dept. of Physics, Central University of Jammu, J&K, India*
[4]*Dept. of Electronic Science, Kurukshetra University, Haryana, India*

Abstract

Recent technological advancements in material science domain, led to rapid improvement in photovoltaic material synthesis and device fabrication. Development of material and photovoltaic device performance requires evaluation and validation via various characterization systems. The focus of this chapter is to provide detailed insight of characterizations method employed for measurement of critical electrical parameters. Initially overview of currently available photovoltaic measuring technologies are presented followed by some assessments and forecasts of the direction that research and characterization. Measurements and characterization must keep pace with cutting-edge developments and technologies as photovoltaics pushes the boundaries of materials research, device engineering, and performance. Different electrical measurement methods like current-voltage, quantum efficiency, Hall effect, light beam induced current technique, photoluminescence spectroscopy, electroluminescence imaging, electron impedance spectroscopy, and ellipsometry that essential to identify the problems and optimization of output parameters such as photoconversion efficiency (PCE), short circuit current (J_{sc}), open-circuit voltage (V_{oc}) and fill factor (FF). These measurement approaches stand out as effective performance metrics for producing outcomes for reliable device performance as well as for improving the device with higher levels of repeatability.

Keywords: Electrical parameters, hall effect, PCE, metal oxide solar cells

---

*Corresponding author: saleem.k21@gmail.com

Rayees Ahmad Zargar (ed.) Metal Oxide Nanocomposite Thin Films for Optoelectronic Device Applications, (239–262) © 2023 Scrivener Publishing LLC

10.1 Introduction

Because it is affordable and endless availability, photovoltaic energy is one of the renewable, green, and clean energies that draw researchers from all over the world [1]. Even though there is a large demand for solar-generated electricity, it can only meet 0.1% of that need due to the lack of technology that is now available to fully utilize solar energy. According to estimates from the International Energy Agency (IEA), developing nations are consuming energy at a rate that is quicker than that of wealthy nations. Additionally, a 44% increase in global energy use is predicted between 2006 and 2030. Increasing demand of energy encouraged researchers to explore new approaches to develop photovoltaic devices (PV) [2–4]. The advancements in fabrication of high density and high efficiency prompted the necessity for measurement and characterization studied of photovoltaic devices. Utilizing advance methods and tools for photovoltaic device characterization, researchers may evaluate device performance, comprehend performance-affecting variables, and define material attributes [5–7]. Figure 10.1 shows the extensively used advance tools for various physical and chemical characterizations. The analysis performed with these systems provides the insight of

Figure 10.1 Classification of PV characterization domain.

CHARACTERIZATION OF PHOTOVOLTAIC DEVICES 241

material compositions, chemical state, morphological conditions (structural, topography, and surface roughness), and thin film growth status [8–10]. In-order to estimate the defects in the PV devices, characterizations needed are depicted in Figure 10.2. Besides these characterizations electrical efficiency and device performance parameter estimations are very critical. The current and voltage produced by the PV device as a result of the incidence light are used to gauge its performance [11–15].

A compilation of characterization techniques that routinely support photovoltaic materials, device, and component evaluations are included in Figure 10.1. All aspects and the entirety of measurements of even this selected set cannot be adequately addressed in single chapter, which serves only as a gateway to the operations and need of these methods in the PV technology [16–18]. The importance, strengths, and limitations of these techniques are stressed, especially their significance to photovoltaics. Included are several techniques that have been developed specifically to address problems and requirements for photovoltaics. The absorbent surface and devices have been thoroughly needs to comprehend for establishing the fundamental factors that lead to materials' high efficiency, low scalability, and degradation mechanisms. Various characterizations utilized for topology and morphological measurements are mentioned in

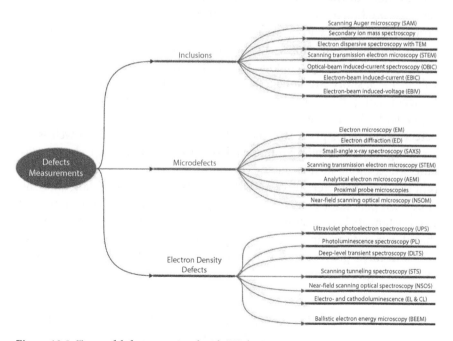

Figure 10.2 Types of defects associated with PV devices.

Figure 10.1. In order examine the chemical and structural details of the material, X-ray and electron microscopy techniques are abundantly used. The focus of this chapter is on the electrical and optical measurements of the PV devices.

10.2 Electrical and Optical Measurements

Studies on electrical measurements and its assessment aid in the construction of effective PV devices with low power dissipation. Understanding its numerous optical aspects and factors, which are almost certainly impacting it, would also be beneficial. Electrical measurements have been used up to this point to evaluate the performance of photovoltaic systems and the factors that affect it, including interfacial recombination losses, carrier lifetime, maximum output power, power conversion efficiency, resistivity, fill factor, Hall voltage, trap density, and others [19–22].

10.3 Current–Voltage (I-V) Characterization

Photovoltaic device performance is assessed by the current and voltage produced in the device by illumination. Several factors influence the produced electrical energy, including the nature of material, structure of device, ambient conditions, and the actual electrical load applied to the device. The I-V measurements provide parameters such as open circuit voltage (V_{oc}), short circuit current (I_{sc}), and fill factor. The efficiency of PV devices is computed from these parameters. Furthermore, the temperature dependent I-V characteristics measurement in the dark are very useful for detecting the redox processes that restricts the device capacity as well as for assessing internal device parameters like series resistance (R_s), shunt resistance (R_{sh}), diode ideality factor (n), and dark or bucking saturation current (I_0). I-V characteristic of a PV device acts like a diode in dark ambient. A current generator is introduced in parallel to the diode during lighted circumstances. It is customary to utilize negative currents, which means that all currents are applied in the opposite direction of their natural flow, which in turns provides positive generated power which can be described by following equation:

$$I = I_{Ph} - I_0 \left[\exp\left(\frac{qV}{nKT}\right) - 1 \right] \qquad (10.1)$$

where
- I_0: diode saturation current
- q: charge
- n: ideality factor
- K: Boltzmann's constant
- T: Absolute temperature (Kelvin)
- I_{Ph}: photocurrent

Also, the short circuit current (I_{sc}) is obtained at zero voltage. Short circuit current in a typical linear solar cell is inversely correlated with solar radiation intensity. The voltage potential at which there is no current flowing is known as the open circuit voltage (V_{oc}). The open circuit voltage behaves as logarithmic function of current as given in below equation

$$V_{oc} = \frac{nkT}{q} \ln\left(\frac{I_{PH}}{I_0} + 1\right) \quad (10.2)$$

In PV devices maximum power point (P_{PM}) is defined as peak power at specific voltage and current. Also, first derivative of product of voltage and current equals to zero refers to as P_{PM}. The specific voltage and current are known as maximum power point voltage (V_{MP}) and current (I_{MP}). Fill factor is the ration of P_{PM} and V_{oc} and I_{sc} product give below. Unity fill factor value signifies high quality PV device.

$$FF = \frac{I_{MP} \cdot V_{MP}}{I_{SC} \cdot V_{OC}} = \frac{P_{MP}}{I_{SC} \cdot V_{OC}} \quad (10.3)$$

Efficiency is likely one of most significant PV device evaluation metric. It is the ratio of output electrical energy leaving the device to photon striking the device, or, to put it another way, the ratio of electrical energy to input power. PV device efficiency is calculated as

$$\eta = \frac{P_{MP}}{G.A} \quad (10.4)$$

where power density of incident light and A is the device area.

Figure 10.3 depicts the standard setup for laboratory I-V curve measurements, which includes a light source, reference device for calibration

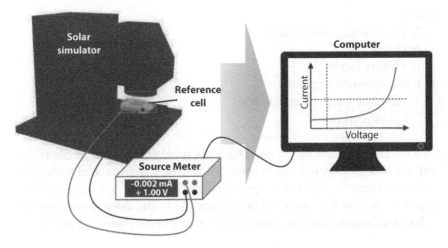

Figure 10.3 Schematics of current-voltage measurement system for PV device. Reprinted with permission from [23]. Copyright (2022) Elsevier Joule.

of irradiance during the measurement, test bench having device under test (DUT) unit, and temperature sensors. Recent measuring equipment also contain software capabilities for processing the obtained data, most importantly for standard testing conditions (STC) adjustment and I-V curve parameter computation. DTU is pre-installed and position vertical between the light source and PV devices. Photovoltaic simulators are light

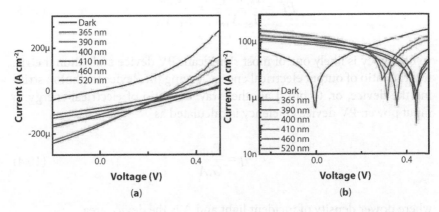

Figure 10.4 I–V profiles of transparent solar cell (NiO/TiO2/FTO/Glass) for broadband of illumination under (a) Linear scale and (b) Log-scale. Reprinted with permission from [24]. Copyright (2022) Elsevier Solar Energy Materials and Solar Cells.

sources (synthetic sunshine) created in a research lab to mimic solar radiation and whose irradiation intensity is calibrated to the brightness of the natural light. In order to examine the total efficiency of the solar, it may be utilized as a test system for solar cell material devices. They offer the ability to adjust electrical parameters so that PV device accurately characterized. Nguyen et al. (2020) fabricated p-NiO/n-TiO$_2$ based PV device capable of converting highest wavelength of light. The maximum photovoltage converted by the device is 600 mV for UV light. Also, under the sunlight the device capability of producing 153.9 mV photovoltage and 0.4 mA of

Table 10.1 Comparison of metal oxide-based PV devices.

Material for electron transport layer	PV device layers	Current (J_{sc})	Voltage (V_{oc})	Ref.
TiO2	FTO/TiO2/MAPbI3/spiro-OMeTAD/Ag	18.35 mA/cm^2	1.650 V	[25]
TiO2–SnO2	FTO/TiO2-SnO2/MAPbI3/spiro-OMeTAD/Ag	22.52 mA/cm^2	1.100 V	[26]
Ag:TiO2	FTO/Ag:TiO2/MAPbI3/spiro-OMeTAD/Ag	21.00 mA/cm^2	0.970 V	[27]
Graphene:TiO2	FTO/graphene:TiO2/MAPbI3/spiro-OMeTAD/Au	25.05 mA/cm^2	1.060 V	[28]
La:TiO2	FTO/La:TiO2/MAPbI3/spiro-OMeTAD/Au	21.30 mA/cm^2	1.050 V	[29]
Eu3+/Sm3+:TiO2	FTO/Eu3+/Sm3+:TiO2/MAPbI3/spiro-OMeTAD/Au	22.47 mA/cm^2	1.100 V	[30]
Yb3+/Er3+:TiO2	FTO/ Yb3+/Er3+:TiO2/MAPbI3/spiro-OMeTAD/Au	20.31 mA/cm^2	0.630 V	[31]
Nb:TiO2	FTO/Nb:TiO2/MAPbI3/spiro-OMeTAD/Au	16.26 mA/cm^2	0.714 V	[32]
Cd:TiO2	FTO/Cd:TiO2/MAPbI3/spiro-OMeTAD/Au	16.90 mA/cm^2	0.784 V	[33]

photocurrent shown in Figure 10.4 [24]. The compared photocurrent and photovoltage by various metal oxide PV device given in Table 10.1.

10.4 Quantum Efficiency

Understanding phenomenon of current generation, recombination, and diffusion dynamics in PV device requires knowledge of the spectrum responsiveness also known as quantum efficiency (QE). The QE is frequently used as a spectrum conversion factor for PV cell and calibration modules labeled in Figure 10.5. The observed spectrum responsivity is measured in Amps/watt as a function of wavelength is used to calculate the QE as electron-hole pairs gathered per incident photon. In order to allow the solar cell to function in intended purpose circumstances while being measured, the majority of QE systematically uses a predefined set a bias light with an intensity equivalent to one sun. Additionally, they employ a mechanical oscillator to allow measuring electronics to differentiate between the current produced by monochromatic light and that produced by bias light or noise. The majority of QE system monochromators use an interferometer to divide the beam's wavelengths. The light from specific wavelength can pass via a set outlet slit owing to the grating's rotation. To effectively cover the wavelength range that solar cells response to, the majority of QE system monochromators include numerous, adjustable gratings. Occasionally is the extra incident light filtering provided by multiple monochromators necessary. Using optical filters, QE system monochromators block light at harmonics of the specified wavelength. Furthermore, a portion of the monochromatic pulse deviate to a screen detector that is connected to a separate channel of the measuring electronics, allowing correction for variations in luminosity between the calibration and testing times. The lock-in approach with a synchronization signal from the electromechanical chopper is frequently used in current measuring electronics. Voltage on the test device is kept at zero.

Typically, semiconductor materials do not absorb all the photons incident over it, due to electron transport phenomena caused by energy bandgap. Restriction in absorption occurs because of surface recombination process at low wavelength and low diffusion at higher wavelength. QE can be classified as external quantum efficiency which includes transmission and losses due to reflection and internal quantum efficiency which represents only absorbed photons. QE of PV devices is represented by following equation

$$QE = \frac{hc.I_{sc}}{e\lambda.P_{in}} \times 100 \qquad (10.5)$$

where
  h: Plank's constant
  c: speed of light
  e: charge of electron
  P_{in}: power of incident wave
  λ: wavelength

Factor which quantifies QE is known as spectral response (SR) which specifically indicates current generation with respect to incident power and is represented as

$$SR = \frac{e\lambda}{hc} \times QE \qquad (10.6)$$

Assessments of spectral responsiveness require determining the photocurrent generated by light of a certain wavelength and power. Since the device may be nonlinear, the quantum efficiency is often tested using bias illumination that simulates baseline circumstances. The QE measurement at zero voltage gives the spectral correction factor.

Sagar and Rao (2020) demonstrated the efficiency variation in silicon-based PV device with various metal oxide such as zinc oxide, magnesium oxide, and aluminum oxide as antireflection layer and their

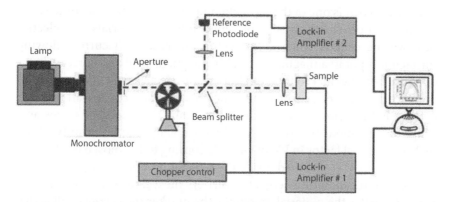

Figure 10.5 Setup of quantum efficiency measurement system. Reprinted with permission from [34]. Copyright (2020) AIP Publishing, Journal of Applied Physics.

Table 10.2 Comparison of metal oxide-based PV devices efficiency and fill factor.

Material for electron transport layer	Efficiency %	Current (J_{sc}) mA/cm^2	Fill factor	Ref.
ZnO	11.13	20.08	0.560	[36]
AZO–ZnO	10.70	19.84	0.600	[37]
ZnCsO	4.06	7.03	0.578	[38]
ZnLiO	3.91	6.77	0.571	[38]
ZnO-Li	17.80	22.00	0.733	[38]
ZnO-Na	18.90	22.50	0.754	[39]
ZnO-K	19.90	23.00	0.771	[39]
Al:ZnO	17.60	21.70	0.756	[40]
In:ZnO	16.10	23.00	0.700	[41]

influence on conversion efficiency. The fabricated device ZnO, MgO, and Al2O3 give 8.380%, 9.864%, and 9.125% efficiency respectively [35]. Efficiency and fill factor of oxide-based PV devices tabulated Table 10.2.

10.5 Hall Effect Measurements

Hall Effect was discovered by E.T. Hall in 1879 and outline view is shown in Figure 10.6. It is one of the most abundantly used technique for to study the conduction characteristics of semiconductor materials. An electric field will be created when a conducting material with current flowing through it is put in a magnetic field. The magnetic field and current directions are both normal to the direction of this electric field. The resistivity, mobility, and carrier density of the studied semiconductor are all revealed by the Hall measurement, which is significant. The mobility serves as a gauge for the impurity level, epitaxial layer quality, and uniformity which makes its crucial factor in PV devices. According to the most basic form of a solar cell, light is absorbed inside the cell and produces charged carriers. To produce meaningful energy, the carriers must flow to the cell's terminals. The better the material's mobility, the quicker the carriers will arrive at the terminals.

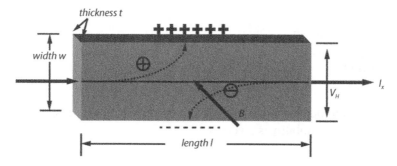

Figure 10.6 Hall effect outline view.

The fundamental theory underlying the Hall Effect derives from the fundamental properties of conduction, according to which current is composed of charged particles (electrons) flowing while drifting under the influence of an applied electric field. These carriers will move at a speed known as the drift velocity (v), and they can move either in the direction of the applied electric field (holes) or in the opposite way (electrons). The applied electric field strength E has a direct relationship with the carriers' drift velocity in terms of magnitude.

The conventional method of Hall measurement uses DC magnetic field. This approach has a long history of accurate measurements with various types of materials, especially semiconductors. However, it is very difficult to analyze low mobility materials using DC Hall effect methods, such as organic electronics, thermoelectric technology, and solar cell technology materials.

The Hall effect occurs when a magnetic field is introduced perpendicular to a current that is flowing in a semiconductor device. The current and magnetic field work together to exert a force on the carriers known as the Lorentz force. The magnetic field lines are circumvented by the carriers' circular motion as a result of this force. Some of the charges will make contact with the sample's sides. The charge accumulation produces an electric field that is perpendicular to the magnetic and current fields. Voltage produced by this electric field is known as Hall voltage and at steady state situation the Hall voltage cancels the Lorentz force given as

$$V_H = \frac{R_H IB}{t} \qquad (10.7)$$

where
R$_H$: Hall coefficient
B: magnetic field
I: current
t: thickness

The relation between Hall coefficient and resistivity with respect to carrier density and mobility is given as

$$n = \frac{1}{R_H e} \ \& \ \mu = \frac{R_H}{\rho} \qquad (10.8)$$

where
n: carrier density
μ: mobility
e: charge of the carrier

Hall voltage is zero in ideal geometry, when applied field is zero. Nevertheless, a mismatch voltage (V$_0$) and a thermoelectric voltage (V$_{TE}$) are also included in the voltage detected (V$_d$) in a field test. The mismatch voltage is equivalent to the material's n resistivity, the current, and a geometry-dependent component (mobility). This factor changes the resistance between the two Hall voltage probes from resistivity. The metal contact between two different semiconductor materials gives rise to thermoelectric voltage. Although it is not reliant on the current, it does depend on the presence of temperature gradients. The detected voltage is given as

$$V_d = \frac{R_H i B}{t} + V_0 + V_{TE} \qquad (10.9)$$

$$V_d = \frac{R_H i B}{t} + \alpha \frac{\rho}{t} i + V_{TE} \qquad (10.10)$$

$$V_d = \frac{\rho i}{t}(\mu B + \alpha) + V_{TE} \qquad (10.11)$$

where α is the factor called the misalignment factor. The typical value is unity and can be as small as zero. The impacts of the thermoelectric voltage and misalignment voltage can be eliminated by using utilizing precise DC magnetic fields. While using DC magnetic field measurement might quickly pick noise and outweigh the real quantity, leading to inaccurate outcomes. Hall measurements on low mobility materials frequently detect the wrong carrier types and generate erroneous mobility values due to this reason. In order to overcome this issue, AC magnetic field is employed. The detected voltage is given as

$$V_d = \frac{\rho i}{t}(\mu B \sin(\omega t) + \alpha) \quad (10.12)$$

The measuring circuitry can accurately distinguish between the intended AC signal and the unwanted DC signal by utilizing a lock-in amplifier. However, the measured voltage have new factor associated with it, which is derivative of magnetic field and proportional to combined inductance of samples and leads and given as

$$V_d = \frac{\rho i}{t}(\mu B \sin(\omega t) + \beta \frac{dB}{dt}) \quad (10.13)$$

Archana *et al.* (2019) synthesized morphology controlled TiO$_2$ nanostructures for enhanced electrical characteristics measured using Hall effect. The measurements were carried out at 300 k temperature and the mobility in TiO2 nanotube showed extremely enhanced electrical characteristics such as high mobility of 3.68x10^3 cm^2/Vs in comparison with TiO$_2$

Table 10.3 Metal oxide mobility parameter.

Material for electron transport layer (ETL)	Mobility	Bandgap	Ref.
Zinc sulfate (Zn$_2$SO$_4$)	10-30 cm^{2}V^{-1}s^{-1}	3.8 eV	[43]
Zinc oxide (ZnO)	200 cm^{2}V^{-1}s^{-1}	3.3 eV	[44]
Tin oxide (SnO$_2$)	250 cm^{2}V^{-1}s^{-1}	3.6-4.0 eV	[45]
Tungsten oxide (WO$_3$)	10-20 cm^{2}V^{-1}s^{-1}	2.6-3.1 eV	[46]
Cerium oxide (CeO$_2$)	0.01 cm^{2}V^{-1}s^{-1}	3.5 eV	[47]
Strontium titanate (SrTiO$_3$)	5-8 cm^{2}V^{-1}s^{-1}	3.25 eV	[48]
Barium stannate (BaSnO$_3$)	150 cm^{2}V^{-1}s^{-1}	3.1 eV	[49]

nanoparticles [42]. Table 10.3 presents the metal oxide materials with bulk mobility.

10.6 Photoluminescence Spectroscopy and Imaging

Luminescence has been proven to be a very effective approach for the characterization of silicon and photovoltaic applications demonstrated by Peter Wurfel's research group theoretically and practically in the late 1980s and 1990s. Later in 2005 photoluminescence (PL) imaging was used for characterization of PV devices. PL can provide maximum open circuit voltage value and optimum diode factor of the material under observation without final PV device being built. Calibrated PL spectra taken at one solar excitation gives the quasi-Fermi level splitting (QFLS). The PL spectra may be used to obtain absorption spectrum of PV device, which enables accurate identification of tail states. Both radiative and non-radiative losses in the QFLS are caused by tail states. In ideal scenario QFLS provides the open circuit voltage given by following equation

$$V_{OC} = \frac{\Delta E_F}{q} \qquad (10.14)$$

where
ΔE_F: QFLS
q: elementary charge

The above equation represents the ideal PV device, however in practical situation metal contacts causes gradients in QFLS which in turns reduces open circuit voltage. The capability of absorber material reflects in QFLS, thus giving the maximum open circuit voltage. Semiconductor material at equilibrium emits back body radiation, thus the emission must be included in absorptions. The accumulative photon emission flux is represented as

$$\Phi_{eq}(E) = A(E)\Phi_{BB}(E) \approx A(E)\frac{E^2}{4\pi^2\hbar^3 c^2}e^{-E/K_BT} \qquad (10.15)$$

Semiconductor material used for the fabrication of PV devices have certain bandgap, thus K_BT is much smaller than emission energies. The emission flux is given as

$$F_{eq} = \int \Phi_{eq}(E)dE = B \times np = B \times n_i^2 \quad (10.16)$$

where

The emission flux depends upon the thickness of the thin film and depends upon hole and electron concentration. The charge concentration increases when subjected to excitation by solar or laser and flux due to excitations is given as

$$F_{PL} = B \times np = B \times n_i^2 e^{\Delta E_F/k_B T} \quad (10.17)$$

Assuming the QFLS remains constant throughout the absorber under excitation photon flux is given as

$$\Phi_{PL} \approx A(E)\Phi_{BB}(E)e^{\Delta E_F/k_B T} \approx A(E)\frac{E^2}{4\pi^2 \hbar^3 c^2} e^{-(E-\Delta E_F)/k_B T} \quad (10.18)$$

Linear fit of above equation to PL spectra can be used to determine QFLS.

There are significant ramifications to the rate of spontaneous emission's proportionality to the charge carrier product. Unlike the majority of other quantification, minority carrier entrapment and artefacts brought on by extra carriers stored in space charge areas have little impact on luminescence measurements. This is significant because the surplus carrier density (i.e., of the injection level) strongly influences the minority carrier lifespan. Therefore, lifetime data that is important for the operation of the solar cell should be measured under circumstances similar to those in which the cell operates in the sun. Whereas other approaches seem to be either insensitive or impacted by the aforementioned artefacts, PL enables carrier lifetime assessments at these additional carrier concentrations and also much lower ones. Appropriate minority carrier lifetime graphics on passivated or diffused semiconductors can be obtained using PL imaging with megapixel resolution in a fraction of a second, which is orders of magnitude quicker than what is conceivable, for example, with electromagnetic photoconductance decay mapping tools which are frequently used for this objective.

10.7 Electroluminescence Spectroscopy and Imaging

Electroluminescence imaging (EL) is classified as high-speed nondestructive characterization method providing spatial features of PV devices. The charge carries moves free path until they recombine in a forward biased material. The recombination of charges releases energy in the form of EL radiations. The minority carrier lifespan, diffusion length, and series resistance that are linked to material effectiveness may all be measured quantitatively using electroluminescence imaging. Additionally, it may be utilized to identify fractures, intergranular, and damaged connections that lead to cell misalignment in the device. EL imaging system consists of image capturing device, power system, and surrounding light shielding system where as EL spectroscopy consists of monochromator and detector. Typically, Si or InGaAs photodiode array are used as cooled detector. Charge-coupled device (CCD) and photodiode arrays are frequently combined for broadening the wavelength range which is important for PV devices. EL intensity at any position is given as

$$\phi_{em}(E,r) = [1-R(E,r)]Q_i(E,r)\phi_{bb}(E)\exp\left(\frac{qV(r)}{kT}\right)$$

$$\phi_{em}(E,r) = Q_e(E,r)\phi_{bb}(E)\exp\left(\frac{qV(r)}{kT}\right)$$

where
 $V(r)$: kT/q
 $Q_e(r)$: $[1-R(r)]Q_i(r)$
The EL signal recorded by the CCD recorder is given as

$$S_{cam}(r) = \int Q_{cam}(E)Q_e(E,r)\phi_{bb}(E)dE\exp\left(\frac{qV(r)}{kT}\right)$$

Where Q_{cam}: sensitivity of the detecting digital recorder and it is independent of position of the device. Thus, the variations in the intensity of EL are from the EQE which is due to recombination and optical losses and internal potential difference which is due to resistive losses. Since the exponential voltage-dependent term predominates in the graphic, EL analysis is a particularly useful technique for analyzing resistive losses. Hence it is promising method for characterization of PV devices.

CHARACTERIZATION OF PHOTOVOLTAIC DEVICES 255

10.8 Light Beam-Induced Current Technique (LBIC)

Measurement of the current caused by light is used to assess the efficiency of a full solar cell. High resolution imaging makes it possible to see and identify shunts and other flaws that reduce cell effectiveness and lifespan. Because LBIC equipment use lasers as their light source because they produce coherent, interferometer light, they are recognized as high-resolution imaging techniques. The schematic of LBIC is depicted in Figure 10.7. Measurements carried out using LBIC, PV devices short-circuit current is distributed laterally. The assessment of the short-circuit current produced by a PV devices step-wise spatially limited irradiation is the foundation of this scanning technique. A measurement and histogram of the photoinduced current are made in relation to the location of the study PV device. Fundamental benefits of LBIC includes such as ability of inspecting both the materials and the end products, assessments of both qualitative and quantitative, and non-invasive operating technique. Figure shows the schematic of the measurement system. Different signals emerge as the concentrated laser beam utilized in the LBIC approach scans the surface of the specimen under examination point by point. The information gleaned from the mapping of spatial pattern enables the whole PV device behavior to be replicated for any set of efficiency-affecting factors.

10.9 Electron Impedance Spectroscopy (EIS)

The essential element energy conversion is to analyze the electromagnetic and photo/electro-active material interactions. The immittance

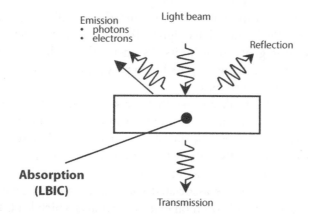

Figure 10.7 Schematic of LBIC measurement.

spectroscopy comprises of impedance, admittance, capacitance, and dielectric measurements which are insightful of chemical and physical properties. An alternating voltage is applied to measure the current response, so that impedance can be computed. In order to comprehend the charge transfer processes, EIS enables the decoupling of a complex system into multiple simpler components. The impedance can be given as

$$Z(t) = \frac{V(t)}{I(t)} = \frac{V_0 \cos(\omega t)}{I_0 \cos(\omega t - \phi)} = Z_0 \frac{\cos(\omega t)}{\cos(\omega t - \phi)}$$

EIS spectroscopy utilizes two or four-point configuration for solid state PV devices. Detected current is dependent upon material's conductivity and device geometry. The time scales of the phase transition under study dictate the necessary frequency range. The sample's structure and conductivity will affect the detection range for current. Impedance spectroscopy is generally carried out in megahertz to subhertz frequency range i.e., from microsecond to second time scale. Although it is technically possible to measure the impedance of PV device at lower or higher frequencies, which may frequently cause generation of unwanted signal. The measurement requires long time span and noise generally accompanies the signal at low frequencies, whereas at very high frequencies some inductive signal from measurement cables may be present in the device output. The standard requirement for getting trustworthy transfer function of Z (impedance spectra) must be linear, causal, stabile, and finite. Since current often fluctuates nonlinearly with applied voltage, the requirement for linearity is evidently not achieved across the entire current-voltage characteristic of PV devices. Consequently, the magnitude of voltage signal is set to be extremely small, so the current varies linearly as well as pseudolinearly with respect to the input voltage signal. Thus, impedance spectra are independently of voltage in linear limits. As a general guideline, voltage signal amplitude is chosen as low as possible to prevent nonlinear distortions in the impedance spectrum [50].

The concept of causality states that the system only reacts to the Vac input rather than to exterior causes. Observations that do not fulfill causality include artefacts for realistic samples. It can be difficult to recognize and remove measurement distortions from impedance spectra because they can provide frequency-dependent patterns that correlate with authentic transition. The inductance associated with measurements cables at high frequency is the cause of common external artifacts which leads to voltage clipping and distortion due to system limitations. These artifacts at high frequency can be overcome by using short measurement cables, calibration measurements with reference device over full frequency spectrum, and accounting the measurement cable inductance [51, 52].

Since impedance is a nonlinear measurement method, no permanent modifications should be made to the sample during the experiment. As a result, stability is a requirement for gathering accurate impedance spectrum data. Impedance measurements might require a few minutes or even many hours, according to the frequency spectrum of the measurement, the quantity of measurement points, and the amount of time necessary to integrate the current response, which is a practical issue. Finally, the impedance should have finite over a wide range of frequency spectrum [53–55].

10.10 Characterization by Ellipsometry Spectroscopy

A particularly powerful optical method for concurrently determining the optical characteristic and thickness of thin films is spectroscopic ellipsometry. When polarized light is reflected or transmitted on a material, the ellipsometer detects the variation in polarization. The incident light beam surface and defines the incident plane. The expression of reflectance ratio in spectroscopic ellipsometry is given as

$$\rho \equiv \tan(\psi)e^{i\Delta} = r_p / r_s = \left(\frac{E_{rp}/E_{ip}}{E_{rp}/E_{is}} \right)$$

where
  ψ: amplitude ratio
  Δ: phase difference of reflected p to s-polarized light
  rp: Fresnel reflection coefficients of p polarized light
  rs: Fresnel reflection coefficients s polarized light
  Ep: electric field for p polarized light
  Es: electric field for s polarized light

Ellipsometry spectroscopy is used to convert measured values into material's characteristics which relates to optical properties and thickness. Refractive index of the material defines the value of ψ and extinction coefficient outline the light absorption. Spectroscopy establishes the amplitude ratio and light absorption. The interaction of a polarized light beam with a material's surface at a predetermined angle of incidence is necessary for ellipsometry measurements. Since this measurement results are acutely susceptible to these factors, it is important to correctly establish the measurement mode, spectral area, and incidence angle [56–58].

10.11 Conclusion

Almost all characterization methods that have been used in semiconductor material and device analysis which directly or indirectly used in solar research. Higher resolution, great accuracy, precision, improved repeatability, simplicity of use, and identification of the artifacts associated which are essential to the optimize the materials and PV devices. To meet the needs of the photovoltaic community, the measurements and characterization community will keep working. The understanding of operating principles of measurements and removal of anomalies which provides the significant results. The benefits have been significant, and it is reasonable to anticipate that they will continue as photovoltaic device advancements.

References

1. Hernández-Balaguera, E., Muñoz-Díaz, L., Pereyra, C., Lira-Cantú, M., Najafi, M., Galagan, Y., Universal control strategy for anomalous ionic-electronic phenomenology in perovskite solar cells efficiency measurements. *Mater. Today Energy*, 27, 101031, Jul. 2022.
2. Irace, A. et al., Transverse probe optical lifetime measurement as a tool for in-line characterization of the fabrication process of a silicon solar cell. *Solid State Electron.*, 43, 12, 2235–2242, Dec. 1999.
3. Karaoğlan, G.K., Hışır, A., Maden, Y.E., Karakuş, M. Ö., Koca, A., Synthesis, characterization, electrochemical, spectroelectrochemical and dye-sensitized solar cell properties of phthalocyanines containing carboxylic acid anchoring groups as photosensitizer. *Dyes Pigm.*, 204, 110390, Aug. 2022.
4. El Radaf, I.M. and El-Bana, M.S., Synthesis and characterization of the $CuSbSe_2$/n-Si heterojunction: Electrical and photovoltaic characterizations. *Phys. B Condens. Matter*, 584, 412067, May 2020.
5. Ghorab, M., Fattah, A., Joodaki, M., Fundamentals of organic solar cells: A review on mobility issues and measurement methods. *Optik (Stuttg)*, 267, 169730, Oct. 2022.
6. Joseph, J.D., Jasmin, M., Sidharth, S.R., Fabrication and characterization of silicon solar cells towards improvement of power efficiency. *Mater. Today Proc.*, 62, 2050–2055, Jan. 2022.

7. Ojha, V., Jansen, G., Patanè, A., La Magna, A., Romano, V., Nicosia, G., Design and characterization of effective solar cells. *Energy Syst.*, 13, 2, 355–382, May 2022.
8. Kusuma, J. and Geetha Balakrishna, R., A review on electrical characterization techniques performed to study the device performance of quantum dot sensitized solar cells. *Sol. Energy*, 159, 682–696, Jan. 2018.
9. Ning, S., Yang, C., Li, S., Bai, J., Wang, H., Ma, S., An analytical investigation on characterizing the properties of glass composition for crystalline silicon solar cells based on the digital microsystem measurement technology. *Sol. Energy Mater. Sol. Cells*, 244, 111814, Aug. 2022.
10. Geisz, J.F. *et al.*, Characterization of multiterminal tandem photovoltaic devices and their subcell coupling. *Cell Rep. Phys. Sci.*, 2, 12, 100677, Dec. 2021.
11. Anspaugh, B., Space solar cell performance measurements and characterization. *Sol. Cells*, 29, 2–3, 245–256, Aug. 1990.
12. Mikulik, D. *et al.*, Conductive-probe atomic force microscopy as a characterization tool for nanowire-based solar cells. *Nano Energy*, 41, 566–572, Nov. 2017.
13. Kim, J., Lee, T., Kwak, J., Lee, C., Photovoltaic characterizing method of degradation of polymer light-emitting diodes based on ideality factor and density of states. *Appl. Phys. Lett.*, 119, 12, 123301, Sep. 2021.
14. Gudovskikh, A.S. *et al.*, New method for interface characterization in heterojunction solar cells based on diffusion capacitance measurements. *Thin Solid Films*, 516, 20, 6786–6790, Aug. 2008.
15. Wulles, C., Rafhay, Q., Desrues, T., Kaminski, A., Theodorou, C., Parasitic oscillation in the low-frequency noise characterization of solar cells. *Solid State Electron.*, 194, 108327, Aug. 2022.
16. Sargent, E.H., Colloidal quantum dot solar cells. *Nat. Photonics*, 6, 3, 133–135, Feb. 2012.
17. Brown, G.F. and Wu, J., Third generation photovoltaics. *Laser Photonics Rev.*, 3, 4, 394–405, Jul. 2009.
18. Green, M.A. *et al.*, Solar cell efficiency tables (version 50). *Prog. Photovolt.: Res. Appl.*, 25, 7, 668–676, Jul. 2017.
19. Kramer, I.J. and Sargent, E.H., The architecture of colloidal quantum dot solar cells: Materials to devices. *Chem. Rev.*, 114, 1, 863–882, Jan. 2014.
20. Rath, A.K., Lasanta, T., Bernechea, M., Diedenhofen, S.L., Konstantatos, G., Determination of carrier lifetime and mobility in colloidal quantum dot films via impedance spectroscopy. *Appl. Phys. Lett.*, 104, 6, 063504, Feb. 2014.
21. Kim, H.J., Ko, B., Gopi, C.V.V.M., Venkata-Haritha, M., Lee, Y.S., Facile synthesis of morphology dependent CuS nanoparticle thin film as a highly efficient counter electrode for quantum dot-sensitized solar cells. *J. Electroanal. Chem.*, 791, 95–102, Apr. 2017.

22. Ren, F., Li, S., He, C., Electrolyte for quantum dot-sensitized solar cells assessed with cyclic voltammetry. *Sci. China Mater.*, 58, 6, 490–495, Jun. 2015.
23. Jeong, S.H. *et al.*, Characterizing the efficiency of perovskite solar cells and light-emitting diodes. *Joule*, 4, 6, 1206–1235, Jun. 2020.
24. Nguyen, T.T., Patel, M., Kim, J., All-inorganic metal oxide transparent solar cells. *Sol. Energy Mater. Sol. Cells*, 217, 110708, Nov. 2020.
25. Xie, F., Zhu, J., Li, Y., Shen, D., Abate, A., Wei, M., TiO2-B as an electron transporting material for highly efficient perovskite solar cells. *J. Power Sources*, 415, 8–14, Mar. 2019.
26. Xie, H. *et al.*, Low temperature solution-derived TiO2-SnO2 bilayered electron transport layer for high performance perovskite solar cells. *Appl. Surf. Sci.*, 464, 700–707, Jan. 2019.
27. Wu, M.C., Liao, Y.H., Chan, S.H., Lu, C.F., Su, W.F., Enhancing organo lead halide perovskite solar cells performance through interfacial engineering using Ag-doped TiO2 hole blocking layer. *Sol. RRL*, 2, 8, 1800072, Aug. 2018.
28. Shen, D., Zhang, W., Xie, F., Li, Y., Abate, A., Wei, M., Graphene quantum dots decorated TiO2 mesoporous film as an efficient electron transport layer for high-performance perovskite solar cells. *J. Power Sources*, 402, 320–326, Oct. 2018.
29. Chen, S.H., Chan, S.H., Lin, Y.T., Wu, M.C., Enhanced power conversion efficiency of perovskite solar cells based on mesoscopic Ag-doped TiO2 electron transport layer. *Appl. Surf. Sci.*, 469, 18–26, Mar. 2019.
30. Gao, X.X. *et al.*, Tuning the Fermi-level of TiO2 mesoporous layer by lanthanum doping towards efficient perovskite solar cells. *Nanoscale*, 8, 38, 16881–16885, Sep. 2016.
31. Wang, L. *et al.*, Inverted pyramid Er3+ and Yb3+ Co-doped TiO2 nanorod arrays based perovskite solar cell: Infrared response and improved current density. *Ceram. Int.*, 46, 8, 12073–12079, Jun. 2020.
32. Lee, S. *et al.*, Nb-doped tio 2: A new compact layer material for TiO2 dye-sensitized solar cells. *J. Phys. Chem. C*, 113, 16, 6878–6882, Apr. 2009.
33. Zaragr, R.A., Micro size developed TiO_2-CdO composite film: Exhibits diode characteristics. *Mater. Lett.*, 335, 15, 133813, 2023.
34. Hierrezuelo-Cardet, P. *et al.*, External quantum efficiency measurements used to study the stability of differently deposited perovskite solar cells. *J. Appl. Phys.*, 127, 23, 235501, Jun. 2020.
35. Sagar, R. and Rao, A., Increasing the silicon solar cell efficiency with transition metal oxide nano-thin films as anti-reflection coatings. *Mater. Res. Express*, 7, 1, 016433, Jan. 2020.
36. Zargar, R.A., Arora, M., Bhat, R.A., Study of nanosized copper-doped ZnO dilute magnetic semiconductor thick films for spintronic device applications. *Appl. Phys. A*, 124, 1–9, 2018.

37. Dong, J. et al., Impressive enhancement in the cell performance of ZnO nanorod-based perovskite solar cells with Al-doped ZnO interfacial modification. *Chem. Commun.*, 50, 87, 13381–13384, Oct. 2014.
38. An, Q. et al., High performance planar perovskite solar cells by ZnO electron transport layer engineering. *Nano Energy*, 39, 400–408, Sep. 2017.
39. Azmi, R., Hwang, S., Yin, W., Kim, T.W., Ahn, T.K., Jang, S.Y., High efficiency low-temperature processed perovskite solar cells integrated with alkali metal doped ZnO electron transport layers. *ACS Energy Lett.*, 3, 6, 1241–1246, Jun. 2018.
40. Tseng, Z.L., Chiang, C.H., Chang, S.H., Wu, C.G., Surface engineering of ZnO electron transporting layer via Al doping for high efficiency planar perovskite solar cells. *Nano Energy*, 28, 311–318, Oct. 2016.
41. Mahmood, K., Khalid, A., Ahmad, S.W., Mehran, M.T., Indium-doped ZnO mesoporous nanofibers as efficient electron transporting materials for perovskite solar cells. *Surf. Coat. Technol.*, 352, 231–237, Oct. 2018.
42. Archana, T., Vijayakumar, K., Arivanandhan, M., Jayavel, R., TiO2 nanostructures with controlled morphology for improved electrical properties of photoanodes and quantum dot sensitized solar cell characteristics. *Surf. Interfaces*, 17, 100350, Dec. 2019.
43. Elseman, A.M., Sajid, S., Shalan, A.E., Mohamed, S.A., Rashad, M.M., Recent progress concerning inorganic hole transport layers for efficient perovskite solar cells. *Appl. Phys. A*, 125, 7, 1–12, Jun. 2019.
44. Zargar, R.A., Arora, M., Alshahrani, T., Shkir, M., Screen printed novel ZnO/MWCNTs nanocomposite thick films. *Ceram. Int.*, 47, 6084–6093, 2020.
45. Lin, S. et al., Efficient and stable planar hole-transport-material-free perovskite solar cells using low temperature processed SnO2 as electron transport material. *Org. Electron.*, 53, 235–241, Feb. 2018.
46. Gheno, A., Thu Pham, T.T., di Bin, C., Bouclé, J., Ratier, B., Vedraine, S., Printable WO3 electron transporting layer for perovskite solar cells: Influence on device performance and stability. *Sol. Energy Mater. Sol. Cells*, 161, 347–354, Mar. 2017.
47. Pandey, R., Saini, A.P., Chaujar, R., Numerical simulations: Toward the design of 18.6% efficient and stable perovskite solar cell using reduced cerium oxide based ETL. *Vacuum*, 159, 173–181, Jan. 2019.
48. Guo, H. et al., Low-temperature processed yttrium-doped SrSnO3 perovskite electron transport layer for planar heterojunction perovskite solar cells with high efficiency. *Nano Energy*, 59, 1–9, May 2019.
49. Zhou, J., Hou, J., Tao, X., Meng, X., Yang, S., Solution-processed electron transport layer of n-doped fullerene for efficient and stable all carbon based perovskite solar cells. *J. Mater. Chem. A Mater.*, 7, 13, 7710–7716, Mar. 2019.
50. Sacco, A., Electrochemical impedance spectroscopy: Fundamentals and application in dye-sensitized solar cells. *Renewable Sustainable Energy Rev.*, 79, 814–829, Nov. 2017.

51. Fabregat-Santiago, F., Garcia-Belmonte, G., Mora-Seró, I., Bisquert, J., Characterization of nanostructured hybrid and organic solar cells by impedance spectroscopy. *Phys. Chem. Chem. Phys.*, 13, 20, 9083–9118, May 2011.
52. Ecker, B., Egelhaaf, H.J., Steim, R., Parisi, J., von Hauff, E., Understanding S-shaped current-voltage characteristics in organic solar cells containing a TiOx interlayer with impedance spectroscopy and equivalent circuit analysis. *J. Phys. Chem. C*, 116, 31, 16333–16337, Aug. 2012.
53. Veal, B.W., Baldo, P.M., Paulikas, A.P., Eastman, J.A., Understanding artifacts in impedance spectroscopy. *J. Electrochem. Soc.*, 162, 1, H47–H57, Nov. 2015.
54. Fasmin, F. and Srinivasan, R., Review—Nonlinear electrochemical impedance spectroscopy. *J. Electrochem. Soc.*, 164, 7, H443–H455, May 2017.
55. Naumann, R.L.C., Electrochemical impedance spectroscopy (EIS), in: *Functional Polymer Films*, vol. 2, pp. 791–807, Jun. 2011.
56. Li, H. et al., A review of characterization of perovskite film in solar cells by spectroscopic ellipsometry. *Sol. Energy*, 212, 48–61, Dec. 2020.
57. Tejada, A., Braunger, S., Korte, L., Albrecht, S., Rech, B., Guerra, J.A., Optical characterization and bandgap engineering of flat and wrinkle-textured FA0.83Cs0.17Pb(I1−xBrx)3 perovskite thin films. *J. Appl. Phys.*, 123, 17, 175302, May 2018.
58. Bailey, C.G., Piana, G.M., Lagoudakis, P.G., High-energy optical transitions and optical constants of CH3NH3PbI3 measured by spectroscopic ellipsometry and spectrophotometry. *J. Phys. Chem. C*, 123, 47, 28795–28801, 2019.

11
Theoretical and Experimental Results of Nanomaterial Thin Films for Solar Cell Applications

Muzaffar Iqbal Khan[1,2]*, Rayees Ahmad Zargar[1†], Showkat Ahmad Dar[3] and Trilok Chandra Upadhyay[2]

[1]*Department of Physics, Baba Ghulam Shah Badshah University Rajouri (J & K), India*
[2]*Department of Physics, Hemvati Nandan Bahuguna Garhwal University (A Central University) Srinagar (Garhwal), Uttarakhand, India*
[3]*Department of Computer Science and Engineering, Annamalai University, Chennai, India*

Abstract

The metal oxide thin films have attracted more and more interest, due to excellent physical, optical, and electric properties. These thin films were employed in a wide range of applications including the laser devices, supercapacitor, solar cells, biosensors, biomedical, photovoltaic, luminescent materials applications. Besides synthesis of metal oxide films have been displayed enhanced functional properties in terms of film microstructure by using different deposition techniques. The current chapter describes the literature regarding metal oxides and their calculation of optical properties by using very simple model known as the Sellmeier model, whereas experimental results of ZnCdO thin films and their important results have been incorporated.

Keywords: Semiconductor, sintering, band gap, nanotechnology, optical properties

11.1 Introduction

Nanotechnology is the branch of science of understanding the structure and behavior of materials at the atomic or molecular level. It is the world of

---

*Corresponding author: muzaffariqbalkhan786@gmail.com
†Corresponding author: rayeesphy12@gmail.com

Rayees Ahmad Zargar (ed.) Metal Oxide Nanocomposite Thin Films for Optoelectronic Device Applications, (263–294) © 2023 Scrivener Publishing LLC

the atoms, molecules, macromolecules, quantum dots, and macromolecular assemblies. The prefix denoting "10^{-9}" is known nano. The symbol used is "n". Therefore, one nanosecond (ns) = 10^{-9} s. Technology involving sizes of only a few nanometers is nanotechnology. Nanoscience is the study of things at the nanometer scale. Nanotechnology is the name for the resulting technology. The ability to control matter, atom by atom, is a component of nanotechnology [1]. At this size, the tasks performed by the components of matter are distinct from those performed by the components or bulk materials [2, 3]. Metal oxide thin films have received a lot of attention in the field of nanotechnology due to its distinctive physical [4], compositional, morphological, electrical, and optical features [5], non-toxic nature, and large band gap value. These films hold promise as components for solar cells [10–13], supercapacitors [10–13], photocatalysis [10–13], luminous materials [9–13], laser devices [10–13], biosensors [6–13], and biomedical devices. Many different deposition techniques, including DC magnetron sputtering [14], chemical bath deposition [15], spray pyrolysis [16], sol-gel spin coating [17], pulsed laser deposition [18], electro deposition [19], thermal evaporation deposition [20], molecular beam epitaxy [21], and chemical vapor deposition [21], have been reported to be used to produce thin films. In particular, researchers concluded vacuum-based deposition method has restriction because of high process temperature, and high manufacturing costs. Solar energy is one of the renewable energy sources [18]. The solar energy is free and solar modules were operated without emission of any toxic gases [19]. Biomass [20], solar [21], hydropower [22], Wind and geothermal [23], and hydropower are a few examples of renewable energy. Sunlight can be converted into power via solar cells. Numerous benefits of solar energy include the fact that it is free and that solar panels work without emitting any harmful gases [24]. Table 11.1 showed first, second, and third generation solar cells [25]. Generally, working principle of these types of solar cells was established on the photo electric effect. Currently, silicon remained the dominant photovoltaic material due to silicon is non-toxic material, abundant and produce higher power conversion efficacy. However, the main drawback was too expensive. Later, development of thin film based solar cell was reported. Researcher highlighted that thin film technology can decrease the amount of active material in cell, and cheaper than crystalline silicon.

In this chapter we focusing on the development or progress of various types of solar cells is the study's main purpose and its main goal (thin film corresponding to solar cells (SC), perovskite solar cells (PSC), and dye sensitized solar cell (DSSC)). The performance of solar cells has been studied. Research findings on open circuit voltage (V_{os}), short current density (J_{sc}),

Table 11.1 Generations of solar cells of various types.

S. no.	Generations	Solar cells description
1	First Generation	1. The crystalline and poly-crystalline solar cells are examples of the oldest commercially available solar cells technologies and made of crystalline silicon [26]. 2. These types of solar cells contributed about 90% world solar cell market. 3. Silicon showed indirect band gap semiconductor material with band gap of 1.12 eV [27].
2	Second Generation	1. Several types of materials including copper indium gallium Di selenide(CIGDS), copper zinc tin sulphide (CZT), cadmium telluride (CT) thin films, and amorphous silicon were produced onto glass and flexible substrate via various deposition methods. 2. Advantages of these solar cells such as cheaper production cost, less material needed, and low temperature large area deposition [28]. 3. The solar cells need photoactive materials (film thickness <1 μm), showed direct band gap, high absorption coefficient to absorb incident sunlight [29].
3	Third Generation	1. In dye sensitized type solar cells, the organic and inorganic dye molecules were used (sensitizers). Production of solar cells could be carried out on rigid and flexible substrates [30]. 2. The C60 and conjugated conductive polymers were employed as the acceptor and the donor materials, subsequently in the bulk heterojunction solar cells [31]. 3. In the perovskite solar cells, the light harvesting active layer was produced by using Lead or tin halide-based hybrid organic inorganic compounds. Perovskite has specific crystal structure, can produce highly efficient in a very short time [32]. 4. The inorganic quantum dots were employed (sensitizers) in quantum dot solar cells due to tuneable nature of the band gap value [33].

fill factor, and efficiency were highlighted under various conditions. Open circuit voltage (V_{os}), short current density (J_{sc}), fill factor, and efficiency were reported under various conditions as part of an investigation into the solar cell's performance.

11.2 Literature Survey

11.2.1 Zinc Oxide (Z_nO)

The zinc oxide symbolically represented by (Z_nO) showed n-type semiconductor materials and has wide band gap. The Z_nO films indicated easy crystallization, high carrier mobility, and technological flexibility. The zinc oxide is based on the type an n-type semiconductor material in a broad band gap in the II-VI range [34]. The Z_nO films demonstrated simple crystallisation [35], high carrier mobility, and adaptability in terms of technology [36]. Carbon fiber, Z_nO, epoxy resin, C_uO, and epoxy resin made up the solar cell [37]. The unplugged wire values of the short circuit current density (Jsc) (20.69 to 32.4) and voltage (Voc) range from 404 to 674.6 mV and efficiency (5.88–13.57%), fill factor (52.33 to 62.12), and mA/cm² were emphasized in various samples (under various composite film thickness, percentage of Z_nO and PVA). The present provider solar cell performance is reduced by mobility. The photo electrode provided the electron with energy, a light-absorbing material, (Z_nO/PVA composite) is exposed to sunlight. Zinc oxide sheets and a counter electrode constructed of lead sulphide were used to create the solar cell. Due to a rise in the surface density of quantum dots, the short circuit current density rose when the growth period was extended [38]. It was reported on the analysis of photovoltaic cell metrics including Jsc (12.6 mA/cm²), fill factor (0.61), Voc (676 mV), and efficiency (5.19%). C_u2O/Z_nO heterojunction solar cell and its power conversion efficiency were displayed in Table 11.2. The solar cell used the polyaniline-zinc oxide nanocomposites. Various samples were used to examine the photovoltaic characteristics [39–44], such as the fill factor (48.57 to 65.36%), efficiency (1.683–2.69%), short circuit current density (7.7 to 8.1 mA/cm²), and open circuit voltage (0.45 to 0.51V). Due to their extremely smooth, compact, and uniform shape, Z_nO films were shown to be a potential electron transport structure for perovskite based solar cells by Sudheer and colleagues [39–45].

Table 11.2 The efficiency of the power conversion solar cells made amidst Z_nO/Cu_2O heterojunctions using different deposition processes.

S. no.	Deposition technique	Description
1	Ionic Layer Adsorption and Reaction in Successive Steps	1. A solar type cell comprised of ITO, NiO, Cu_2O, ZnO, SnO_2, and Al was created, according to Chatterjee and colleagues [39]. 2. Power conversion efficiency reached 1.12%.
2	Electro Deposition Method	1. FTO/ZnO/Cu_2O/Au solar cell was fabricated. 2. 1.43% power conversion efficiency was attained, as stated by Fujimoto and colleagues [40].
3	Thermal Oxidation/ PLD Method	1. Cu/Cu_2O/ZnO/AZO solar cells were constructed. 2. Approximately 3.85% of power is converted, according to Minami and colleagues [41].
4	Galvanostatic Electro Deposition	1. Solar cell consisted of FTO/ZnO/Cu_2O/Au. 2. According to Jeong and colleagues' research, the power conversion efficiency is roughly 0.41 percent [42].
5	Magnetron Sputtering	1. Solar cell was made from IGZO/ZnO/Cu_2O/Au. 2. According to Ke and colleagues' findings [43], the power conversion efficiency was 1.68%.

11.2.2 Tin Oxide ($Sn O_2$)

Rameshkumar and co-researchers [46] reported that tin oxide (symbolically represented as $Sn O_2$) films are suitable for solar cell applications. These materials can absorb maximum radiation in ultraviolet (UV), and then drops in visible region. The optical transmittance showed maximum transmittance about 70- 80%, cut-off frequency is estimated to be 287 nm. The glass/FTO/SnO$_2$/TiO$_2$/P$_3$HT:PCBM/MoO$_3$/Ag layer solar cell was made [47]. The photoactive layers produced electron hole pairs with a maximum transmittance of 78%. According to experimental results, semiconductor materials have great uniformity, low surface roughness, and high transparency. Jsc (10.4 mA/cm^2), Voc (0.78V), fill factor (35) and efficiency (2.87%) of the photovoltaic system were reported. Solvothermal methods were used to create the SnO$_2$ microsphere (diameter = 75 nm) [48]. These samples demonstrated 16.85% power conversion efficiency and the largest amount of light absorption. 66.4%, 1.12 V, and 21.8 mA/cm^2 were the fill factor, Voc, and Jsc values, respectively. Wang and colleagues' [49] fabrication of 2.2 m diameter SnO$_2$ microspheres was highlighted. According to experimental findings, dye-sensitive solar cells could employ a photo electrode with a power efficiency 6.25% (fill factor = 0.55, Voc = 803 mV, and Jsc = 14.11 mA/cm^2). According to Govindhasamy and colleagues [50], SnO$_2$ films can be employed as an electron conducting medium because of their improved anti-reflective properties, increased electron mobility, wider band gaps, and better band edges. The solar cell's power efficiency was 8.38%, by a fill factor of 45%, a Voc of 0.96 mV, and a Jsc of 18.99 mA/cm^2 for the CH$_3$NH$_3$PbI$_3$ perovskite-infiltrated SnO$_2$ material. SnO$_2$ can be used in perovskite type solar made cells rate to its deep conduction band and valence band, according to Xiong and colleagues [51]. The use of low temperature SnO$_2$ films generated from sol increased the power conversion efficiency to 17%. The electron transport layer's (SnO$_2$) shortcomings, such as its propensity for deterioration at high temperatures and its significantly smaller conduction band, could result in voltage loss in the solar cell. By utilizing SnO$_2$ films, Jung and colleagues [52] have revealed hysteresis free planar perovskite type based solar made cells. The potential efficiency of the solar type cell built from (HC(NH$_2$)$_2$PbI$_3$)$_{0.875}$(CsPbBr$_3$)$_{0.125}$) was 19.17% (fill factor: 0.74, Voc: 1.144 V, Jsc: 22.64 mA/cm^2). The absorbent substance was created on SnO$_2$ films that were 40 nm wide, 0.1 M, and 250°C temper. According to other researchers, Table 11.3 displayed the potential conversion of tin oxide type solar cells.

11.2.3 Cadmium Oxide (*CdO*)

Cadmium oxide (symbolically represented as *CdO*) is an inorganic compound. This chemical is one of the important precursors to other cadmium compounds. Similar to sodium chloride, it crystallizes in a cubic rocksalt lattice with octahedral cation and anion centers [59]. It naturally

Table 11.3 Efficiency of solar cells made of tin oxide in converting energy.

S. no.	Experimental findings	References
1	1. The smoother and more compact flexible perovskite type madesolar based cells have a potential efficiency of 19.80% and can sustain a potential efficiency of 94.5% even after 3000 cycles of bending. 2. The potential efficiency of perovskitetype solar made cells with annealing-free electron transport layers was greater (20.36%) than that of annealed SnO_2 ETLs (17.42%).	Zhihao and co-researchers [53]
2	1. A perovskite solar cell made of p-SnO_2 had a 19.56% potential efficiency. 2. Tin oxide films were created using atmospheric Ar/O_2 plasma energy at high speed and ambient temperature. 3. Results showed that p-SnO_2 thin films created more uniform surface, high electrical conductivity, and good electron mobility than SnO_2 films that had been thermally annealed.	Haejun and co-researchers [54]
3	1. On rigid substrates, SnO_2 thin films with low power nitrogen plasma treatment had a power conversion efficiency of 19.1%. 2. On polyethylene terephthalate substrates coated with indium tin oxide, flexible perovskite solar cells were created, achieving a power conversion efficiency of roughly 18.1%. According to investigations, after 1000 bending cycles, 90% of its initial performance will still be there.	Anand and co-researchers [55]
4	1. Organic type and inorganic type lead halide perovskite based solar made cells exhibited power conversion efficiencies of 16.46%, 17.92%, and 21.09%, respectively, using compact SnO_2, ground SnO_2 and composite G-SnO_2/C-SnO_2 as the electron transporting layer. 2. A commixture of G-SnO_2/C-SnO_2 solar cell demonstrated good long-term stability. According to studies, these materials maintained 89% of their power conversion efficiency with encapsulation after 105 days in dry ambient air.	Singh and co-researchers [56]

(*Continued*)

Table 11.3 Efficiency of solar cells made of tin oxide in converting energy. (*Continued*)

S. no.	Experimental findings	References
5	1. Low temperature atomic layer deposition of SnO_2 resulted in films with outstanding electron extraction and hole-blocking properties. 2. The potential power efficiency of SnO_2 film-based perovskite type solar made cell was roughly 20%. 3. Experimentation outcomes reduce charge recombination at boundary bounded by perovskite type and electron transporting structure, these materials' surfaces must be passivated.	Yonghui and co-researchers [57]
6	1. Photovoltaic parameters were studied. Power conversion efficiency = 18.8%, fill factor = 72%, V_{os} = 1.11V, J_{sc} = 23.6 mA/cm^2. 2. Studies showed that steady power output for perovskite solar cells with two inkjet-printed layers is highly exceptional.	Valentina and co-researchers [58]

manifests as the odd mineral monteponite. In addition to brown or red crystals, cadmium oxide can also be found as a white amorphous powder [60]. At ambient temperature, band gap of the n-type semiconductor cadmium oxide as 2.18 eV (2.31 eV) (298 K) [61]. Thin films made of cadmium oxide are crucial components for solar cell applications. The n-type semiconductor *CdO* films exhibit outstanding conductivity at ambient temperature, a greater mobility of 130 cm^2/Vs, a high electron concentration of more than 1019/cm^3, a small direct band gap between 2.2 and 2.5 eV and high transparency in the visible region. Al/p-Si/n-*CdO*/Al solar cell as described by Murugasamy and colleagues [62]. The absorbing substance was created at 400°C and displayed well-developed tiny grains (102 nm) scattered across the *CdO* film's whole surface. Transmission decreased in the UV area, with the maximum transmission measured at 650 nm at 60% (less than 400 nm). Researchers discovered that, particularly for materials with high carrier concentrations and a significant degree of non-stoichiometry, light absorption is moved towards the lower wavelength side. For the formation of films made at 350, 400, and 450°C correspondingly, fill factor (0.31 to 0.32) and efficiency (0.6 to 0.8%) values improved, then decreased to 0.3 and 0.7%. According to Gaurav and colleagues [63], $CdO:TiO_2$ has the best conductivity and the lowest resistivity, making it an effective anti-reflecting layer that also increases the efficiency of heterojunction solar cells. Conclusions were drawn on the photovoltaic performance, including efficiency (3.23%), open circuit voltage (0.26V), and short circuit current (7.6 mA). A solar cell made of Al, p-Si, n-CdO, and Al was created by Fahrettin [64]. As illumination intensity rose, the photocurrent increased (more produced photo carriers). Under AM 1.5 illumination, the greatest values of Voc (0.41) and Jsc (2.19 mA/cm^2) were found. According to an analysis using atomic force microscopy, the high degree of roughness (58.54 nm) increases light absorption and decreases light reflection in the visible zone [65]. It was stated that n-CdO/p-Si potential efficiency of 5.5% and the fill factor value of 0.652. After the addition of impurity atoms, the efficiency of Si/CdO solar cells improved, according to Hasan and colleagues [66]. Muneer and co-researchers [67] reported the synthesis of cadmium oxide (CdO) films via chemical technique and drop casting method. The obtained samples were cubic structure, polycrystalline and no trace of the other material based on the XRD results. Photovoltaic parameters were studied (Voc = 4.1V, Jsc = 1.44mA, fill factor = 36.2%, efficiency = 6.8%).

11.2.4 Nickel Oxide (*NiO*)

Nickel oxide (symbolically represented as *NiO*) is an intrinsic p-type semiconductor with a large band gap, best temperature, and chemical stability. Nickel oxide has been used as a photocathode material in tandem DSSCs and organic bulk heterojunction solar cells in addition to being used as a hole collector in perovskite heterojunction solar cells and p-type DSSCs. Due to its suitable band gap value, lower price, small hysteresis, excellent chemical stability, excellent transmittance, strong photo stability, low electrode polarization, and lower processing temperature, nickel oxide has been cited by numerous researchers as playing a significant role as a hole transport material [68]. According to Rui and colleagues [69], nickel oxide films made using the sol gel process can be used to create high performance p-i-n perovskite solar cells. Jsc of 17.95 mA/cm^2, Voc of 1.12 V, and approximately 14.85% of maximum efficiency. *NiO* films have a broad band gap (3.19 eV), which Monika and colleagues [70] identified as having the potential to improve ion transfer. The solar cell was created, and Isc was investigated for 24 hours (1.5 mA), 48 hours (1.2 mA), 96 hours (0.9 mA), and 144 hours (0.8 mA). The results showed that 0.45 and 14.04%, respectively, were the best fill factors and power conversion efficiencies. According to Shuangshuang and colleagues [71], solar cells with doped transport layers exhibit greater power conversion efficiency than solar cells without them (due to improved charge separation and hole mobility). According to Mudgal and colleagues [72], the solar cell was created by Ag/ITO/NiO/n-Si/Al carrier selective contact. Studying both with and without the SiO_2 interlayer allowed researchers to compare the photovoltaic metrics Voc (509, 573 mV), Jsc (34.2 and 36.5 mA/cm^2), fill factor (712, 71%), and efficiency (12.4 and 14.9%). Fei and colleagues [73] came to the conclusion that the use of nickel oxide in the manufacture of solar cells showed low fill factor and undesirable Voc values in photovoltaic properties. Under various experimental setups, the nickel oxide and silicon heterojunction solar cell was studied [74]. When the NiO thickness was increased up to 0.1 m, the V_{oc} (572.8 mV), J_{sc} (21.4 mA/cm^2), the fill factor (54.37), and the efficiency (6.67%) all improved. On the other hand, a decrease in valence band offset led to an increase in Voc when the buffer layer band gap was reduced. Boyd and co-researchers [75] discussion of photovoltaic performance in various hole transport layers. Utilizing poly TPD, the Voc, Jsc, fill factor, and efficiency were 1.12 V, 20.42 mA/cm^2, 80.8%, and 18.55%, respectively. By adding poly (trial amine), the solar cell was able to reach Voc of 1.11 V, Jsc of 20.53 mA/cm^2, fill factor of 74%, and efficiency of 16.63% (PTAA). Finally, for the solar cell made using nickel oxide films, the Voc, Jsc, fill factor, and efficiency were 0.89V, 20.72 mA/cm^2, 79.2%, and 14.57%, respectively. The solar cell built from NiO and PTAA under varied concentrations was reported by Yawen and colleagues [76]. For concentrations of 0.3 mg/ml PTAA/NiO, 0.5 mg/ml PTAA/NiO, 0.7 mg/ml PTAA/NiO, and 1 mg/ml PTAA/NiO, respectively, the power conversion efficiency was 14.5%, 16.7%, 15.5%, and 14.8%. Table 11.4 represented the power conversion efficiency for nickel oxide solar cells.

11.2.5 Magnesium Oxide (*MgO*)

Magnesium oxide symbolically represented by *MgO* is a naturally occurring colourless, crystalline mineral with a high melting point that is produced on a huge scale and is used in many industries. Magnesium oxide has minimum electrical conductivity and large thermal conductivity. The superior stability and potent antibacterial activity of MgO in compared to

Table 11.4 Efficiency of nickel oxide-based solar cells in converting power conversion.

S. no.	Experimental findings	References
1	1. The hydrothermal method used to create nickel oxide resulted in a vertical charge transport channel at the base of perovskite sheets. 2. The fill factor and open circuit voltage of perovskite solar cells based on PEA and FPEA were enhanced, and the power conversion efficiency reached 15.2%.	Jianghu and co-researchers [77]
2	1. In order to decrease Voc loss and increase the power conversion efficiency (20.2%), coumarin 343 was added to perovskite. 2. According to experimental findings, after ageing at 85°C for 500 hours in a nitrogen glovebox, C343 doped solar cells maintained over 94% of their initial power conversion efficiency.	Sanwan and co-researchers [78]
3	1. At 250°C, the solution combustion technique was used to create the NiO film. 2. The solar cell, which was made from pure NiO, had potential efficiency of about 13.6%.	Selvan and co-researchers [79]
4	1. The hole concentration, hole mobility, and electrical conductivity of the NiO films were successfully improved by Li-Co-Mg doping. 2. Mg-Li-Co doped NiO based perovskite solar cells depicted the enhanced short circuit current density (17.3 to 23.9 mA/cm^2), open circuit voltage (0.932 to 1.06 V), fill factor (0.46 to 0.52), and power conversion efficiencies (7.47 to 13.22%).	Fatma and co-researchers [80]
5	1. Perovskite solar cell photovoltaic properties of ITO/NiOx/FA0.9MA0.1PbI2.7Br0.3/C60/ZnSe/Ag. The highest Voc (1.05V) was achieved by a solar cell with a NiO film that had been annealed at 250°C as a result of complexes in the layer that were seen in the results of ultraviolet photoelectron spectroscopy. 2. The solar cell's maximum power conversion efficiency (15.55%) was achieved with a 300°C-annealed NiO layer.	Imran and coresearchers [81]

organic antimicrobial agents has been an interesting field of investigation in recent years. Cu_2O/SnO_2 solar cell power conversion efficiency could be increased by adding an extremely thin MgO layer (prepared by using electron beam evaporation method). According to Chao and colleagues' research, this layer reduces the density of CuO defects [82]. By adding a MgO/ZnO bilayer, the polymer solar cell constructed from PTB7-Th:PC71BM achieved 11% efficiency. Black holes and surface flaws in fluorine-doped tin oxide glass were reduced because MgO played a significant role [83]. The open circuit voltage (Voc) in a perovskite solar cell was unaffected by the compact layer of titanium dioxide (TiO_2) (50–60 nm) coated with a thin layer of magnesium oxide (MgO), while MgO incorporated into TiO_2 showed a significant improvement in Voc from 0.86V to 0.98V [84]. Different processes were used to create the perovskite solar cell based on CH3NH3PbI3. When compared to solar cells with SnO_2 present just (19%), the maximum power conversion efficiencies could be seen at 200lx (25%) and 400lx (26.9%) interior illumination [85].

11.2.6 Aluminium Oxide (Al_2O_3)

Aluminum oxide symbolically represented with chemical formula Al_2O_3 is an amphoteric oxide of aluminum. It is also frequently known as alumina. Aluminum oxide has a rather good heat conductivity for a ceramic material while being an electrical insulator (30 Wm1 K1). The efficiency of the solar cell may be improved by covering the transparent conductive oxide (TCO) layer with a dielectric coating, such as Al_2O_3, according to scientists. High dielectric strength, strong stability, excellent transparency, and durability in adverse conditions were all displayed by Al_2O_3 [86]. Al_2O_3/ITO (double-layered anti-reflection coating) was used to create the solar cell, which shown improvements in efficiency (from 20.95–21.6%), external quantum efficiency (from 76.89–84.34%), Jsc (from 39.9 to 41.13 mA/cm²), and average reflectance (from 9.3–4.74%). Researchers looked at the effects of double- and triple-layer anti-reflection (DLAR and TLAR) coatings on solar cell performance [87]. Comparing TLAR coatings to DLAR coatings, the TLAR coatings showed greater fill factor (88.98%), efficiency (32.71%), Jsc (12.15 mA/cm²), and Voc (2.74). Adoption perovskite type solar made cells and hole transporter and light absorber was highlighted by the researcher. Al_2O_3 was reportedly used as a mesoporous scaffold for theformation of solar based cells, according to Naser and colleagues [88]. According to research findings, the average roughness and scaffold thickness both boosted the collecting efficiency. Between TiO_2 films and the perovskite layer, Liang and colleagues [89] described an insulating ultra-thin Al_2O_3

interface layer. According to experimental findings, excessively thick thin coatings caused an increase in the electron transfer barrier. With different Al_2O_3 soaking times, the solar cell's performance was investigated. With longer soaking times (from 0 to 60 minutes), the values of Voc (0.86 to 0.93 V) and Jsc (17.21 to 21.16 mA/cm²) rose. The power conversion efficiency was 9.36%, 10.2%, 12.79%, and 11.32% for 0, 10, 30, and 60 minutes of soaking, respectively. Bernd and colleagues developed the novel Al_2O_3 and SiC:B passivation stack [90]. In these experiments, Voc, Jsc, fill factor, and efficiency values in these studies range from 670 to 681.8 mV, 37.9 to 40.5 mA/cm², 75.9–82.5%, and accordingly 20.5–21.4%. Utilizing plasma accelerated atomic layer deposition, the Al_2O_3 films (used as a moisture barrier layer) were created. During the experiment, cells that were not enclosed (low shunt resistance) and those that were encased in Al_2O_3 (difficult to degrade) were created. These findings demonstrated that a 50 nm Al_2O_3 layer can shield a GaInP/GaAs/ge-Al_2O_3 solar cell from moisture over 1000 hours of testing at 85°C and 85% relative humidity (RH) [91]. Al_2O_3 was created by Choi and colleagues [92] using the atomic layer deposition technique, and it can be employed as a protective coating for perovskite solar cells. After 7500 hours of 50% RH at room temperature, experimental data showed that PTAA-based solar cells exhibited a less than 4% reduction in performance.

11.2.7 Cobalt Oxide (*CoO*)

Cobalt and oxygen atoms make up the class of chemical compounds known as cobalt oxide. Cobalt(II) oxide (cobaltous oxide), symbol CoO, and cobalt(III) oxide are examples of compounds in the cobalt oxide family (cobaltic oxide), represented by symbol $C_{o2}O_3$ and Cobalt(II, III) oxide, represented by symbol $C_{o2}O_4$. The $C_{o2}O_3$/mesoporous carbon composites were created using the oxidation and ammonia nitridation processes, which also acted as counter electrodes. In comparison to the counter electrode constructed of platinum (4.88%), the power conversion efficiency rose (5.26%) when these composites were applied to dye-sensitive solar cells [93]. Thioglycolic acid was used to hydrothermally produce the 2-mm cobalt oxide. X-ray diffraction (XRD) and scanning electron microscopy (SEM) investigations reveal good crystalline and outstanding agglomeration. A few of the photovoltaic metrics that have been published [94] are fill factor (0.69), power conversion efficiency (9.05%), open circuit voltage (0.8), and short circuit density (19.02 mA/cm²). Using the direct current magnetron sputtering deposition method, ultrathin CoO films (10 nm) were applied to the fluorine-doped tin oxide glass substrate. The solar cell, which had a maximum power conversion efficiency of 10%, was made using cobalt-copper binary oxide. The solar cell's initial power conversion efficiency of 90% persisted for 12 days (in dark and humid settings) according to experimental data, which demonstrated good stability [95]. As a hole transport layer, p-type ultrafine cobalt oxide was treated in a solution. The resulting CoO films displayed great hole mobility, smooth shape, and excellent conductivity [96]. In perovskite solar cells, a number of hole extraction layers have been researched. The ITO/CoO/CH3NH3PbI3/PCBM/Ag solar made cell has the highest potential efficiency (14.5%) when compared to conventional hole extracting layers like NiO (10.2%), CuO (9.4%), and PEDOT:PSS (12.2%). After 1000 hours, the solar cell's stable CoO layer still had a 12% power conversion efficiency [97]. Spin-coating CoO layers onto a perovskite solar cell allowed for the creation of a stable device. The potential efficiency of this kind of solar made cell was 20.72% [98]. When a CoO nanorod was used as the counter electrode, the CdS/CdSe/ZnS quantum dot sensitized solar cell attained a fill factor of 0.45, potential efficiency of 6.02%, a short circuit current density of 20.42 mA/cm^2, and an open circuit voltage of 0.655V [99]. Reduced graphene oxide, tungsten carbide, and co3o4 were used in the construction of the counter electrode of dye-sensitive solar cells. According to experimental findings [100], this counter electrode generated solar cells with a greater potential efficiency (7.38%) than platinum (6.85%).

11.2.8 Tungsten Oxide (WO_3)

Tungsten oxide symbolically represented with chemical formula WO_3, also called as tungsten trioxide is a the chemical composite compound of oxygen and the transition metal tungsten. Due to its close kinship to tungstic acid H_2WO_4, the substance is sometimes known as tungstic anhydride. It is a crystalline solid that is light yellow. Hetero junction solar made cells in bulk, polymer solar cells, and organic solar type cells, tungsten oxide may be employed as a hole extraction and transport layer. Tungsten oxide films that have undergone solution processing could be used in polymer solar cells (active anode buffer layer). According to Dip and colleagues' research [101], the tungsten oxide semiconductors had a broad band gap (3.5 ev) and direct transition material. A WO3/PEDOT:PSS/WO3 solar cell was used in the PCDTBT:PC71BM (as hole transport layer). There was research done on the Voc (933 mV), fill factor (53%), Jsc (11.1 mA/cm^2), and power conversion efficiency (58.9 percent) [102]. One benefit of the poly is that it can be processed as a solution at room temperature. It does, however, take into consideration device deterioration brought on by unintended interactions with photoactive substances and indium tin oxide [103, 104]. According to experimental findings, the low-temperature S-WO3 film had a high conductivity, matching energy level, and low roughness, making it suitable for use in organic type solar made cells [105]. The best power conversion (3.63%) was found at a concentration of 100 mg/L when the performance of solar cells using various concentrations of ammonium tungstate was evaluated. 10.45 mA/cm^2, 0.63 V, or 55.48% for Voc, Jsc, and fill factor are all acceptable values. The WO3 films could be used as an anode buffer layer in polymer solar cells instead of PEDOT:PSS. The spin coating of tungsten (VI) isopropoxide solution on an ITO electrode and 10 minutes of air annealing at 150°C were used to create the solution processed WO3 (s-WO3) layer [106]. The band gap of tungsten oxide films made of N-type semiconductors is wide, and they have favorable physicochemical properties. In comparison to WO3 electrode-based devices (4.36%), dye-sensitized solar cells using WO3/Pt electrodes had a power conversion efficiency of 8.1%. To aid in the decrease of in I 3-dye-sensitized solar cells, WO3 was coated on platinum nanoparticles, which possessed a 3-dimensional structure and a sizable surface area [107].

11.3 Theoretical and Experimental Results

11.3.1 Theoretical Model

Figure 11.1 illustrates a straightforward model of the deposition of a ZnO thin film on a transparent glass substrate in terms of the film's thickness (d) and refractive index (n). Contrarily, it is thought that substrate thickness exceeds film thickness by several orders of magnitude (d). The substrate's refractive index is shown in the figure as s, and the air's is shown as $n_0 = 1$. The vertical arrow depicts the incident light's partial reflection from the substrate's surface (R_1) and partial reflection from the top surface (R_1 and R_2). After being reflected numerous times, the light from the ZnO thin film on the transparent substrate is transferred in the final ray (indicated by T). This passage [108] claims that the reflection from the interface between the substrate and higher air is insignificant.

11.3.2 Sellmeier Model

The variation of the refractive index in the transparent area [109] with wavelength is described by the empirical model known as the Sellmeier model. The extinction coefficient, k, in this model is set to 0, implying that

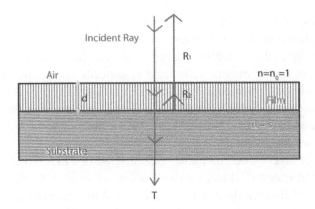

Figure 11.1 Model of the transparent thick glass substrate with thin absorbent sheets [108].

the sub-band absorption is also zero. The following equation describes a thin film with a refractive index of n [108]:

$$n^2(\lambda) = A + \frac{B\lambda^2}{(\lambda^2 - C)} + \frac{D\lambda^2}{(\lambda^2 - E)} \qquad (11.1)$$

The symbols A, B, C, D, & E in this case stand for fitting parameters defined for theparticular material of interest. The expression of the transmission of all the thin film/substrate defined [110]:

$$T(\lambda) = T_0(\lambda) - \sqrt{R_1 R_2} \cos[\delta(\lambda)] \qquad (11.2)$$

where $T(\lambda)$ is the final transmittance brought on by numerous reflections and $T_0(\lambda)$ is the absolute value of transmittance with interference effect from thin film. The expression shows that the values of R_1, R_2 and the phase $\delta(\lambda)$ have an impact on the $T(\lambda)$ value. These parameters are given by the following expressions (R_1, R_2 and phase $\delta(\lambda)$ defined by):

$$\delta(\lambda) = 2\pi \times \left(\frac{2n(\lambda)}{\lambda}\right) d + \pi \qquad (11.3)$$

where the phase shift d denotes the change in the optical field as it moves through the layer.

$$R_1 = \frac{\{(n-1)^2 + k_2\}}{\{(n+1)^2 + k_2\}} \quad (11.4)$$

and

$$R_2 = \frac{\{(s-n)^2\}}{\{(n+1)^2\}} \quad (11.5)$$

where, k represents the thin film's extinction coefficient. Equation (11.1) provides the Sellmeier equation for the ZnO thin film's refractive index (n) as a mapping of the wavelength and is once more expressed as:

$$S^2(\lambda) = A + \frac{B\lambda^2}{(\lambda^2 - C)} + \frac{D\lambda^2}{(\lambda^2 - E)} \quad (11.6)$$

where the constants A, B, C, D and E are fitted modeled parameters. The values of these parameters for ZnO are tabulated in the Table 11.5 [111].

From Equation (11.4), it is clear that the value of k is necessary to calculate R_1, and since Sellmeier's model assumes that extinction coefficient k is equal to zero, the value of k for ZnO is provided by Cauchy's relation, which is given as

$$k(\lambda) = F_k \lambda exp^{-G_k\left(\frac{1}{H_k} - \frac{1}{\lambda}\right)} \quad (11.7)$$

The values of various Cauchy's constraints are given below Table 11.6.

Table 11.5 Fitting of Sellmeier's connection model parameters for zinc oxide [108, 111].

A	B	C (nm)	D	E (nm)
2.0065	1.5748 × 10⁶	1 × 10⁶	1.5868	260.63

Table 11.6 The values of Cauchy's constraints for ZnO [108, 110].

$F_{k(nm^{-1})}$	$H_{k(nm)}$	$G_{k(nm)}$
0.0178	7327.1	337.87

11.3.3 Optical Properties Derive from the Above Equations

Figure 11.2(a) for ZnO thin films shows the absorbance spectra, which were calculated from transmission data by applying the formula (11.8) [112]. It is also referred to as optical density and is calculated as a logarithmic function of T (%).

$$A = log_{10}\left\{\frac{1}{T}\right\} = log_{10}\left\{\frac{I_0}{I}\right\} \tag{11.8}$$

The disclosed spectra show that absorbance rises as film thickness grows, which makes it the best indicator of a material that is appropriate for an optoelectronic application. The absorbance (A) and the absorption coefficient (α) are defined as follows by Beer-law: Lambert's

$$log_{10}\left\{\frac{I_0}{I}\right\} = (2.303 \times A) = \alpha t \tag{11.9}$$

where the value of incident light is given by the quantity $\left\{\frac{I_0}{I}\right\}$ and linked with t = thickness formation film. Therefore, the expression (11.9) is given by

$$\left\{\frac{2.303 \times A}{t}\right\} = \alpha \tag{11.10}$$

According to Figure(11.1a), the behavior of the transmittance spectra and the absorbance spectra are utterly at odds with one another. Figure (11.2b), which plots the absorption coefficient against wavelength, shows a striking increase in absorption coefficients as film thickness increases. The rising of grain boundaries is what causes the change in absorption coefficient of films based on thickness. Therefore, the estimation of energy (E_g) can be obtained using the Tauc's relation [113] defined as

$$h\nu\alpha = P[h\nu - E_g]^n \tag{11.11}$$

where, P is constant, while E_g is band gap energy. The assessment of the exponent "n" determines the kind of change for materials with a direct

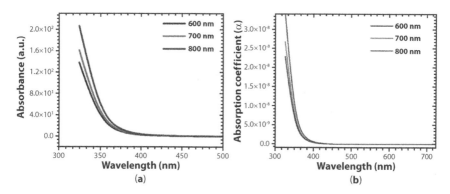

Figure 11.2 (a) Absorbance spectra and (b) Absorption coefficient with thickness (600 nm, 700 nm and 800 nm) [108].

band gap. The figure of $(h\nu\alpha)^2$ vs photon energy $(h\nu)$ in Figure (11.3a) depicts a linear curve, which is a direct transition and corresponds to ZnO films' maximum absorption. This type of plotting, known as a Tauc plot, is depicted along varying thicknesses (600, 700, and 800 nm), and the band gap energies were discovered to be between 3.38 and 3.33 eV. This decrease in E_g may be related to changes in crystallinity, variations in the distribution of dopants in the film, and larger particle size [114]. When individual atoms are moved closer to one another, the free atoms have a tendency to produce overlapping levels, which could account for the band gap narrowing with increasing thickness [115]. In addition, as shown in Figure (11.3b), the band-edge absorption of the formation films was

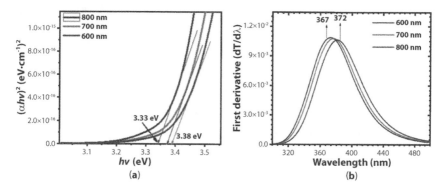

Figure 11.3 (a) Tauc's plots for determining the energy band-gap (b) 1st derivative of thickness-dependent transmittance spectra (600 nm, 700 nm, 800 nm) [108].

determined by using first derivative of the transmittance. The $\frac{dT}{d\lambda}$ verses λ curves exhibit a peak that is indicative of band edge absorption. As can be seen in Figure (11.3b), the band location in the figures is clearly changing approaching larger wavelengths, which suggests that the band gap is narrowing as thickness increases.

11.4 Experimental Results of Optical Properties

Information regarding the electron transition occurring in the optical domain has been deduced using the UV-visible absorption spectroscopy approach. Using established relations, the optical parameters, band gap, refractive index, extinction coefficient, and dielectric constants obtained in distinction to the observed optical absorption spectra. The absorbance spectra of developed and sintered films are shown in Figure11.4(a). The average absorbance may increase with increasing sintering temperature because of a localized energy level between the valence and conduction bands [116].

The red shift in the absorption edge for 600°C sintered film is caused by an increase in disordering in the film structure. The SEM images also lend credence to this view. The following relationship [118] is used to calculate

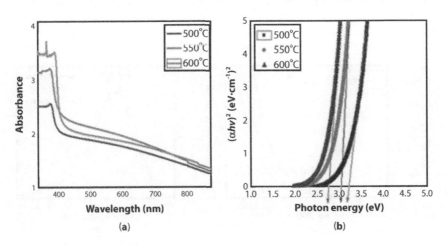

Figure 11.4 (a) ZnCdO film absorbance spectra (b) Tauc's Plot ZnCdO films were sintered at 500°C, 550°C, and 600°C [117].

the fluctuation of the absorption coefficient for the direct electronic transition in a semiconductor.

$$h v \alpha = [h v - E_g]^{\frac{1}{2}} \quad (11.12)$$

The optical band gap, absorption coefficient and incident photon energy are represented here E_g, α and hv. Figures $[hv]^2$ vs photon energy $[hv]$ variation are shown in Figure11.4 (b). When extrapolated to the X-axis, Tauc's figure, which represents the linear curve and respectively to phase change, provides optical band gap (E_g) data for observation. Band gaps of films sintered at temperatures of for 500°C, for 550°C, and for 600°C are estimated 3.21 eV, 2.97 eV, and 2.81 eV. This 0.40 eV decrease in band gap energy is the result of defect density states that emerged during the sintering process. The structural deformations in the lattice chain are brought on by changes in link length and bond angle in conjunction with interstitial space between the atoms during the high-temperature sintering of films [119]. Refractive index (n), extinction coefficient, and dielectric constants should be given additional consideration. These parameters, which were generated from reflectance data, primarily provided information on the material compatibility for manufacturing different optoelectronic devices. The formulas for calculating the extinction coefficient (k) and refractive index (n) are given below [120, 121].

$$n = \left[\frac{1+R}{1-R}\right] + \left[\left\{\frac{4R}{(1-R)^2}\right\} - k^2\right]^{\frac{1}{2}} \quad (11.13)$$

and

$$k = \left(\frac{\alpha \lambda}{4\pi}\right) \quad (11.14)$$

The wavelength of the incident photon, the absorption coefficient, and the word reflectance are all represented by the symbol R. The refractive index changes of ZnCdO coated films are shown in Figure 11.5(a) as a function of photon energy and as a function of rising sintering temperature. It is clear that the refractive index drops with rising photon energy and sintering energy up to a point of saturation at 3.37 eV. Therefore, it is

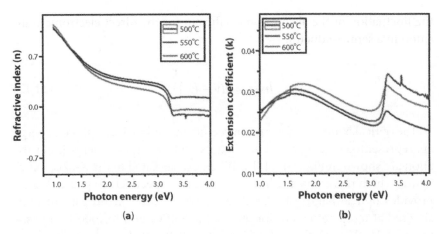

Figure 11.5 Figure shows the ZnCdO films sintered at 500°C, at 550°C, and at 600°C temperatures (a) Refractive index (n) versus photon energy (eV) and (b) Extinction coefficient (k) versus photon energy (eV) plots [117].

recommended to use band gap regime refractive index normal dispersion. Figure 11.5(b) displays the films' extinction coefficients (k), and these k fundamentally represent the semiconductor's inelastic scattering of electromagnetic waves (EMW), including the photoelectric effect, Compton effect, and other effects [122].

The behavior k initially showed an ascending trend as the sintering aid increased up to 1.5 eV, after which it started to decline with photon energy until 3.37 eV, and then it stayed almost constant at higher photon energies. This shift of k and n inside the band gap regime upon sintering temperature is due to the local structural modification that places the Cd atoms near the Zn atoms and, as a result, results in a change in the bond length. As a result, localized conditions are what ultimately lead to illnesses and surface flaws [123]. Due to their higher k and lower n in the UV-visible region, they are skilled at creating optoelectronic devices [124]. To better understand the problem, the electrical polarizability was investigated by considering the dielectric constants of these sintered films. The real (ε') and imaginary (ε'') dielectric constants with respect to the photon energy spectra depicted in Figures 11.6(a) and 11.6(b) have been calculated using the following equations [121]. The real dielectric constant obtained by

$$\varepsilon' = n^2 - k^2 \qquad (11.15)$$

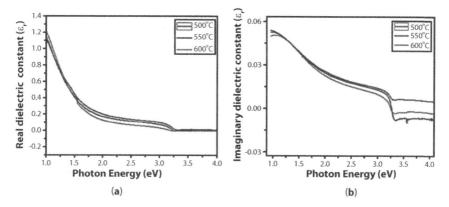

Figure 11.6 Shows the variations (a) real (ε') and (b) imaginary (ε'') components of the dielectric constants of ZnCdO films sintered at 500°C, at 550°C, and at 600°C [117].

and the imaginary dielectric constant obtained by

$$\varepsilon'' = 2nk \tag{11.16}$$

These graphs demonstrate that the real and imaginary sections of the dielectric constants vary with the photon energy as a result of minute interactions between the photons and the electrons in the material. A shift in reflectance and absorbance is what causes this perceptible fluctuation in actual and hypothetical parameters with photon energy and sintering assistance. Because a refractive index has a smaller value than a constant, the behavior of both constants is similar to that of a refractive index. The complex dielectric constant's real component is greater than its imaginary component for 550°C and 600°C. These elements are taken into account while calculating the power loss rate and dissipation factor for an oscillation mode in films. Applications involving optical communication call for these parameters.

11.5 Conclusions

This chapter reveals basic information of metal oxide thin films regarding optical properties at nano and bulk level respectively. To control these optical properties a particular wavelength of electromagnetic radiation and ultimately improve the optoelectronic properties, it is crucial to design an appropriate thin film. Selimer, a theoretical model seems to be important to study under different thickness. Optical properties to study experimentally has been investigated vastly. An overall analysis of MO thin film solar cells is provided in this chapter. The effects of material properties and structures are optimized for the best performances using a variety of MO semiconductors to investigate various structures and combinations for solar cells.

Acknowledgement

The authors are highly thankful to the BGSBU administration for providing peaceful environment and facility to carry out this sort of research for the visibility of University.

References

1. Pauling, L., *General chemistry*, Dover, New York, 1988.
2. Moritarty, P., Nanostructured materials. *Rep. Prog. Phys.*, 64, 297, 2001.
3. Kakani, S.L. and Kakani, A., *Materials science*, New Age International (P) Ltd., New Delhi, 2006.
4. Wang, S. *et al.*, Recycling of zinc oxide dust using ChCl-urea deep eutectic solvent with nitrilotriacetic acid as complexing agents. *Miner. Eng.*, 175, 107295, 2022.
5. Alkalbani, A.M., Chala, G.T., Myint, M.T.Z., Experimental investigation of rheological properties of water-base mud with zinc oxide nanoparticles using response surface methodology. *J. Pet. Sci. Eng.*, 208, 109781, 2022.
6. Tripathy, N. and Kim, D.H., Metal oxide modified ZnO nanomaterials for biosensor applications. *Nano Convergence*, 5, 1–10, 2018.
7. Samuel, E. *et al.*, ZnO/MnO x nanoflowers for high-performance supercapacitor electrodes. *ACS Sustain. Chem. Eng.*, 8, 9, 3697–3708, 2020.
8. Mageshwari, K. and Sathyamoorthy, R., Studies on photocatalytic performance of MgO nanoparticles prepared by wet chemical method. *Trans. Indian Inst. Met.*, 65, 49–55, 2012.
9. Devaraja, P.B. *et al.*, Luminescence properties of MgO: Fe3+ nanopowders for WLEDs under NUV excitation prepared via propellant combustion route. *J. Radiat. Res. Appl. Sci.*, 8, 3, 362–373, 2015.
10. Ho, S.M., Synthesis and properties of cadmium oxide thin films: A review. *Int. J. Curr. Adv. Res.*, 5, 7, 1038–1041, 2016.
11. Ho, S.M., Preparation and characterization of tungsten oxide thin films. *J. Chem. Pharm. Res.*, 8, 7, 414–416, 2016.
12. Ho, S.M., Study of the growth of magnesium oxide thin films using X-ray diffraction technique: Mini review. *Recent Adv. Petrochem. Sci.*, 1, 555558, 2017.

13. Jung, H.S. et al., Preparation of nanoporous MgO-coated TiO2 nanoparticles and their application to the electrode of dye-sensitized solar cells. *Langmuir*, 21, 23, 10332–10335, 2005.
14. Khan, M.I., and Upadhyay, T.C., Phase transition study of thermal dependence of soft mode frequency, dielectric constant and dielectric tangent loss properties in CsH 2 PO 4 (CDP) and CsD 2 PO 4 (DCDP) crystals. *J. Low Temp. Phys.*, 203, 401–418, 2021.
15. Ho, S.M., Synthesis and characterization of tin oxide thin films: A review. *Der Pharma Chem.*, 8, 3, 20–23, 2016.
16. Ho, S.M., Preparation of nanocrystalline aluminum oxide thin films: A review. *Int. J. Chem. Sci.*, 15, 115–121, 2017.
17. Joni, H. et al., Pulsed laser deposition of metal oxide nanoparticles, agglomerates, and nanotrees for chemical sensors. *Proc. Eng.*, 120, 1158–1161, 2015.
18. Ho, S.M. and Emmanuel, A., A short review of recent advances in copper oxide nanostructured thin films. *Res. J. Chem. Environ.*, 23, 138–145, 2019.
19. Ibrahim, I.M., Characterization and gas sensitivity of cadmium oxide thin films prepared by thermal evaporation technique. *Int. J. Thin Film Sci. Technol.*, 1, 1, 1–8, 2012.
20. Arafat, M.M. et al., Gas sensors based on one dimensional nanostructured metal-oxides: A review. *Sensors*, 12, 6, 7207–7258, 2012.
21. Vallejos, S. et al., Chemical vapour deposition of gas sensitive metal oxides. *Chemosensors*, 4, 1, 4, 2016.
22. Zhou, Y., Zhao, J., Zhai, Q., 100% renewable energy: A multi-stage robust scheduling approach for cascade hydropower system with wind and photovoltaic power. *Appl. Energy*, 301, 117441, 2021.
23. Barbara, T. et al., Utilization of renewable energy sources in desalination of geothermal water for agriculture. *Desalination*, 513, 115151, 2021.
24. Ho, S.M. et al., Advanced research in solar energy: Malaysia, UAE and Nigeria. *Eurasian J. Anal. Chem.*, 13, 312–331, 2018.
25. Ho, S.M. et al., Review on dye-sensitized solar cells based on polymer electrolytes. *Int. J. Eng. Technol.*, 7, 3001–3006, 2018.
26. Girtan, M., *Future solar energy devices*, pp. 1–101, Springer International Publishing, Switzerland, 2018.
27. Hashmi, G. et al., Band gap measurement of P-type monocrystalline silicon wafer. *Bangladesh J. Sci. Ind. Res.*, 53, 3, 179–184, 2018.
28. Ramanujam, J. et al., Flexible CIGS, CdTe and a-Si: H based thin film solar cells: A review. *Prog. Mater. Sci.*, 110, 100619, 2020.
29. Amalraj, P.A. and Alkaisi, M.M., Nanostructures for light trapping in thin film solar cells. *Micromachines*, 10, 9, 619, 2019.
30. Rho, W.Y. et al., Enhancement of power conversion efficiency with TiO2 nanoparticles/nanotubes-silver nanoparticles composites in dye-sensitized solar cells. *Appl. Surf. Sci.*, 429, 23–28, 2018.

31. Zhen, J. et al., Pyridine-functionalized fullerene additive enabling coordination interactions with CH 3 NH 3 PbI 3 perovskite towards highly efficient bulk heterojunction solar cells. *J. Mater. Chem. A*, 7, 6, 2754–2763, 2019.
32. Wei, J. et al., Enhanced light harvesting in perovskite solar cells by a bio-inspired nanostructured back electrode. *Adv. Energy Mater.*, 7, 20, 700492, 2017.
33. Ghosh, D. et al., Dependence of halide composition on the stability of highly efficient all-inorganic cesium lead halide perovskite quantum dot solar cells. *Sol. Energy Mater. Sol. Cells*, 185, 28–35, 2018.
34. Ahmadpour, G. et al., Effect of substrate surface treatment on the hydrothermal synthesis of zinc oxide nanostructures. *Ceram. Int.*, 48, 2, 2323–2329, 2022.
35. Sanchez-Lopez, A.L. et al., Influence of erbium doping on zinc oxide nanoparticles: Structural, optical and antimicrobial activity. *Appl. Surf. Sci.*, 575, 151764, 2022.
36. Abdullah, F.H., Bakar, N.A., Bakar, M.A., Current advancements on the fabrication, modification, and industrial application of zinc oxide as photocatalyst in the removal of organic and inorganic contaminants in aquatic systems. *J. Hazard. Mater.*, 424, 127416, 2022.
37. Khan, S.A. et al., Performance investigation of ZnO/PVA nanocomposite film for organic solar cell. *Mater. Today: Proc.*, 47, 2615–2621, 2021.
38. Dong, S. et al., Preparation of tree-structured ZnO films for solar cells application. *Mater. Lett.*, 305, 130782, 2021.
39. Chatterjee, S., Saha, S.K., Pal, A.J., Formation of all-oxide solar cells in atmospheric condition based on Cu2O thin-films grown through SILAR technique. *Sol. Energy Mater. Sol. Cells*, 147, 17–26, 2016.
40. Fujimoto, K., Oku, T., Akiyama, T., Fabrication and characterization of ZnO/Cu2O solar cells prepared by electrodeposition. *Appl. Phys. Express*, 6, 8, 086503, 2013.
41. Minami, T. et al., High-efficiency oxide solar cells with ZnO/Cu2O heterojunction fabricated on thermally oxidized Cu2O sheets. *Appl. Phys. Express*, 4, 6, 062301, 2011.
42. Jeong, S.S. et al., Electrodeposited ZnO/Cu2O heterojunction solar cells. *Electrochim. Acta*, 53, 5, 2226–2231, 2018.
43. Ke, N.H. et al., The characteristics of IGZO/ZnO/Cu2O: Na thin film solar cells fabricated by DC magnetron sputtering method. *J. Photochem. Photobiol. A: Chem.*, 349, 100–107, 2017.
44. Prasanna, A.E.P. et al., Investigation on zinc oxide nanoparticle incorporated polyaniline nano composites for solar cell applications. *Mater. Today: Proc.*, 48, 245–252, 2022.
45. Sudheer, M. et al., Synergic effect of Zinc (Zn) precursors on the microstructural, optical and electrical properties of Zinc oxide thin films for solar cell applications. *Optik*, 206, 163750, 2020.

46. Rameshkumar, C. et al., An investigation of SnO2 nanofilm for solar cell application by spin coating technique. *AIP Conf. Proc.*, 1, 23411, 2021.
47. Yunus, O. and Candan, I., SnO2 interlayer effects on the inverted polymer solar cells. *Chem. Phys. Lett.*, 740, 137078, 2020.
48. Fan, X. et al., Spray-coated monodispersed SnO2 microsphere films as scaffold layers for efficient mesoscopic perovskite solar cells. *J. Power Sources*, 448, 227405, 2020.
49. Wang, Y.F. et al., Controllable synthesis of hierarchical SnO2 microspheres for dye-sensitized solar cells. *J. Power Sources*, 280, 476–482, 2015.
50. Murugadoss, G. et al., An efficient electron transport material of tin oxide for planar structure perovskite solar cells. *J. Power Sources*, 307, 891–897, 2016.
51. Xiong, L. et al., Review on the application of SnO2 in perovskite solar cells. *Adv. Funct. Mater.*, 28, 35, 1802757, 2018.
52. Jung, K.H. et al., Solution-processed SnO2 thin film for a hysteresis-free planar perovskite solar cell with a power conversion efficiency of 19.2%. *J. Mater. Chem. A*, 5, 47, 24790–24803, 2017.
53. Li, Z. et al., Annealing free tin oxide electron transport layers for flexible perovskite solar cells. *Nano Energy*, 94, 106919, 2022.
54. Haejun, Y. et al., Superfast room-temperature activation of SnO2 thin films via atmospheric plasma oxidation and their application in planar perovskite photovoltaics. *Adv. Mater.*, 30, 10, 1704825, 2018.
55. Subbiah, A.S. et al., Novel plasma-assisted low-temperature-processed SnO2 thin films for efficient flexible perovskite photovoltaics. *ACS Energy Lett.*, 3, 7, 1482–1491, 2018.
56. Singh, M. et al., Facile synthesis of composite tin oxide nanostructures for high-performance planar perovskite solar cells. *Nano Energy*, 60, 275–284, 2019.
57. Yonghui, L., Lee, S., Peak, S., Cho, T., Seo, G., Efficient planar perovskite solar cells using passivated tin oxide as an electron transport layer. *Adv. Sci.*, 5, 1-6, 1800130, 2018.
58. Valentina, R. et al., Analytical study of solution-processed tin oxide as electron transport layer in printed perovskite solar cells. *Adv. Mater. Technol.*, 6, 2, 2000282, 2021.
59. Calligaris, M., Structural inorganic chemistry by AF Wells. *Acta Crystallogr. Sect. B: Struct. Sci.*, 41, 3, 208–208, 1985.
60. Chu, T.L. and Chu, S.S., Degenerate cadmium oxide films for electronic devices. *J. Electron. Mater.*, 19, 1003–1005, 1990.
61. Al-Taie, H. et al., Erratum: Cryogenic on-chip multiplexer for the study of quantum transport in 256 split-gate devices. *Appl. Phys. Lett.*, 102, 243102, 2013.
62. Ramamurthy, M., Balaji, M., Thirunavukkarasu, P., Characterization of jet nebulizer sprayed CdO thin films for solar cell application. *Optik*, 127, 8, 3809–3819, 2016.

63. Upadhyay, G.K., Kumar, V., Purohit, L.P., Optimized CdO: TiO2 nanocomposites for heterojunction solar cell applications. *J. Alloys Compd.*, 856, 157453, 2021.
64. Yakuphanoglu, F., Nanocluster n-CdO thin film by sol–gel for solar cell applications. *Appl. Surf. Sci.*, 257, 5, 1413–1419, 2010.
65. Zaien, M., Ahmed, N.M., Hassan, Z., Fabrication and characterization of nanocrystalline n-CdO/p-Si as a solar cell. *Superlattices Microstruct.*, 52, 4, 800–806, 2012.
66. Hussein, H.I., Shaban, A.H., Khudayer, I.H., Enhancements of p-Si/CdO thin films solar cells with doping (Sb, Sn, Se). *Energy Proc.*, 157, 150–157, 2019.
67. Jadduaa, M.H., Habubi, N.F., Ckal, A.Z., Preparation and study of CdO-CdO2 nanoparticles for solar cells applications. *Int. Lett. Chem. Phys. Astron.*, 69, 34–41, 2016.
68. Khan, M.I., Singh, P., Upadhyay, T.C., Dielectric properties of hydrogen-bonded cesium dihydrogen phosphate-type ferroelectric crystals. *Ferroelectrics*, 587, 1, 198–206, 2022.
69. Liu, R. *et al.*, Pristine inorganic nickel oxide as desirable hole transporting material for efficient quasi two-dimensional perovskite solar cells. *J. Power Sources*, 512, 230452, 2021.
70. Srivastava, M. *et al.*, Fabrication of room ambient perovskite solar cell using nickel oxide HTM. *Mater. Today: Proc.*, 34, 748–751, 2021.
71. Zhao, S. *et al.*, F4-TCNQ doped strategy of nickel oxide as high-efficient hole transporting materials for invert perovskite solar cell. *Mater. Sci. Semicond. Process.*, 121, 105458, 2021.
72. Mudgal, S. *et al.*, Carrier transport mechanisms of nickel oxide-based carrier selective contact silicon heterojunction solar cells: Role of wet chemical silicon oxide passivation interlayer. *Solid State Commun.*, 334, 114391, 2021.
73. Ma, F. *et al.*, Nickel oxide for inverted structure perovskite solar cells. *J. Energy Chem.*, 52, 393–411, 2021.
74. Labed, M. *et al.*, Study on the improvement of the open-circuit voltage of NiOx/Si heterojunction solar cell. *Opt. Mater.*, 120, 111453, 2021.
75. Boyd, C.C. *et al.*, Overcoming redox reactions at perovskite-nickel oxide interfaces to boost voltages in perovskite solar cells. *Joule*, 4, 8, 1759–1775, 2020.
76. Du, Y. *et al.*, Polymeric surface modification of NiOx-based inverted planar perovskite solar cells with enhanced performance. *ACS Sustain. Chem. Eng.*, 6, 12, 16806–16812, 2018.
77. Liang, J. *et al.*, Overcoming the carrier transport limitation in Ruddlesden-Popper perovskite films by using lamellar nickel oxide substrates. *J. Mater. Chem. A*, 9, 19, 11741–11752, 2021.
78. Liu, S. *et al.*, Boost the efficiency of nickel oxide-based formamidinium-cesium perovskite solar cells to 21% by using coumarin 343 dye as defect passivator. *Nano Energy*, 94, 106935, 2022.

79. Thiruchelvan, P.S., Lai, C.C., Tsai, C.H., Combustion processed nickel oxide and zinc doped nickel oxide thin films as a hole transport layer for perovskite solar cells. *Coatings, 11*, 6, 627, 2021.
80. Choi, F.P.G. et al., First demonstration of lithium, cobalt and magnesium introduced nickel oxide hole transporters for inverted methylammonium lead triiodide based perovskite solar cells. *Sol. Energy*, 215, 434–442, 2021.
81. Imran, M. et al., Role of annealing temperature of nickel oxide (NiOx) as hole transport layer in work function alignment with perovskite. *Appl. Phys. A*, 127, 1–8, 2021.
82. Qin, C. et al., Surface modification and stoichiometry control of Cu2O/SnO2 heterojunction solar cell by an ultrathin MgO tunneling layer. *J. Alloys Compd.*, 779, 387–393, 2019.
83. Huang, S. et al., Highly efficient inverted polymer solar cells by using solution processed MgO/ZnO composite interfacial layers. *J. Colloid Interface Sci.*, 583, 178–187, 2021.
84. Kulkarni, A. et al., Revealing and reducing the possible recombination loss within TiO2 compact layer by incorporating MgO layer in perovskite solar cells. *Sol. Energy*, 136, 379–384, 2016.
85. Dagar, J. et al., Highly efficient perovskite solar cells for light harvesting under indoor illumination via solution processed SnO2/MgO composite electron transport layers. *Nano Energy*, 49, 290–299, 2018.
86. Zahid, M.A. et al., Improved optical and electrical properties for heterojunction solar cell using Al2O3/ITO double-layer anti-reflective coating. *Results Phys.*, 28, 104640, 2021.
87. Khan, M.I., and Upadhyay, T.C., General introduction to ferroelectrics, in: *Multifunctional Ferroelectric Materials*, pp. 1–23, IntechOpen Publisher, London, UK, 2021.
88. Abdi, N., Abdi, Y., Alemipour, Z., Effects of morphology and thickness of Al2O3 scaffold on charge transport in Perovskite-based solar cells. *Sol. Energy*, 153, 379–382, 2017.
89. Liang, Z. et al., Interface modification via Al2O3 with retarded charge recombinations for mesoscopic perovskite solar cells fabricated with spray deposition process in the air. *Appl. Surf. Sci.*, 463, 939–946, 2019.
90. Steinhauser, B. et al., PassDop rear side passivation based on Al2O3/a-SiCx: B stacks for p-type PERL solar cells. *Sol. Energy Mater. Sol. Cells*, 131, 129–133, 2014.
91. Kim, J.H. et al., Stability enhancement of GaInP/GaAs/Ge triple-junction solar cells using Al2O3 moisture-barrier layer. *Vacuum*, 162, 47–53, 2019.
92. Choi, E.Y. et al., Enhancing stability for organic-inorganic perovskite solar cells by atomic layer deposited Al2O3 encapsulation. *Sol. Energy Mater. Sol. Cells*, 188, 37–45, 2018.
93. Chen, M. et al., Cobalt oxide and nitride particles supported on mesoporous carbons as composite electrocatalysts for dye-sensitized solar cells. *J. Power Sources*, 286, 82–90, 2015.

94. Ambika, S. *et al.*, Structural, morphological and optical properties and solar cell applications of thioglycolic routed nano cobalt oxide material. *Energy Rep.*, 5, 305–309, 2019.
95. Huang, A. *et al.*, Fast fabrication of a stable perovskite solar cell with an ultrathin effective novel inorganic hole transport layer. *Langmuir*, 33, 15, 3624–3634, 2017.
96. Li, B. *et al.*, Solution-processed p-type nanocrystalline CoO films for inverted mixed perovskite solar cells. *J. Colloid Interface Sci.*, 573, 78–86, 2020.
97. Shalan, A.E. *et al.*, Cobalt oxide (CoO x) as an efficient hole-extracting layer for high-performance inverted planar perovskite solar cells. *ACS Appl. Mater. Interfaces*, 8, 49, 33592–33600, 2016.
98. Dou, Y. *et al.*, Toward highly reproducible, efficient, and stable perovskite solar cells via interface engineering with CoO nanoplates. *ACS Appl. Mater. Interfaces*, 11, 35, 32159–32168, 2019.
99. Zhang, Q. *et al.*, First application of CoO nanorods as efficient counter electrode for quantum dots-sensitized solar cells. *Sol. Energy Mater. Sol. Cells*, 206, 110307, 2020.
100. Chen, L., Chen, W., Wang, E., Graphene with cobalt oxide and tungsten carbide as a low-cost counter electrode catalyst applied in Pt-free dye-sensitized solar cells. *J. Power Sources*, 380, 18–25, 2018.
101. Nandi, D.K. and Sarkar, S.K., Atomic layer deposition of tungsten oxide for solar cell application. *Energy Proc.*, 54, 782–788, 2014.
102. Sokeng D., A. *et al.*, Improved hole extraction selectivity of polymer solar cells by combining PEDOT: PSS with WO3. *Energy Technol.*, 9, 12, 2100474, 2021.
103. Sun, K. *et al.*, Review on application of PEDOTs and PEDOT: PSS in energy conversion and storage devices. *J. Mater. Sci.: Mater. Electron.*, 26, 4438–4462, 2015.
104. Bießmann, L. *et al.*, Monitoring the swelling behavior of PEDOT: PSS electrodes under high humidity conditions. *ACS Appl. Mater. Interfaces*, 10, 11, 9865–9872, 2018.
105. Khan, M.I., and Upadhyay, T.C., Phase transitions in H-bonded deuterated Rochelle salt crystal. *Eur. Phys. J. Plus*, 136, 1, 1–14, 2021.
106. Tan, Z. *et al.*, Solution-processed tungsten oxide as an effective anode buffer layer for high-performance polymer solar cells. *J. Phys. Chem. C*, 116, 35, 18626–18632, 2012.
107. Dang, H.L.T. *et al.*, Micro-wheels composed of self-assembled tungsten oxide nanorods supported platinum counter electrode for highly efficient liquid-junction photovoltaic devices. *Sol. Energy*, 214, 214–219, 2021.
108. Zargar, R.A. *et al.*, Optical characteristics of ZnO films under different thickness: A MATLAB-based computer calculation for photovoltaic applications. *Phys. B: Condens. Matter*, 631, 413614, 2022.
109. Palik, E.D., *Handbook of optical constants of solids*, vol. 3, Academic Press, University of Maryland, USA, 1998.

110. Khan, M.I., and Upadhyay, T.C., Phase transition thermal dependence of ferroelectric and dielectric properties in H-bonded PbHPO 4 (LHP) crystal. *Appl. Phys. A*, *126*, 11, 881, 2020.
111. Sun, X.W. and Kwok, H.S., Optical properties of epitaxially grown zinc oxide films on sapphire by pulsed laser deposition. *J. Appl. Phys.*, *86*, 1, 408–411, 1999.
112. Zargar, R.A. *et al.*, Screen printed TiO2 film: A candidate for photovoltaic applications. *Mater. Res. Express*, 7, 6, 065904, 2020.
113. Hu, Y. *et al.*, Hydrothermal preparation of ZnS: Mn quantum dots and the effects of reaction temperature on its structural and optical properties. *J. Mater. Sci.: Mater. Electron.*, 29, 16715–16720, 2018.
114. Mariappan, R. *et al.*, Influence of film thickness on the properties of sprayed ZnO thin films for gas sensor applications. *Superlattices Microstruct.*, 71, 238–249, 2014.
115. Khan, M.I., and Upadhyay, T.C., Ferroelectric phase transitions in PbHPO4-type crystals. *J. Mater. Sci.: Mater. Electron.*, 32, 11, 14569–14583, 2021.
116. Chandramohan, S. *et al.*, Swift heavy ion beam irradiation induced modifications in structural, morphological and optical properties of CdS thin films. *Nucl. Instrum. Methods Phys. Res. Sect. B: Beam Interact. Mater. At.*, 254, 2, 236–242, 2007.
117. Zargar, R.A. *et al.*, Structural, optical, photoluminescence, and EPR behaviour of novel Zn0·80Cd0·20O thick films: An effect of different sintering temperatures. *J. Lumin.*, 245, 118769, 2022.
118. Zargar, R.A. *et al.*, Screen printed TiO2 film: A candidate for photovoltaic applications. *Mater. Res. Express*, 7, 6, 065904, 2022.
119. Soylu, M. and Yazici, T., CdO thin films based on the annealing temperature differences prepared by sol–gel method and their heterojunction devices. *Mater. Res. Express*, 4, 12, 126307, 2017.
120. Shkir, M. *et al.*, Enhanced opto-non-linear properties of low cost deposited pure and Ni@ PbI2/glass nanostructured thin films for higher order nonlinear applications. *J. Phys. Chem. Solids*, 157, 110197, 2021.
121. Chackrabarti, S. *et al.*, Realization of structural and optical properties of CdZnO composite coated films for photovoltaic cell applications. *Optik*, 127, 20, 9966–9973, 2016.
122. Czichos, H. and Saito, T., *Springer handbook of materials measurement methods*, vol. 978, L. Smith (Ed.), pp. 399–429, Springer, Berlin, 2006.
123. Behera, M. *et al.*, Role of Te on the spectroscopic properties of As50Se40Te10 thin films: An extensive study by FTIR and Raman spectroscopy. *Opt. Mater.*, 66, 616–622, 2017.
124. Boucle, J., Ravirajan, P. and Nelson, J., Hybrid polymer–metal oxide thin films for photovoltaic applications, *J. Mater. Chem.*, 17.30, 3141–3153, 2007.

12

Metal Oxide-Based Light-Emitting Diodes

Shabir Ahmad Bhat[1]*, Sneha Wankar[2]†, Jyoti Rawat[3] and Rayees Ahmad Zargar[4]‡

[1]Department of Chemistry, Government Degree College for Women, Anantnag, Jammu and Kashmir, India
[2]Post Graduate Department of Chemistry, S. K. Porwal College of Arts, Science and Commerce, Kamptee, Nagpur, Maharashtra, India
[3]Department of Physical Science, Rabindranath Tagore University, Bhopal, MP, India
[4]Department of Physics, BGS University, Rajouri, India

Abstract

An advancement in designing and developing desired electronics and optically active devices emerged as an important field of research. In this context, construction of new material as LEDs with superior quality as compared to existing one is an intriguing part of discussion. Light emitting diodes, owing to their potential application in crafting displays, construction of mobile phone accessories, and many more enticed much inquisitiveness in research from last few decades and considered as promising entrants for replacing current technologies to reshape the future world. The most abundant and promising component adopted for designing of such material is metal oxide. The promising characteristics of metal oxides are optimum band gap (efficient for short wavelength photonics), high emission ranging from UV to IR region. This chapter will deal with general outline of transition and lanthanide metal oxide-based LEDs. It covers the synthetic strategies, characterization, and luminescent properties of various metal-based LEDs and their applications. The present chapter is appraised to be a vital contribution for future research strategies.

Keywords: Nanoparticles, ZnO, electroluminescence, LEDs, optoelectronic properties

---

*Corresponding author: shabirjmi0671@gmail.com
†Corresponding author: snehawankar21@gmail.com
‡Corresponding author: rayeesphy12@gmail.com

12.1 Introduction

Because of countless applications like in energy, fiber optics, medicine, and communications, the demand of Light-based technologies grew enormously from the last few decades. The world has commemorated year 2015 as the International Year of light. In 1962, Nick Holonyak, Jr., practically invented LED for the first time. These LEDs were used as better replacements for incandescent indicators, seven-segment displays, and then LEDs were employed in other electronic appliances as well. The advancement in the materials utilized in the LED construction were selected in order to enhance its efficiency.

The LEDs, due to an appealing and remarkable features, they have firmly bested artificial light sources, therefore, are potential candidates to future lighting. In comparison to incandescent bulbs, the LEDs have efficient lumen power and therefore find their use in energy-saving devices. Furthermore, the lifespan of these LEDs bulbs is much more (one lakh hours) than usual fluorescent (ten thousand hours) and incandescent bulbs (one thousand) and do not require any maintenance, therefore, they are more reliable for use.

Unlike fluorescent and incandescent bulbs, they are even highly resistant to external shock and starts operating within microseconds even under weak current or low input voltage. They are considered as eco-friendly as they lack mercury unlike fluorescent lamps and are recyclable. They emit different color emissions without use of any filter. The light-based technologies, presently, Light-emitting diodes (LEDs) are most substantial devices with efficient conversion of electrical energy to light and have outstripped the most conventional sources [1]. Lot of efforts has been taken for the development of lightning devices from earlier to present multicolor LEDs. The blue-green LEDs prepared from aluminum gallium indium nitride (AlGaInN) make the possible use of lightening devices in industries. The semiconductor-based LED chips used in solid state lightning are based on GaN [2, 3], InGaN [4, 5], AlGaN [6, 7], and AlInGaP [8] that produces emission of different wavelengths. Presently, the blend of LEDs with organic inorganic perovskite materials is also being explored.

However, there are certain considerations that must be taken into account before an LED can be used for specific purpose in the global market. The LEDs fabrication materials must be tested for both their safety and outcomes on human health. Though they do not contain mercury however concentration of elements like Cu, Pb and As must be low. They must possess overlapping excitation and the emission spectra, efficient luminescent quality (efficient quantum efficiency) high physical and chemical stability.

The LEDs introduced to for the first time were semiconductor diodes which are highly accompanied with emitting light. More explicitly, the electric current convert into an optical radiation. The diodes are pen junction and that drive the transformation of electric energy into radiation. The metal generally adopted for pen junction are silicon or germanium as they certified for competent electric flow through them without any mutilation as they are lesser subtle to temperature. The intrinsic semiconductors tampered with pentavalent ions like phosphorus, arsenic, antimony, or bismuth as impurity. Due to some shortcomings stumble upon in the use of Si and Ge based LEDs, elements like Ga, As came into glare of publicity for their higher quantum efficiency [9]. The application of materials like AlGaP, GaP, and GaAsP are acceptable in LEDs are remarkable owing to their befitting the energy gap and exhibiting efficient phenomenon of electroluminescence. The burst through juncture in the LED industry came with the origination of InGaN chips that gave an opportunity to cover the blue region of the electromagnetic spectrum [10].

12.2 Structure of LEDs

The framing of LED is akin pen junction diode; furthermore, the structure of semiconductor is ranked in terms of three different layers positioned on the substrate. The layers stipulated as p-type region at the top along with holes, the middle layer as the operational region and charged carriers, whereas the bottom layer is constituted of n-type region with electrons.

The bottom of the substrate straightway allied with anode while the top layer is club together with p-layer via wire band. The reflector cup consociate with the bottom aiming to reflect the earned light towards the upward direction. The integrated set up accommodated inside a hemispherical shaped capsule made up of epoxy resin; this capsule also acts as a defensive shielding from any expedient electric shocks or any vibrational turmoil from external periphery. It is apparent to ensure that some parameters like temperature stability, refractive index, and chemical inertness while adopting any polymer material for designing of capsule. The dome shaped is constructed with intent of receiving reflected light from the cup for generation of maximum illumination. For better understanding, an anode part kept elongated as compared to the cathode terminal. The design of top portion is very important for light reflection from the reflector cup should

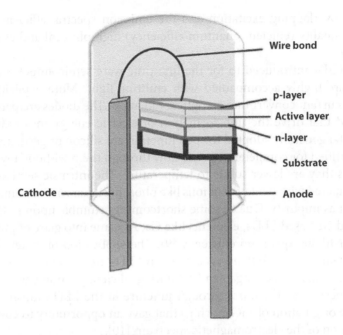

Figure 12.1 Basic construction of LEDs.

be domed shape in order to give maximum illumination. In order to distinguish between anode and cathode in the constructed LEDs, the anode is bit longer in comparison to the cathode terminal. The basic construction of LED is represented in Figure 12.1.

12.3 Working Principle of LEDs

The phenomenon of electroluminescence was first discovered by H. J. Round while working with electric current passing through the carborundum [11]. The suitable condition for earning electroluminescence is forward biased settings in LEDs, wherein the p-type terminal is joined to the positive terminal of the battery whereas the negative terminal is linked to n-type part. The mechanism of forward biased is evident as the free holes released from the p-region while the free electrons released from the n-region and flow towards the pen junction. It is witnessed that p-type and n-type regions are conducting individually whereas their juncture is in nonconductive. The region close to the junction denoted as depletion or active region (Figure 12.2). Furthermore, the electrostatic forces between the positive holes and negatively charged electrons over power each other

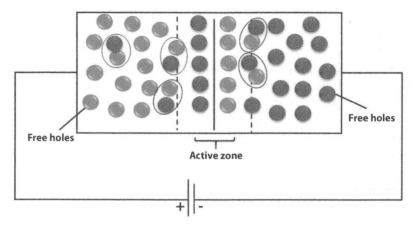

Figure 12.2 Schematic representation of working of LED.

and phenomenon termed as recombination. The continuation process of recombination results in the depletion or narrowing of junction. Because of depletion at junction the potential barriers sinks and thus facilitates the electrons to cross the junction. Generally, when electrons flow from n-type to p-type it leaves holes behind it. However, in the local region existing free electrons upshots the emission of radiative energy. The release of energy or emitted photon is a characteristic exemplification of semiconducting material embraced for designing of diode.

12.4 Selection of Material for Construction of LEDs

An assortment of chief material for assembling various fragments of LED, the material like silicon carbide, gallium nitride, or silicon is cast off. Amid these materials, GaN received more appreciation in comparison to other material in view of its ideal band gap value. The size of the band gap is amounted to 3.39 eV and considered ideal for designing of colored LEDs *viz.* blue, violet. These LEDs are also reported for their temperature tolerance up to 600°C [12].

As an alternate selection for the substrate sapphire explored much, however silicon carbide (SiC) is structurally similar to the gallium nitride. The use of SiC exhibits decreased in defect density while enhance both durability and efficiency. In addition to this, SiC failed to manage its applicability for designing LEDs owing to their high cost. It is rather challenging to accomplish the high demands of LEDs in substantial numbers using sapphire or silicon carbide substrates, as they need much more amendments in order to ensure their applicability. An effective way to expand the light extraction part is by reducing the epilayer defect densities to create an appropriate patterning on the substrate. Thus, patterned sapphire substrate (PSS) in nanoscale range proven as better performer with dropped cost and manufacturing at ease [13–15]. The adoption of Zn doped in GaN to yield p-type GaN witnessed to be important contribution in the field of LEDs production. Somehow doping of Zn in an appropriate concentration with GaN resulted in the formation of effective radiative center for emitting blue light efficiently at 2.86-0.02 eV. With increasing doping concentration of zinc in GaN the photoconductivity peak shifted towards higher energies up to 70 meV [16]. Furthermore, Mg-doped GaN film also reported for p-type conduction proposed for low energy electron-beam irradiation (LEEBI) treatment [17]. The photoluminescence efficiency of Mg-doped GaN enriched while resistivity fall significantly. The weak blue emission measurement and high resistivity observed in GaN film due to formation of acceptor-H neutral complexes [18]. But dropping of resistivity is also observed in such system due to thermal annealing in the nitrogen-ambient atmosphere referred as the removal of atomic hydrogen from the acceptor-H neutral complexes.

Moreover, the selection of packing material for covering the entire set up is an important part, as the electronic circuits are susceptible to damage under the thermal and mechanical stress. For such packing material phosphor material is dispersed over matrix followed by deposition of die. As the nature of epoxy or silicon-based material are highly stable under heat and resistant to the high-energy blue radiation and therefore most widely used as encapsulant for LEDs. Therefore, usage of silicon-based encapsulators are good enough for long-term stability however very expensive and challenging to consider for mass production and their usability to the society [19].

12.5 Basic Terminology Involved in Fabrication of LEDs

12.5.1 Color Rendering Index (CRI)

The efficiency of light source estimated by means of its color rendition property denoted as Ra. The light source reflects different colors under suitable illumination. The quality of light is assessed by CRI value amounted by any luminaries. The LEDs phosphors designed with doped rare earth reduces the cost complexities. Thus, high quality phosphor hold high CRI values that facilitates the performance of luminaire in preferred color region.

In the latest, CIE make a speech about inadequacies of color rendering index that every so often led to a unjust opinion for solid-state lighting sources [20]. CRI has miscarried to state truthfully the color gamut and failed to describe the color fidelity of a light source. Consequently, the scientific society was obliged to implement a new and more precise metric to enumerate the color fidelity of a light source. CIE released a technical report to introduce the color fidelity index (Rf) as the new metric to fill the gaps left out by CRI.

12.5.2 CIE Color Coordinates

In 1931, International Commission on Illumination (or Commission International de l'Eclairage CIE) established the concept of basic colors red, green and blue (RGB) colors. The color space model practices combination of three primary color values (tristimulus) in order to produce color which may be perceived by human eye. Thus, tristimulus values X, Y, and Z used to characterize the composition of colors. The tristimulus values X, Y, and Z used to characterize the composition of colors. The CIE chromaticity coordinates (x, y, z) are calculated from the tristimulus values by the ratio of X, Y or Z to their total sum. Since the sum of these coordinates is always unity, therefore, color is suitably characterized using only (x, y) coordinates. The CIE chromaticity coordinates x, y and z are calculated from the tristimulus values by the ratio of X, Y or Z to their total sum. Since the sum of these coordinates is always unity, therefore, color is suitably characterized using only (x, y) coordinates [21, 22].

The efficiency of a phosphor is considered as prime factor before selecting it in the fabrication of LED devices. Sometimes, those phosphors that show unmatched efficiency out of the package can be slightly inefficient within the LED package. Hence, different terminologies were quoted to describe the phosphor efficiency in different stages of fabrication. The term quantum efficiency or the quantum yield is often used to describe the phosphor's efficiency. Quantum yield can be defined as the ratio of the number of photons emitted to the number of photons absorbed by the phosphor. On the other hand, the efficiency of an LED chip expressed in terms of the external quantum efficiency (EQE). This term signifies the efficiency of the chip that converts the electrons into photons.

The injected electrons in the active region of LED chip tunnel through the device that undergoes recombination with holes and results in photon formation. Nevertheless, it is not necessary that all the electron-hole recombination will lead to the generation of photons. The photons, which are produced from the radiative electron-hole recombination in the active region, escape from the LED chip. The segment of such photons produced in the active region and that are able to discharge from the device is called extraction efficiency.

12.5.3 Forster Resonance Energy Transfer (FRET)

The FRET is an interesting process, which designates an energy transfer takes place from one chromophore to another. In mechanistic way, a donor chromophore located in its excited state transfer its energy *via* non-radiative pathway. A long-range dipole-dipole coupling mechanism to an acceptor chromophore in close proximity of the donor chromophore, typically energy transfer mechanism is termed Förster resonance energy transfer, which was named after the German scientist de: Theodor Förster.

12.6 LEDs Based on ZnO (Zinc Oxide)

The advancement in electronic and optoelectronic devices are often seek the new materials with desired properties superior to the existing one. An exploration of important element like germanium, replaces silicon, it modernizes the semiconductor industry. In addition to this, gallium arsenide and zinc oxide (Figure 12.3) are aroused as a significant module for designing optoelectronics for vigorous working devices. The performance of ZnO is promising as compared to existing one as gallium nitride employed for short-wavelength optoelectronic devices. The ZnO appears to be potential entrant to challenge the supremacy of III-nitrides, but applicability of ZnO remain as a challenge in usability at industrial level. However, working devices have been built from ZnO and related materials that show the potential inherent in this material system and received attention. In case of ZnO, the obtained binding energy, thermal energy explicitly at room temperature makes ZnO LEDs and diode more efficient for bright light emission. The dramatic consequences for near- and far-ultraviolet (UV) emitters where GaN failed to performed sound in terms of cost and efficiency [23].

The increasing importance of solid-state lighting during recent years, GaN-based LEDs found to be an important entrant, but ZnO surpass owing to its controlled doping system and become technically impactful. The work have been reported by several groups globally to validate the working of ZnO LEDs in both heterojunction and homojunction.

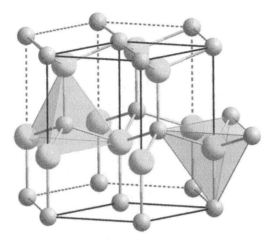

Figure 12.3 Crystal structure of ZnO.

The studies on commercial quality, blue-emitting diodes based on III-nitride technology first developed in Japan during the 1980s and 1990s. The early pioneering demonstrations gave rise to the entire solid-state lighting industry of today. In year 2014, Nobel Prize was awarded jointly to Isamu Akasaki, Hiroshi Amano, and Shuji Nakamura in Physics for "the invention of efficient blue light-emitting diodes, which were permitted for bright and energy-saving white light sources."

The GaN-family semiconductors possess favorable properties to enable their use for making blue and UV LEDs and laser diodes. Of the several possibilities, ZnO clearly viewpoints out because of its many similarities to GaN with some superior features. Thus, it is not surprising that despite continuing difficulties in ZnO technology many attempts have been made at constructing working LEDs from this material [24].

There are certain salient features, which are important to list out; ZnO is well-recognized crystalline structure from prehistoric time. The fascinating electronic properties of ZnO pawed in semiconductor age in the late 1940s. In this regard material science researchers, crystallographers and various physicists are advantageously considering this feature of ZnO. The crystal of ZnO was explored in 1936 as Wurtzite form hexagonal crystal structure. It also involves binary sulphides and oxides. The remarkable piezoelectric properties of ZnO are imperative to understand the behavior of ZnO-based devices as this material is even more active piezoelectric than GaN. The tetrahedral coordination seen in ZnO is also indicates the sp3 bonding and to be sure ZnO exhibits mixed ionic-covalent character with nearly equal contribution from each bonding type. This feature gives rise to some interesting chemical reactions that are possible with this oxide [25].

As a semiconductor, ZnO exhibited ideal separation between the valence band and conduction band edges amounted to 3.37 eV at room temperature. The physical appearance of ZnO do vary with band gap and they seems to be different. The larger band gap appears to be glass like and found to be transparent to visible radiation. ZnO also found to be superior to GaN in terms of adopting in shaping them in optoelectronic application. ZnO are considered as ecofriendly and nontoxic to humans. The biocompatibility of ZnO has led to its use in many medical treatments, while GaN is relatively consider as heavy metal and toxic to humans. ZnO requires no special disposal as compared GaN.

The bulk ZnO is more comfortable for the fabrication of device as compared to GaN because it is gladly acquire for hydrothermal growth. The density of hydrothermally grown ZnO is close to 5.6 g/cm^3, corresponding to 4.2 × 1022 ZnO U/cm^3. The ZnO based substrate provide an ideal matrix for subsequent epitaxial growth, and without any thread dislocation that was observed in case of GaN-based devices. It is an attention-grabbing point that, state that similar chemical structures between GaN and ZnO lattice. The mismatch is amounted to 1.8% and complement each other to grow on each other epitaxially [26, 27]. This observation is advances the application of combined GaN–ZnO hybrid devices and indeed most heterojunction ZnO LEDs demonstrated so far.

Furthermore, the wet chemical processes are certainly available for to modulate the nature of ZnO, however, dry etch chemistries are broadly available for ultra-small features [28, 29]. It has been found in many cases, even submicron features can be etched with acidic etchants, such as dilute orthophosphoric acid, containing appropriate type and concentration of surfactants. Generally ferric chloride and oxalic acid are adopted for etching of ZnO while GaN, is very difficult to etch through wet chemistries [30, 31].

There are some processes reported in literature regarding dry etching of ZnO. Some chlorine and hydrocarbon based dry etch chemistries reported by various research groups. The results obtained from inductive coupled plasma study exhibited that under the controlled conditions of high quality bulk single crystal ZnO sample was studied. At low ion energies, $CH_4/H_2/Ar$ causes faster etching while Cl_2/Ar plasmas required higher ion energy. In the case of methane-based chemistry, low ion energies are not very effective, as breaking of Zn–O is high bond strength, and thus requires high-energy ion impacts. However, once the bonds are broken,

ZnO can volatilize as the dimethyl zinc species. The high vapor pressure of this compound (301 mTorr at 20°C) is because of the high-ultimate etch rate seen with methane-based chemistries. On the other hand, for $ZnCl_2$ formed with chlorine-containing chemistries, the etch product has a lower vapor pressure of about 1 mTorr at 428°C consistent with the slower etch rates seen with this chemistry. This ion assisted nature of ZnO dry etching creates nice vertical side-wall profiles. Also because the near-surface stoichiometry is maintained, so the etching process produces only a small degree of surface roughening. Kim et al. [32] investigated ZnO etch rates as a function of $BCl_3/Cl_2/Ar$, BCl_3/Ar, and $BCl_3/CH_4/H_2$ chemistries. They found that compared with Ar, Cl_2/Ar, and CH_4/H_2-based gas mixtures, pure BCl_3 gas results in substantially high etch rates; indicating that B and BCl radicals react with ZnO and form volatile compounds [33]. The oxygen is thus removed by boron, whereas zinc is removed concomitantly by chlorine, with the formation of $ZnCl_2$. It should also be noted here that dry etching results in both mechanical and radiation damage to most materials but ZnO is exceptionally radiation-hard and can endure high amounts of irradiation without any noteworthy damage [34].

The observed refractive index of ZnO is lower than that of GaN, this is an essential element that means light can be easily extracted more easily from confines of ZnO based light emitting device. To enhance light extraction more easily, ZnO photonic crystal structure can be produced on ZnO based light emitters. ZnO shows exciton binding energy close to 60 meV, which is more than that of the value of GaN. Thus, light emitting devices constructed from ZnO should be brighter than comparable to GaN-based devices at the same operating current. In other words, ZnO-based light emitters should be more efficient when compared with similar GaN based devices. This also means that laser diodes made out of the ZnO system should have lower threshold current requirement when compared with GaN-based laser diodes.

The electrical contacts are an essential element required for functioning of all type of electronic and optoelectronic devices. Thus, a contact technology for ZnO family materials has made substantial advancement in recent years [35, 36]. Various metallization schemes have been reported for n- and p-type contacts to ZnO. As expected, n-type contacts display lower specific contact resistance as compared to p-type contacts.

Considering ZnO also has some features that are not as striking for making light-emitting devices. For one, ZnO is considerably softer than GaN and more easily attacked by a variety of chemicals, making it a less robust material. Then ZnO is also thermodynamically less stable than GaN and related compounds. A sample of ZnO turns yellow on heating-a strong thermochromic effect due to the formation of in situ oxygen ion vacancies. Furthermore, ZnO is also much harder to dope with optically active ions, such as the rare earths. GaN, in contrast, is easy to dope with both transition metal and rare-earth ions [37].

The ZnO exhibits as n-type and ultimately established due to a large number of native defects, such as oxygen vacancies, zinc vacancies, oxygen interstitials, and zinc interstitials [38]. Among these, oxygen vacancies have the lowest formation energy and are thus the easiest to form. This feature makes it difficult to induce significant amount of p-type character through known doping techniques. The major enduring problem with ZnO is unavailability of suitably highly doped p-type material. This problem is also observed with other III–V and II–VI wide bandgap semiconductors-almost all of which are easy to dope n-type but much harder to dope p-type. The only major exception appears to be ZnTe, which is quite easy to dope p-type [39]. Over the past few years, substantial progress has been made in this area and now material with hole concentration in excess of 3×1019 cm^{-3} has been demonstrated. However, the various processes developed so far have either suffered from lack of reproducibility, low achievable hole concentration, and/or gradual dwindling of p-type character over time. Indeed, further developments are obligatory in this track so that controllably doped p-type material of practically high hole concentration and mobility becomes more often than not. Most of the work is been conceded on developing good quality p-type ZnO [40–42]. The substitution of group-I elements for Zn sites or the substitution of group-V elements for O sites lead to p-type ZnO. However, both theoretical and experimental studies revealed that the situation for ZnO doping is much more complicated as compared to silicon. The relative sizes of substituting ions, charge compensation effects, defect complexes, and bond strains all plays an important role in the case of p-type ZnO doping. The photonic crystal light extraction structures on ZnO devices are shallower and easier to fabricate. This should lead to brighter devices that have convincing external quantum efficiency and benefit over GaN based devices. Perhaps the most often cited advantage of ZnO over GaN for the solubility, placement, and activation of a range of potential dopants in ZnO.

Recently, Reynolds et al. [43] have shown that sufficiently high nitrogen doping followed by appropriate thermal annealing that leads to significantly high (~1018 cm^{-3}) p-type behavior in ZnO, at room temperature. Through secondary ion mass spectrometry, Raman-scattering, photoluminescence (PL), and Hall-effect studies, they concluded that the observed p-type electrical activity is a result of doped N atoms evolving from their initial incorporation on Zn sites to a final shallow acceptor complex $V_{Zn}-N_O-H_+$, with an ionization energy of around 130 meV. This complex, they determined, is responsible for the p-type character of ZnO thus doped. It should be mentioned here that very recent work has also reported on ferromagnetism in N-doped ZnO thin films, which may have interesting implications for future magneto-optic devices. Further work is proceeding on finding the best route for p-type doping of ZnO, which is stable and reproducible [44].

12.7 Transition Metal Oxide-Based LEDs

Transition metal oxides (TMO) have been investigated from last few years back to the late 1990's, the significant reports are available in the field of organic electronic that are developed by holes injected using thin films of vanadium, molybdenum and ruthenium oxide as interlayers between the anode and organic material. The initial work was followed by a series of reports on the use of such compounds in OLEDs and organic photovoltaic (OPV) devices, as hole-injection and hole-extraction interlayers, and in charge generation and charge recombination layers [45].

Transition metals are elements with partially filled d orbitals, with one to nine electrons in the outer shell. In a solid, these d orbitals form relatively narrow d bands. In Transition metal oxides, the 2p orbitals, which originate from the oxygen anion, completely filled and form the valence band of the material. The transition metals are used as functional components in catalysis, useful in designing of high temperature superconductors, employed in magnetic recording devices based on the giant magnetoresistance effect [46] or in gas sensors and electro-/photo-/thermochromic devices [47, 48]. TMOs have also been extensively studied for their fundamental properties. They were turned out to be an exceptional, since classical solid state physics theory was not able to describe experimental observations made on some of these materials. However, a TMO such as MnO with partially filled 3d states exhibits good insulating properties with a conductivity of the order of 10 − 11 S/cm [49]. Such TMOs are now described as Mott-Hubbard, or charge transfer, insulators because a strong Coulomb interaction inhibits the formation of a 3d band and localizes the electrons at the transition metal ions [50]. But TMOs can also have metallic properties, as demonstrated by MoO_2, which has a conductivity on the order of 104 S/cm [51]. Some predominant metal oxides such as MoO_3, V_2O_5 and WO_3, which are the most widely used in organic electronics because of their favorable electronic properties, low optical absorption in the visible spectrum and a high level of technological compatibility. The electronic structure of TMOs as determined via combinations of direct and inverse photoemission spectroscopy and charge carrier transport measurements. The interface energetics, i.e., energy level alignment between TMOs and organic semiconductors, and the resulting hole-injection and extraction properties of these materials are important.

12.7.1 Electronic Structure of TMO Films

Some reports presented on TMOs such as MoO_3, V_2O_5, and WO_3 as conducting p-type materials [52, 53]. Ultra-violet photoemission spectroscopy (UPS) and inverse photoemission spectroscopy (IPES) are generally used techniques that provide appropriate measurement of the density of electronically filled and empty states of materials, here the valence band and conduction band states for the above-mentioned metal oxide is shown in Figure 12.4. Using a combination of UPS and PES, measurements of the VB and CB states of TMO films were established and the alignment of energy levels at interfaces between (organic) materials and TMO films was determined [54].

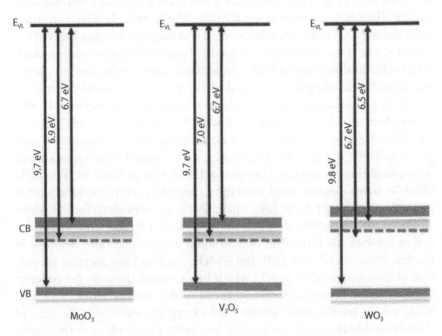

Figure 12.4 The arrangement of conduction band (CB) and valence band (VB) in MoO_3, V_2O_5 and WO_3.

The work function of each TMO film is directly estimated from the photoemission onset via standard method, the vacuum level position and by comparing this position with the Fermi level determined separately on a metallic electrode. All the observation evidently explain that MoO_3, V_2O_5 and WO_3 exhibit very similar deep lying electronic states, with a valence band edge around 2.5-3.0 eV below the Fermi level and a conduction band edge very close to the Fermi level, indicative of a highly n-type material. The n-type conductivity of these oxides is most likely a result of a slightly non-stoichiometric composition, with some oxygen deficiency, a situation which has been confirmed via X-ray photoemission spectroscopy (XPS) by several groups.

The similarity between the three materials is striking. V_2O_5 exhibits the largest WF (7.0 eV) closely followed by MoO_3 and WO_3 with values of 6.9 eV and 6.7 eV, respectively. Note that WF values in excess of 5.5–6 eV for these transition metal oxide films have been known for over a decade [55, 56]. The valence band edge or hole-transport level, of all transition metal oxide derives from the occupied 2p orbital of oxygen atom. However, the unusually deep valence band (ionization energy IE > 9 eV) makes hole-injection and transport highly unlikely in these materials. On the other hand, the strong n-type nature of these transition metal oxide films makes electron transport via the conduction band far more favorable. The most commonly used transition metal oxide in organic electronics applications is MoO_3 because it get evaporated at relatively low temperature $\sim 400°C$ and can therefore be easily deposited in vacuum from a crucible. In contrast, the evaporation temperatures of V_2O_5 and WO_3 are expressively greater and can lead to a more noteworthy release of oxygen during the evaporation and may furthermore lead to a considerable heating of the substrates that to be coated. In general, reduced transition metal oxide shows an upsurge in conductivity, stronger optical absorption [57, 58]. For instance, Greiner et al. witnessed a metallic valence band structure when MoO_3 strongly reduced from Mo^{5+} to Mo^{4+} states. But the metal-like electrical properties of reduced MoO_3 losses absorption property, which obstruct their applications where high transparency is obligatory [59].

The order of magnitude of thermal conductivity of transition metal oxide based thin films is larger than that of organic semiconductors. The OLEDs are expected to be feature material with specific parameters featuring a 100 nm thick NiO based with resistance amounted to 8 k Ω^2 and transparency of 50% [60]. The conductivity of other transition metal-based OLEDs; MoO_3 and WO_3 OLEDs reported as lesser value of conductivity as compared to NiO. For a 100 nm thick MoO_3 layer the conductivity value obtained as $\sim 1 \times 10^{-7}$ S/cm at room temperature.

12.8 Lanthanide-Based OLEDs

Organic light-emitting diode (OLED) technology fascinates growing interest probably because high surface area light sources and displays could be achieved from these materials [61, 62]. The OLEDs could be either solution-processed or vacuum-deposited. The former are cost effective and permits high resolution and therefore are commercially appealing. There are several examples of these solution-processed OLEDs, however, the efficiency is typically inferior to vacuum-deposited OLEDs. Therefore, these days, the growth of solution-processed OLEDs is a significant task [63].

The considered materials for the design of effective OLED should have radiative relaxation of their triplet excited states as 75% of electro-excitons possess triplet nature [64a,b–66]. The iridium-based complexes and promising thermally activated delayed fluorescence (TADF) materials are effective emitters having triplet radiative relaxations. In spite of high efficiencies of these material-based OLEDs they have two downsides: large cost and large luminescence band-width (over 100 nm) that thereby restricts contrast qualities of diodes. On the other hand, lanthanide coordination compounds are a class of emitters involving triplet levels in the luminescence processes and have narrow emission bands [67]. The most inherent property of lanthanide complexes is that by changing lanthanide ion the different colors of luminescence could easily be obtained. Even the Near-IR OLEDs could be designed from praseodymium, neodymium, holmium, erbium, thulium and ytterbium ions. The lanthanide complexes containing aromatic carboxylates and β-diketones due to their desirable properties are potential materials for OLED applications. The significant challenge is achieving OLEDs functioning in NIR for applications like medical diagnostics, photo-dynamic therapy, plastic-based telecommunications, night vision, optical sensors and information processing.

In general, now-a-days, scientists have intense initiative towards the strategy of lanthanide-based complexes used in organic light emitting devices (OLEDs) [61, 62]. The lanthanide complexes possess distinctive sharp luminescence having bandwidth less than 10 nm with high color purity and long lifetimes that arise due to shielding nature of 4f orbitals by outer 5s and 5p orbitals. Importantly, such a narrow emission is almost unattainable for common organic luminophores. Besides, the emission color could be easily tweaked over the visible region simply by switching lanthanide ions, for example, europium shows red emission, terbium emits green, samarium complexes show characteristic pink/orange luminescence and dysprosium complexes appear yellow/white. However, due to low absorptivity coefficient of 4f - 4f transitions, the direct excitation of lanthanide ions give rise to frail emission. The luminescence from lanthanide complexes is usually achieved by indirect sensitization by making use of organic chromophore with very large molar absorption coefficient. Any organic moiety that has the strong ability to absorb either UV or vis light can potentially sensitize lanthanide (III) ions. This phenomenon of indirect sensitization of lanthanide ions is called Antenna Effect. The more efficient the energy transfer is, the more luminescent will be the lanthanide complex. As stated above, the carboxylates and β-diketones are well known chromophores commonly used to synthesize lanthanide complexes that efficiently sensitize Ln (III) ions [68, 69]. The advantages of using β-diketones as sensitizers or antenna chromophores in lanthanide complexes are; (a) these possess strong absorption property in a wider wavelength range (b) are easily accessible or easy to synthesize and (c) being hard base the oxygen of carbonyl group has strong coordinating ability resulting in stable tris (β-diketonate) lanthanide complexes that possess high thermodynamic stability, volatility and solubility in common organic solvents. The β-diketones based lanthanide complexes are mostly explored as they are highly luminescent which are vital requirements for the fabrication of OLEDs [70]. Based on the nature of lanthanide ion, the different colors like blue (praseodymium and thulium), green (terbium), orange (samarium) and red (europium) OLEDs could be designed. Some lanthanide ions particularly neodymium and ytterbium are known for design of near-infrared OLEDs.

The luminescent lanthanide tris β-diketonates are incipient as budding materials in the field of OLEDs as (a) the atom-like emission is attained from these materials because of 4f - 4f transitions happening from the Ln^{3+} emitting state and (b) the intramolecular energy transfer excitation of lanthanide ion to its singlet state as well as relaxation from the singlet to the triplet state through intersystem crossing outcomes use of both singlet and triplet exciton energy for emission produced by electron-hole recombination. While the intersystem crossing efficiency less than 100%, the electroluminescence efficiency of lanthanide materials might supposedly reach 100% or evidently outstrip generally employed organic fluorescent materials. The electroluminescent devices based on europium, terbium and dysprosium complexes are of excellent importance as they respectively provide sharp atom-like red, green, and white/yellow emission. For improvement of OLED performance, scientists have synthesized lanthanide complexes by integrating some modifications and doping [71]. While various devices are reported in literature, the quest for improved electroluminescent materials comprising of lanthanides for fabricating new devices is still required.

12.9 Conclusion

This chapter covers the importance of LEDs and gives an idea about basic construction and working principle of LEDs. The performance of LEDs are valued by means of various parameters as CRI, CIE coordinates and FRET. Furthermore, it also documented the LEDs based on zinc oxide and GaN, the distinctive properties of these LEDs are well compared. The selective transition metal oxide are listed herewith, and their electronic structure is well described for their implementation, furthermore lanthanide based OLEDs are also found to be an excellent component for designing LEDs. The entire content of this chapter is expected to be useful in understanding the basics of LEDs.

References

1. Holonyak, N., Jr., Is the light emitting diode (LED) an ultimate lamp. *Am. J. Phys.*, 68, 864, 2000.
2. Pankove, J.I., Optical absoption of GaN. *Appl. Phys. Lett.*, 17, 197, 1970.
3. Pankove, J.I., Berkeyheiser, J.E., Miller, E.A., Properties of Zn-doped GaN. I. Photoluminescence. *J. Appl. Phys.*, 45, 1280, 1974.
4. Zhao, Y., Tanaka, S., Pan, C.C., Fujito, K., Feezell, D., Speck, J.S., Den Baars, S.P., Nakamura, S., High-power blue-violet semipolar (2021) InGaN/GaN light-emitting diodes with low efficiency droop at 200 A/cm^2. *Appl. Phys. Express*, 4, 2–5, 2011.
5. Masui, H., Yamada, H., Iso, K., Nakamura, S., Den Baars, S.P., Optical polarization characteristicsof m-oriented InGaN/GaN light-emitting diodes with various indium compositions in single-quantum-well structure. *J. Phys. D Appl. Phys.*, 41, 225, 2008.
6. Nakamura, S., InGaN/AlGaN blue-light-emitting diodes. *J. Vac. Sci. Technol. A*, 13, 705, 1995.
7. Iida, K., Kawashima, T., Miyazaki, A., Kasugai, H., Mishima, S., Honshio, A., Miyake, Y., Iwaya, M., Kamiyama, S., Amano, H., Akasaki, I., 350.9 nm UV laser diode grown on low-dislocation-density AlGaN. *Jpn. J. Appl. Phys.*, 43, 2, 499, 2004.
8. Setlur, A.A., Phosphors for LED-based solid-state lighting. *Electrochem. Soc. Interface*, 18, 32, 2009.
9. Keyes, R.J. and Quist, T.M., Recombination radiation emitted by gallium arsenide. *Proc. IRE*, 50, 1822, 1962.
10. Nakamura, S., Senoh, M., Mukai, T., P-GaN/N-InGaN/N-GaN double-hetero structure blue-light emitting diodes. *Jpn. J. Appl. Phys.*, 32, 8, 1993.
11. Round, H., A note on carborundum. *Electron. World*, 49, 309, 1907.

12. Maruska, H.P. and Tietjen, J.J., The preparation and properties of vapor-deposited single-crystal-line GaN. *Appl. Phys. Lett.*, 15, 327, 1969.
13. Liu, W.J., Hu, X.L., Liu, Y.J., Performance enhancement of GaN-based near-ultraviolet flip-chip light-emitting diodes with two-step insulating layer scheme on patterned sapphire substrate. *J. Mater. Sci. Mater. Electron.*, 30, 3013, 2019.
14. Cho, C.Y., Park, K.H., Park, S.J., Enhanced optical output power of blue light-emitting diode grown on sapphire substrate with patterned distributed Bragg reflector. *ECS J. Solid State Sci. Technol.*, 7, 66, 2018.
15. Xu, Y., Zou, J., Lin, X., Wu, W., Li, W., Yang, B., Shi, M., Quality-improved GaN epitaxial layers grown on striped patterned sapphire substrates ablated by femtosecond laser. *Appl. Sci.*, 8, 1842, 2018.
16. Pankove, J.I., Properties of Zn-doped GaN. II. Photoconductivity. *J. Appl. Phys.*, 45, 3892, 1974.
17. Amano, H., Kito, M., Hiramatsu, K., Akasaki, I., P-type conduction in Mg-doped GaN treated with low-energy electron beam irradiation (LEEBI). *Jpn. J. Appl. Phys.*, 28, 2112, 1989.
18. Nakamura, S., Mukai, T., Senoh, M., Iwasa, N., Thermal annealing effects on P-type Mg-dopedGaN films. *Jpn. J. Appl. Phys.*, 31, 139, 1992.
19. Chang, L.B., Pan, K.W., Yen, C.Y., Jeng, M.J., Wu, C.T., Hu, S.C., Kuo, Y.K., Comparison of silicone and spin-on glass packaging materials for light-emitting diode encapsulation. *Thin Solid Films*, 570, 496, 2014.
20. Yaguchi, H., David, A., Fuchida, T., Hashimoto, K., Heidel, G., Jordan, W., Jost-Boissard, S., Kobayashi, S., Kotani, T., Luo, R., Mizokami, Y., Ohno, Y., Pardo, P., Richter, K., Smet, K., Teunissen, K.N., Tsukitani, A., Wei, M., Whitehead, L., Yano, T., CIE 2017 Color fidelityindex for accurate scientific use, CIE Central Bureau, Jeju Island, 2017.
21. Wright, W.D., A re-determination of the trichromatic coefficients of the spectral colors. *Trans. Opt. Soc.*, 30, 141, 1929.
22. Guild, J., The colorimetric properties of the spectrum. *Philos. Trans. R. Soc. A Math. Phys. Eng. Sci.*, 230, 149, 1932.
23. Zargar, R.A., Arora, M., Bhat, R.A., Study of nanosized copper-doped ZnO dilute magnetic semiconductor thick films for spintronic device applications. *Appl. Phys. A*, 124, 1–9, 2018.
24. Rahman, F., Solid-state lighting with wide band gap semiconductors. *MRS Energy Sustain.*, 1, 1, 6, 2014.
25. Zargar, R.A., Arora, M., Alshahrani, T., Shkir, M., Screen printed novel ZnO/MWCNTs nanocomposite thick films. *Ceram. Int.*, 47, 6084–6093, 2020.
26. Zargar, R.A., ZnCdO thick film: A material for energy conversion devices. *Mater. Res. Express*, 6, 9, 095909, 2019.
27. Kobayashi, A., Room temperature layer by layer growth of GaN on atomically flat ZnO. *Jpn. J. Appl. Phys.*, 43, 53, 2004.

28. Yoo, D.G., Fabrication of the ZnO thin films using wet-chemical etching processes on application for organic light emitting diode (OLED) devices. *Surf. Coat. Technol.*, 202, 22, 5476, 2008.
29. Zargar, R.A., Fabrication and improved response of ZnO-CdO composite flms under diferent laser irradiation dose. *Sci. Rep.*, 12, 10096, 2022.
30. Stocker, D.A., Schubert, E.F., Redwing, J.M., Crystallographic wet chemical etching of GaN. *Appl. Phys. Lett.*, 73, 18, 2654, 1998.
31. Palacios, T., Wet etching of GaN grown by molecular beam epitaxy on Si(III). *Semicond. Sci. Technol.*, 15, 10, 996, 2000.
32. Kim, H.K., Inductively coupled plasma reactive ion etching of ZnO using BCl_3-based plasmas. *J. Vac. Sci. Technol. B*, 21, 4, 1273, 2003.
33. Bae, J.W. et al., High-rate dry etching of ZnO in $BCl_3/CH_4/H_2$ plasmas. *Jpn. J. Appl. Phys.*, 42, 5B, 535, 2003.
34. Coskun, C. et al., Radiation hardness of ZnO at low temperatures. *Semicond. Sci. Technol.*, 19, 6, 752, 2004.
35. Ip, K. et al., Contacts to ZnO. *J. Cryst. Growth*, 287, 1, 149, 2006.
36. Zargar, R.A., Kumar, K. et al., Structural, optical, photoluminescence, and EPR behaviour of novel Zn0·80Cd0·20O thick films: An effect of different sintering temperatures. *J. Lumin.*, 245, 118769, 2022.
37. Park, J.H. and Steckl, A.J., Demonstration of a visible laser on silicon using Eu-doped GaN thin films. *J. Appl. Phys.*, 98, 5, 056108, 2005.
38. Zargar, R.A., Kumar, K., Mahmoud, Z.M.M., Shkir, M., AlFaify, S., Optical characteristics of ZnO films under different thickness: A MATLAB based computer calculation for photovoltaic applications. *Physica B*, 63, 414634, 2022.
39. Flores, M.A., Defect properties of Sn- and Ge-doped ZnTe: Suitability for intermediate-band solar cells. *Semicond. Sci. Technol.*, 33, 015004, 2017.
40. Van de Walle, C.G. et al., First-principles calculations of solubilities and doping limits: Li, Na, and N in ZnSe. *Phys. Rev. B*, 47, 15, 9425, 1993.
41. Ma, Y. et al., Control of conductivity type in undoped ZnO thin films grown by metalorganic vapor phase epitaxy. *J. Appl. Phys.*, 95, 11, 6268, 2004.
42. Özgür, Ü., AlivovYa, I., Liu, C., Teke, A., Reshchikov, M.A., Doğan, S., Avrutin, V., Cho, S.J., Morkoç, H., A comprehensive review of ZnO materials and devices. *J. Appl. Phys.*, 98, 4, 041301, 2005.
43. Reynolds, J.G., Reynolds Jr., C.L., Mohanta, A., Muth, J.F., Rowe, J.E., Everitt, H.O., Aspnes, D.E., Shallow acceptor complexes in p-type ZnO. *Appl. Phys. Lett.*, 102, 15, 152114, 2013.
44. Zargar, R.A., Chackrabarti, Malik, M.H., Sol-gel syringe spray coating: A novel approach for rietveld, optical and electrical analysis of CdO film for optoelectronic applications. *Phys. Open*, 7, 10069, 2021.
45. Reineke, S., Lindner, F., Schwartz, G., Seidler, N., Walzer, K., Leussem, B., Leo, K., White organic light-emitting diodes with fluorescent tube efficiency. *Nature*, 459, 234, 2009.

46. Moritomo, Y., Asamitsu, A., Kuwahara, H., Tokura, Y., Giant magnetoresistance of manganese oxides with a layered perovskite structure. *Nature*, 380, 141, 1996.
47. Barsan, N., Koziej, D., Weimar, U., Metal oxide-based gas sensor research: How to? *Sens. Actuators B*, 12, 18, 2007.
48. Zargar, R.A. et al., ZnCdO thick film films: Screen printed TiO2 film: A candidate for photovoltaic applications. *Mater. Res. Express*, 7, 065904, 2020.
49. Chen, L. and Schoonman, J., MnO thin film cathode for rechargeable microbatteries. *Solid State Ionics*, 60, 227, 1993.
50. Torrance, J.B., Lacorre, P., Asavaroengchai, C., Metzger, R.M., Why are some oxides metallic, while most are insulating? *Physica C*, 182, 351, 1991.
51. Shi, Y., Guo, B., Corr, S.A., Shi, Q., Hu, Y.S., Heier, K.R., Chen, L., Seshadri, R., Stucky, G.D., Ordered mesoporous metallic MoO_2 materials with highly reversible lithium storage capacity. *Nano Lett.*, 9, 4215, 2009.
52. Kröger, M., Hamwi, S., Meyer, J., Riedl, T., Kowalsky, W., Kahn, A., P-type doping of organic wide band gap materials by transition metal oxides: A case-study on Molybdenum trioxide. *Org. Electron.*, 10, 932, 2009.
53. Meyer, J., Zilberberg, K., Riedl, T., Kahn, A., Electronic structure of m vanadium pentoxide: An efficient hole injector for organic electronic materials. *J. Appl. Phys.*, 110, 033710, 2011.
54. Salaneck, W.R., Classical ultraviolet photoelectron spectroscopy of polymers. *J. Electron. Spectrosc.*, 174, 3, 2009.
55. Huang, J.S., Chou, C.Y., Liu, M.Y., Tsai, K.H., Lin, W.H., Lin, C.F., Solution-processed vanadium oxide as an anode interlayer for inverted polymer solar cells hybridized with ZnO nanorods. *Org. Electron.*, 10, 1060, 2009.
56. Gershon, T., Metal oxide applications in organic-based photovoltaics. *Mater. Sci. Technol. Ser.*, 27, 1357, 2011.
57. Granqvist, C.G., Progress in electrochromics: Tungsten oxide revisited. *Electrochim. Acta*, 44, 3005, 1999.
58. Granqvist, C.G., Electrochromic oxides: A unified view. *Solid State Ionics*, 678, 70, 1994.
59. Greiner, M.T., Helander, M.G., Wang, Z.B., Tang, W.M., Qiu, J., Lu, Z.H., A metallic molybdenum suboxide buffer layer for organic electronic devices. *Appl. Phys. Lett.*, 96, 213302, 2010.
60. Park, S.W., Choi, J.M., Kim, E., Im, S., Inverted top-emitting organic light-emitting diodes using transparent conductive NiO electrode. *Appl. Surf. Sci.*, 244, 439, 2005.
61. Koden, M., *OLED displays and lighting*, p. 232, John Wiley & Sons, Ltd., Chichester, UK, 2016.
62. Tsujimura, T., *OLED display fundamentals and applications*, p. 1801536, John Wiley & Sons, Inc., John New Jersey, 2017.
63. Kozlov, M.I., Aslandukov, A.N., Vashchenko, A.A., Medvedko, A.V., Aleksandrov, A.E., Grzibovskis, R., Goloveshkin, A.S., Lepnev, L.S., Tameev, A.R., Vembris, A., Utochnikova, V.V., On the development of a new approach

to the design of lanthanide-based materials for solution-processed OLEDs. *Dalton Trans.*, 48, 17298, 2019.
64a. Ibrahim-Ouali, M. and Dumur, F., Recent advances on metal-based near-infrared and infrared emitting OLEDs. *Molecules*, St. Alban-Anlage 66, CH-4052 Basel, Switzerland 24, 1412, 2019.
64b. Utochnikova, V.V., Lanthanide complexes as OLED emitters, in: *Handbook on the Physics and Chemistry of Rare Earths*, J.-C.G. Bunzli and V. Pecharsky (Eds.), Elsevier B.V., 2021.
65. Ikeda, N., Oda, S., Matsumoto, R., Yoshioka, M., Fukushima, D., Yoshiura, K., Yasuda, N., Hatakeyama, T., Solution-processable pure green thermally activated delayed fluorescence emitter based on the multiple resonance effect. *Adv. Mater.*, 32, 2004072, 2020.
66. Xu, H., Sun, Q., An, Z., Wei, Y., Liu, X., Electroluminescence from europium(III) complexes. *Coord. Chem. Rev.*, 293, 228, 2015.
67. Platt, A.W.G., Lanthanide phosphine oxide complexes. *Coord. Chem. Rev.*, 340, 62, 2017.
68. Yang, L., Gong, Z., Nie, D., Lou, B., Bian, Z., Guan, M., Huang, C., Lee, H.J., Baik, W.P., Promoting near-infrared emission of neodymium complexes by tuning the singlet and triplet energy levels of β-diketonates. *New J. Chem.*, 30, 791, 2006.
69. Uekawa, M., Miyamoto, Y., Ikeda, H., Kaifu, K., Nakaya, T., Synthesis and luminescent properties of europium complexes. *Synth. Met.*, 91, 259, 1997.
70. Quirino, W.G., Legnani, C., Cremona., M., Lima, P.P., Junior, S.A., Malta, O.L., White OLED using β-diketones rare earth binuclear complex as emitting layer. *Thin Solid Films*, 494, 23, 2006.
71. Li, B., Liu, L., Fu., G., Zhang, Z., Li, H., Lü, X., Wong, W.K., Jones, R.A., Color-tunable to direct white-light and application for white polymer light-emitting diode (WPLED) of organo-Eu^{3+}- and organo-Tb^{3+}-doping polymer. *J. Lumin.*, 192, 1089, 2017.

13

Metal Oxide Nanocomposite Thin Films: Optical and Electrical Characterization

Santosh Chackrabarti[1]*, Rayees Ahmad Zargar[2], Tuiba Mearaj[1] and Aurangzeb Khurram Hafiz[1]

[1]Central for Nanoscience and Nanotechnology, JMI, New Delhi, India
[2]Department of Physics, BGSB University, Rajouri, J&K, India

Abstract

Synthesis of metal oxide nanocomposite thin films is the current area of research being explored. These types of materials are being grown in labs to harness their novel properties (mechanical, electrical, optical, thermal, etc.) in applications like photocatalytic generation, solar energy converters, etc. The synthesis is usually done by embedding desired nanoparticles in metal oxide matrix thin films. This is done using laser technology i.e., by choosing a laser irradiation source of desired wavelength. The metal oxide nanocomposite thin films hence formed have enhanced mechanical, optical, electrical, thermal properties. It has been observed that these properties can also be tailored by the choice of the laser irradiation parameters. In current optoelectronic technology, it is highly desirable to have size miniaturization of electronic devices with improved semiconducting properties. Therefore, insight knowledge of metal oxide thin film is paramount to engineer their band gap for a wide range of optical, electronic, catalytic, chemical, and thermal properties. For device miniaturization, focus is on the synthesis of metal oxide thin film. This chapter starts with a brief introduction of the nanocomposite thin films, their formation/growth and applications. A brief discussion is made on the various available methods/processes for the preparation of nanocomposite thin films to give a basic understanding to the readers. There are known examples of carbon nanotube- metal oxide nanocomposites, silicon mixed oxide nanocomposites, graphene based metal oxide nanocomposites, and so on that can be used for the purpose. The parameters that are altered using different laser irradiation sources to achieve metal oxide nanocomposite thin films for specific applications are also discussed. Finally, along with synthesis processes and applications,

---

*Corresponding author: sdev.phy@gmail.com

Rayees Ahmad Zargar (ed.) Metal Oxide Nanocomposite Thin Films for Optoelectronic Device Applications, (317–360) © 2023 Scrivener Publishing LLC

characterization techniques are also shortly discussed. Here, we discuss the optical and electrical characterization techniques. The kind of materials that is used to obtain metal oxide nanocomposite thin films of desired optical and electrical parameters is discussed. Electrical properties like conductivity or resistivity can change upon formation of above mentioned thin films. Likewise, optical properties such as transmittance, reflectance, refractive index, etc. also change upon formation of new nanomaterials. The change in electrical and optical properties of novel nanomaterials finds application in vast variety of fields like energy storage, photonic devices, etc. This is discussed in detail.

Keywords: Thin films, metal oxides, nanocomposites, optical, electrical

13.1 Introduction

Sensing materials is quite involved extensive issue. Truly, all sensing materials are effective in preparing the chemical sensors both natural sensors, as that tracked down in living creatures, and that they commonly respond with electrochemical signs, wherein measurements are transferred and felt in state of electricity.

The branch of science which frames fabrication followed by characterization and application of materials having one dimension less than 100nm is defined as nanotechnology. Thus, a molecule size inside 1–100 nm territory shows radically variations in synthetic properties and actual properties as compared to bulk materials that promise numerous potential and difficulties as well. Nanoscience and Nanotechnology is a multidisciplinary field which coordinates physics, material science, biology and chemistry etc. The application of this field for biomedical purposes, detections, energy storage, water therapy and different applications have altered the scene of many issues of climate, energy, well-being and so on [1–6] (see Figure 13.1).

Nanocomposites have intrigued the consideration of researchers overall by their prospects and multifunctional properties. Nanocomposites possess the base one of the aspects is under 100nm. Researchers all around the world are chipping away at developing new methodologies to synthesize nanocomposites that exhibit unique properties. Research is centered on progress in viability, proficiency, strength of reasonable utilization by synthesizing appropriate constituents for the amalgamation of specific nanocomposites [7–9].

Nanocomposite materials are preferred to surpass obstructions of monolithics by means of controlling of added substances of the composite variables in stoichiometry way. Nanocomposite films are thin layers designed with the aim of utilizing admixing irrelevant substances having

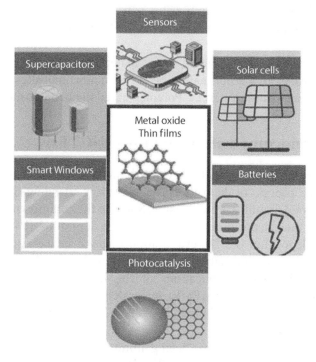

Figure 13.1 Metal oxide thin films and their uses.

nano-layered segment(s) to control and increase new and higher systems. Further, extraordinary nanocomposites might be carried out as exuberant substances for gas sensors, containing consolidated metallic oxides, polymers joined with metals or metallic oxides, or carbon nanotubes joined with metallic oxides or polymers etc. Nanocomposites are placed for their capability to make execution of medications, biomaterials, impetuses and different exorbitant cost presented substances. From the mechanical point of view, nanocomposites consolidate one or more prominent discontinuous levels designated in a single continuous phase, continuous phase is designated as "matrix" while intermittent phase is classified as "reinforcement".

13.1.1 Classification of the Nanocomposites

(A) On the basis of presence of polymer:
It may be divided as polymer based – or non-polymer based-nanocomposites [5–8]. Polymer based nanocomposites may be sub-divided as: I. Polymer/Ceramic nanocomposite; II. Polymer/Layered Silicate Nanocomposites; III. Polymer/Polymer Nanocomposites; IV. Inorganic/Organic Polymer Nanocomposites; V. Biocomposites; VI. Organic/Inorganic Hybrid Nanocomposites.

Non-polymer type nanocomposites may be sub-divided as: I. Metal/Metal Nanocomposite; II. Ceramic/Ceramic Nanocomposites; III. Metal/Ceramic.

(B) In terms of reinforcement fabric material – I. Metal oxide–based nanocomposites; II. Carbon nanotube–based nanocomposites; III. Noble metal based nanocomposites; IV. Polymer-based nanocomposites.

13.2 Nanocomposite Thin Films (NCTFs)

Thin films are the layers of material created through deposition techniques. Actually, the process of deposition is achieved by changing the states of matter i.e. solid, liquid, gas and plasma. Some of the common processes to accomplish their deposition are physical vapor deposition (PVD), chemical vapor deposition (CVD), evaporation through vacuum sublimation, and sometimes even through the combination of these methods [10–13].

Some of the major applications of the thin films include microelectromechanical systems (MEMS), integrated circuit chips, LEDs, microelectronic optical systems, photovoltaic cells, batteries, optical coatings etc. [14, 15].

A class of thin films that has been gaining prominence is that of nanocomposite thin films (NCTFs) are the class of thin films getting higher attention due to the combination of the properties that allow achieving various features for use in optoelectronic applications. Due to their excellent properties like excellent resistance to corrosion, oxidation at high temperatures, high hardness and appropriate tri-biological properties, they are used in chemical, electronics, automotive industries [16, 17].

13.3 Materials Used for Preparation of NCTFs

Several unique characters of metals in quick cited on this segment locate wonderful applications in the field of sensors improvement beginning from electric connectors as electrodes to unique features depending on catalytic interest or work function. Because of those special properties, some particular metals (see Figure 13.2) have extra importance in sensor production and fabrication than others. For example:

13.3.1 Gold

Gold (Au) is chemically inert and very soft element that can withstand oxidation and has excessive biocompatibility. Consequently, it's far taken

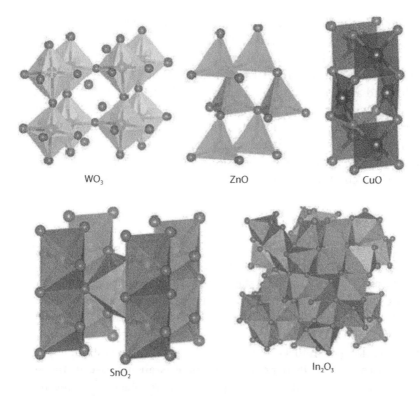

Figure 13.2 Crystal structure of some metal oxide nanocomposites.

into consideration to be one of the exceptionally promising metals within the sensor industry [18]. The development for AuNPs utility is short as a blaze for disclosure of ecological contamination, and biomolecules at the same time [19–22]. In light of the brilliant hydrophily, great scattering and moderate conglomeration of Hg^{+2}, cyclic voltammetry by means of diffusion strategy was applied for its determination in ecological water tests and a couple of natural product using glass carbon terminal changed through chitosan-AuNPs taking in consideration the significant electro-synthetic effects of AuNPs [23]. The AuNPs embedded at the rGO/PANI nanocomposites outfitted a valuable texture for increasing the selectivity of the connection point to serotonin, but furthermore improving the conductivity of the imprinted cathode. Besides, such imprinted cathode was successfully transformed for the discovery of serotonin in human serum and will effectively keep away from the interferences achieved from endogenous compounds [24]. The improvement of new sensor with changes in CPEs with Au NPs transformed into voltammetric analyzation of brexpiprazole with the increase in the sensitivity of the CPEs toward brexpiprazole [25]. The adapted electrode has exceptional extreme reusability, might be recovered for washing in EDTA solution [26]. Likewise, the Au/SnO$_2$/rGO feature an enormous linearity from 1 to 1,000 ppm with phenomenal reproducibility and a low operating temperature [27]. The advances in synthesis of Au-graphene nanocomposites offer the improvement of big surface region, non-enzymatic electrode for determination of glucose with high reliability [28, 29]. Au rGO-PDDA nanocomposites had been executed to make uncommonly tricky electrodes with an immense linearity assortment for nitrite ions commitment in meat and dairy stock examples [30].

13.3.2 Platinum

Platinum is a gray-white, dense, highly unreactive, soft transition metal which is very resistant to corrosion. It may be found freely in nature or may be available in combined states with other elements. Platinum (Pt) has been used as a catalyst while converting methanol to formaldehyde or as a catalytic converter that combines CO and unburned fuel with oxygen to form CO_2 and water vapor. It is also used in many petroleum industries for hydrocracking or in fuel cell devices for electricity generation.

With the use of Pt nanoparticles, there is an increase in the surface area for a reaction and also increase in the percentage of platinum atoms responsible for contact with molecules involved in the reaction. So small quantity of Pt may be used for such purposes.

In order to boom the electron switch efficacy among redox protein/enzyme and electrode, Pt-three-D graphene oxide aerogel (3DGA) nanocomposite became synthesized and used as a progressive biocompatible service for protein/enzyme [31, 32]. For medical purposes, high speed estimation of glucose via fabrication of non-enzymatic sensor with true stability, excessive selectivity demands for synthesis of Pt–NiO nanoplate array decreased graphene oxide nanocomposites which confirmed excessive electrocatalytic hobby for glucose oxidation with a vast linearity [33]. Also, due to smooth fabrication technique of Pt–TiO$_2$-decreased graphene oxide nanocomposites changed electrodes; they show off common software for amperometric dedication of H_2O_2 in milk [34]. Hydrogen fuels made out of degradation of risky natural compounds in air or water may be effortlessly detected with excessive functionality at room temperature via the usage of a hybrid natural PVDF–inorganic M–rGO–TiO$_2$ (M = Ag, Pt) nanocomposites supplying cheaper technology due to their inherent TiO$_2$/Ag self-cleansing properties [35]. Pt nanoparticles have been sensitized with the aid of using mesoporous WO$_3$ semiconductor for the detection of CO in sensible applications due to their catalytic activity [36]. Facile one-pot hydrothermal technique was used in the production of Pt–graphene nanocomposites in which the EG lessened each graphene oxide and H_2PtCl_6 to supply small sized Pt nanoparticles having high electron transfer rate for effective electrocatalytic nitrite sensing [37].

13.3.3 Silver

Silver nanoparticles and the composites based on these particles are gaining too much attention because of their excellent properties like high chemical stability, high thermal conductivity, good catalytic activities, surface-enhanced Raman scattering and nonlinear optical behavior. Here we discuss some literatures explaining the synthesis and development of AgNPs.

Herein, synthesis and alertness of silver nanoparticles (Ag-NPs) decorated nitrogen doped single-walled carbon nanotube electrode changed into confirmed for non-enzymatic electrochemical detection of urea. The proposed electrode suggests excessive catalytic hobby with massive floor location and electric properties [38–40]. An electrochemical sensor primarily based totally on copper porphyrin steel natural frameworks (Cu-MOFs) and Au and Ag core–shell nanoparticles decorated multi-walled carbon nanotubes (MWCNT) changed into assembled acetaminophen and dopamine.

Also, a first-rate cyclic voltammetric and amperometric reaction to H_2O_2 with fast reaction, large linear variety was acquired with the aid of using electrochemical sensor of as prepared Ag/p-g-C_3N_4 composite [41–43].

13.3.4 Palladium

Palladium sensors display a linear relationship activity as a characteristic of H_2 awareness specifically at low concentrations, with the gain that sensor resistance completely recovers upon exposure to air. This sensor makes use of Pd Functionalized Multi-layer graphene nanoribbon networks [44, 45]. An amperometric sensor primarily based totally on palladium nanoparticles (PdNPs) held through graphene oxide nanosheets modified glassy carbon electrode (GCE) was invented for bromate detection [46]. Polystyrene nanosphere lithography as an easy manufacturing method allows the improvement of PdNPs-embellished Si nanomesh shape as a H_2 sensor [47–49]. Depending on the potential of graphene to isolate and maintain PdNPs right into a planar substrate further to provide area for PdNP enlargement at some stage in the absorption and desorption of H_2

molecules, palladium graphene nanocomposites as resistivity-type hydrogen sensing displaying long-time stability high resistance to oxidation [50].

13.3.5 Boron

A bare boron-doped diamond electrode was carried out to estimate ibuprofen in prescription drugs and human urine samples electrochemically with affordable selectivity of feasible endogenous compounds along with reasonable selectivity in prescription drugs and selected agents commonly found in urine (glucose, ascorbic acid, uric acid, dopamine and caffeine) [51]. A current molecular imprinted sensor primarily based totally on dimensional (2D) hexagonal boron nitride (2D-hBN) nanosheets included with core–shell kind nanoparticles (Fe and AuNPs) changed into cypermethrin detection in wastewater samples [52, 53].

13.3.6 Nickel

Nickel oxide is a superb substrate for the producing Pt–NiO/rGO/GCE for non-enzymatic glucose sensing [33, 54]. Furthermore, nickel was incorporated with nitrogen-doped graphene nanoribbons as green electrocatalysts for oxygen evolution reaction [55–57].

13.3.7 Titanium

The corrosion resistance performance and biocompatibility of polypyrrole/chitosan composite coating on Ti has been studied for biomedical applications [58, 59]. The synthesized binary nanocomposite (BNC) prepared from binary Titanium–Tin Oxide Nanocomposite (BNC) through the usage of SnCl2.5H2O and C12H28O4Ti precursors changed into lead ions (Pb2+) elimination from aqueous solution. Moreover, it additionally determined that the binary nanocomposite bears suitable antimicrobial activities against selected stains [60]. Also, TiO2 alter the sensing interest of semiconductor PANi while included in TiO_2-PANI Nanocomposites in which the conductivity of nanocomposite substances changed [61, 62].

13.3.8 ZnO

The overall performance of photo-electrochemical and gas sensing properties of TiO_2 nanotube was improved with thin ZnO coating and increased anodic reaction was discovered with the increase of ZnO coating thickness [63–65]. In addition, integration among Ru-Al_2O_3-ZnO in micro-chip turned into correctly used for SO_2 sensation nano-materials [66] and aggregate of ZnO with SnO_2 nanostructures gives ultrasensitive device with very speedy reaction and less restoration time at quite low running temperature for detection of NO_2 gas [67]. A breakthrough fixing the hygiene issues related to cotton clothing is completed with the aid of using incorporation of zinc oxide nanoparticles (ZnO NPs) as antimicrobial agent. The antibacterial performance in opposition to *S. aureus* is examined with the aid of using disc diffusion test. It determined that expanded ZnO NPs concentration, expanded inhibition area and therefore confirmed excellent antibacterial activity [68–70]. ZnO and Co3O4 grains shape a nanocomposite mixture with practical activity for detection of ethanol and formaldehyde [71]. Graphene covered with ZnO nanocomposites have been synthesized then used as a sorbent for microextraction in packed syringe of carbamate insecticides from juice samples previous to excessive-overall performance liquid chromatography detection [72–74].

13.3.9 Iron Oxide

The formation of PANI/a-Fe_2O_3 hybrid nanocomposite sensor by solid state synthesis method have been monitored for several oxidizing and reducing gases [75]. The a-Fe_2O_3/g-C_3N_4 nanocomposites prepared by hydrothermal and pyrolysis method are promising for the practice of the detection of ethanol [76, 77]. Also cellulose acetate (CA)/iron oxide nanocomposites (NC1 and NC2) were synthesized and utilized as extractors for quantitatively adsorption and the selective separation of fluorene from aqueous samples [78]. Also, Fe_3O_4-reduced graphene oxide (Fe3O4-RGO) binder free counter electrode (CE) using an easy and low cost electrophoretic deposition method and controlling the hydrogen evolution process followed by an electrochemical reduction process for dye-sensitized solar cell [79, 80].

13.3.10 Silicon (Si)

A trustful H_2 gas sensor changed into advanced with PdNPs decorated Si nanomesh shape the usage of too smooth manufacturing system through polystyrene nanosphere lithography [81]. Also, Silicon oxycarbide glasses and glass-ceramics had exquisite community architectures dictating their attractive functional properties like piezoresistive sensing behavior, reversible Li (and Na) uptake/release, drug release behavior, bioactivity, optical houses in addition to gas sensing or photocatalytic houses [82–84].

13.3.11 Cobalt

Co_3O_4 in aggregate with ZnO shape nanocomposites had top applicability for gas detection [68, 69] with appealing responses and selectivity. Sensors depending upon mesoporous Co_3O_4 nanochains confirmed excessive power detection for hydrogen sulfide gas characterized by means of quick reaction and healing time. Results can be associated to structural functions of mesoporous Co_3O_4 nanochains which include big contact surface area and tiny pore size [85].

13.3.12 Molybdenum

To defend the surroundings and meal sources from poisonous pesticide residues (Fenitrothion), the improvement of compatible sensors for the real-time detection of insecticides could be very important. Incidentally, the function of the new, progressive practical substances including niobium carbide (NbC) relies on molybdenum nanoparticles (NbC and Mo). It has drawn extremely good attention in traditional sensory systems [86, 87].

13.3.13 SnO_2 (Tin Oxide)

A polyaniline (PANI)/tin oxide (SnO_2) composite for a CO sensor was fabricated through the sol–gel method. The SnO_2 nanoparticles supplied a high surface area to enhance the alternation in CO awareness at low working temperature. The first rate sensor reaction changed into particularly attributed to the high properties of PANI in the redox reaction during sensing, which produced an outstanding resistance distinction among the air and CO gas [88, 90]. The reaction of the $ZnO-SnO_2$ nanocomposites to NO_2 changed into significantly stricken by the ZnO content. This sensitivity may be defined through chemical interactions related to the adsorption of gases and response with O_2 and additionally to adjustments in digital binding power inside the composite [74–91]. In addition, nanocomposite sensors comprised of a homogeneously dispensed of SnO_2 nanoparticles at the rGO sheets have been applied for NO_2 sensing [92, 93]. Also, the NO_2 gas sensing performance of the SnO_2-SnO hybrid is a lot better than that of natural SnO_2 nanoparticles [94–96].

13.3.14 CuO

A non-enzymatic H_2O2 sensor primarily based on Cu_2O–rGOpa composite was constructed. It was observed that, in Cu_2O–rGO, the inclusion of graphene will increase the catalytic activity of nanocuprous oxide particles for the reduction of H_2O_2 [97]. As a result of the inhibitory impact of ascorbic acid at the peroxidase—like action of copper nanoparticles—carbon nanocomposites, an easy colorimetric approach for ascorbic acid determination was developed. This proposed approach was efficaciously implemented to detect the level of ascorbic acid in drugs and get wider use in bioanalysis [98–100].

13.3.15 Tungsten

Polypyrrole n-tungsten oxide hybrid nanocomposites were studied to verify the detection reaction for NO_2 gas via impedance spectroscopy [101]. Beside this usage, the gas sensing residences of the WS_2/WO_3 sensor have been studied in various concentrations of ethanol gas. The sensor exhibited low detection limit, fantastic selectivity, solid repeatability, extremely good stability, and fast rapid-recovery features. The sensing mechanism of the sensor is attributed to the unique charge carrier transfer properties of the WS_2/WO_3 heterojunction and the synergistic consequences of the aggregate WS_2 and WO_3 [102].

13.3.16 CdS

The promising electrocatalytic overall performance of cadmium sulfide nanorods—decreased graphene oxide composites become examined as an enzyme-unfastened biosensor for H_2O_2 [103, 104]. As a green photocatalyst and powerful antibacterial agent, novel nanocomposites derived from ZnO-CdS QDs embedded cross-linked chitosan become organized. The nanocomposite demonstrates splendid efficiency in the direction of the photocatalytic degradation of cationic dyes [malachite green (MG), and safranin (SF)] and poisonous natural molecule 2,4-dichloro phenol (2,4-DCP) beneath the exposure of sunlight. Moreover, the composite shows excellent antibacterial activity towards *E. coli* and *B. subtilis* [105]. Meanwhile, the addition of CdS augments the degradation ability of tetracycline through widening the variety of absorption spectrum. It also indicates a very good degradation performance of Rhodamine B (RhB) [106].

13.3.17 Graphene Oxide

rGO-Pt-NiO nanocomposites were established for creating enzyme free glucose sensing [33]. Furthermore, an ultra-low detection restriction glucose sensor using decreased graphene oxide concave tetrahedral Pd NCs and CuO composite [107]. TiO_2 nanoparticles embellished decreased graphene oxide nanocomposite may be used for ultrasensitive electrochemical determination of rifampicin [108]. Multi-layer based graphene oxide coated zinc oxide nanoflower (ZnO NFs) as a modified electrode substances was used for high sensitive electrochemical determination of biomarker, 8-hydroxy-2 -deoxyguanosine (8-HDG) that's one of the essential most cancers and oxidative pressure biomarker [109, 110].

13.3.18 Carbon Nanotubes

Carbon nanotubes represent a substrate for several numbers of electrochemical sensors in aggregate with metals and metal oxides. Assembled CNT on graphene outcomes synergistic impact which allows the conductive paths to get higher electric conductivity [111]. In addition, ammonia gas-sensing properties at low temperature had been agreeably more advantageous by comprising carbon nanotubes with tungsten oxide nanobricks sensors as compared to CNT-primarily based and WO_3 based nanobrick-sensors separately [112, 113]. The evaluation of epirubicin anticancer drug in real samples was correctly performed via fabrication of high sensitive electrochemical sensor from Fe_3O_4-SWCNTs/MOCTICl/CPE [114]. Furthermore, the gadgets labored through combining electrochemical structures and nanosensors compromise a quite appropriate gear for immediate evaluation of medicinal drugs and dangerous additives both in environment, food or fitness application [115]. Atorvastatin isomers can be differentiated through creation of stereoselective sensor relying on multiwalled carbon nanotube cross-connected with Hydroxypropyl-b-cyclodextrin primarily based totally chiral nanocomposite [116]. Oil-in water emulsion solvent evaporation technique was deployed for the preparation of PCL-CNT nanocomposites. These nanocomposites discover a sound in drug transport applications [117–120].

13.4 Methods of Preparation of NCTFs

The nanocomposite thin films growth can be carried out through several techniques where they can be deposited as phases or a combination of these phases. Let's discuss some of the techniques for the preparation of nanocomposites thin films.

13.4.1 Cold Spray Approach

Cold spraying allows manufacturing of coating layers at a temperature lower than the sprayed materials melting points. The Cu [121, 122], Al [123, 124], Co [125] metal matrix or alloy matrix [126] nanocomposite coating is produced through this approach.

13.4.2 Sol–Gel Approach

The sol–gel approach is affordable to get extra high-quality films for the order of micron thickness. This approach indicates some disadvantages related to thickness limits and crackability. Sol–gel technique has been a famous technique for higher potential substitutes for the environmentally chromate metal surface pretreatment. Inorganic nanofillers, consisting of the inorganic/inorganic nanocomposite coatings may be used as a 2d segment [127–131].

13.4.3 Dip Coating

The process is used especially as depositing aqueous-based liquid phase coating solutions onto the substrate surface and acquired through dipping a substrate in nanocomposite solution then dragged up at a steady and managed speed. After eliminating the substrates from the solution, it'll be coated with a layer of nanocomposite [132].

13.4.4 Spray Coating and Spin Coating

These strategies are frequently used for formation of polymer nanocomposite coatings. In the case of spray coating approach, nanocomposite coatings with great properties may be acquired through the usage of an atomizer imparting uniform thin films to flat substrates. In the spin-coating procedure, the right thin film nanocomposite coatings have been produced relying on dissolving the substrate in an appropriate solvent then deposited into coating material through centrifugal pressure while turned around at excessive speed [133, 134].

13.4.5 Electroless Deposition

This approach become efficiently and regularly followed within the manufacturing of nanocomposite coating layers with Niken matrix and nanofillers as carbide, nitrite or boride [135–137] with a purpose to develop their qualities consisting of corrosion, abrasion resistance and hardness.

13.4.6 *In Situ* Polymerization Approach

This approach permits the fabrication of conducting polymer which have been nanocomposite coatings with different monomers with initiators or natural matrices [130, 138]. The nanofillers have been metals or metal oxides. The polymerization manner may be done via the usage of electricity, photon [139–141] oxidizing agents [142] or electrodeposition [143].

13.4.7 Chemical Vapor Deposition Approach

Generally, this approach used for production of inorganic/inorganic nanocomposite coatings. With the goal of enhancing the excellent of coating, the aerosol assisted chemical vapor deposition approach may be carried out via atomization and evaporation of precursor droplets [144–147].

13.4.8 Physical Vapor Deposition Approach

This approach can be a famous innovation that's widely applied for the deposition of thin films with respect to many things done at better temperatures, inducing more stresses within the coatings and substrate being applied essentially while the required coating desires to be deposited the usage of this process [148–150].

13.4.9 Thermal Spray Approach

Usually, a matrix of metal or alloy may be coated with nanocomposite coatings through this approach. The spray material composed of a nanosized alloy powder [151] become disbursed in a suspension using thermal spray method [152] or suspension plasma spray process [153].

13.4.10 Solution Dispersion

This approach become efficiently carried out for the preparation of polymer nanocomposite coatings bolstered with nanofillers consisting of carbon nanotube, steel oxides and nanoclay [154, 155]. The ultrasound-assisted stirring approach become used to enhance the dispersion of nanofillers into polymer matrices, similarly to using mechanical stirring methods.

13.5 Applications

Nanocomposites metal oxides show tremendous properties in diverse applications. Few are mentioned below:

13.5.1 Gas Sensors

Sensor technology has been continually growing in both government and private sectors. Gas sensors have been increasingly applied for semiconductor, environmental monitoring, automobiles and health care [156–163].

13.5.2 Batteries

Due to various environmental issues, rechargeable batteries have gained high attention in the recent years. Lithium-ion batteries (LIBs) have played a vital role in our life as it is excessively used in electric vehicles and hybrid vehicles because of long life cycles [164, 165].

13.5.3 Solar Cells

Metal oxides are used to develop metal oxide p–n junctions. Indium tin oxide (ITO) electrodes on glass substrates have proved to be useful candidates for the creation of solar cells that offer the same performance as ITO and have more stable electric and mechanical properties [166–168].

13.5.4 Antennas

Antennas are designed to create electric current from the electromagnetic (EM) waves in space. Since most of the metals are susceptible to oxidation for various reasons, so it is important that some alternatives is to be considered for the antenna material. Vanadium dioxide has the ability to make sudden transformation from insulators to conductors at 66.85°C that could be triggered by light, electric field, mechanical stress etc. So researchers are preparing composites containing VO_2 that showed the probabilities of varying electrical and optical properties or manipulating the EM waves [169, 170].

13.5.5 Optoelectronics

For their peculiar properties, nanocomposites are being excessively used for long standing applications in optoelectronic industries as touch screens, flat panel displays, LEDs, solar cells production etc. [171, 172].

13.6 Examples

13.6.1 Graphene-Based Metal Oxide Nanocomposites

As already mentioned, rGO-Pt-NiO nanocomposites was established for creating enzyme free glucose sensing [33]. Furthermore, an ultra-low detection restriction glucose sensor using decreased graphene oxide concave tetrahedral Pd NCs and CuO composite [107]. TiO_2 nanoparticles embellished decreased graphene oxide nanocomposite may be used for ultrasensitive electrochemical determination of rifampicin [108]. Multilayer based graphene oxide coated zinc oxide nanoflower (ZnO NFs) as a modified electrode substances was used for high sensitive electrochemical determination of biomarker, 8- hydroxy-2'-deoxyguanosine (8-HDG) that's one of the essential most cancers and oxidative pressure biomarker [109, 110].

13.6.2 Carbon-Based Metal Oxide Nanocomposites

Here also, Carbon nanotubes represent a substrate for several numbers of electrochemical sensors in aggregate with metals and metal oxides. Assembled CNT on graphene outcomes synergistic impact which allows the conductive paths to get higher electric conductivity [111]. In addition, ammonia gas-sensing properties at low temperature had been agreeably more advantageous by comprising carbon nanotubes with tungsten oxide nanobricks sensors as compared to CNT-primarily based and WO_3 based nanobrick-sensors separately [112, 113]. The evaluation of epirubicin anticancer drug in real samples was correctly performed via fabrication of high sensitive electrochemical sensor from Fe_3O_4-SWCNTs/MOCTICl/CPE [114]. Furthermore, the gadgets labored through combining electrochemical structures and nanosensors compromise a quite appropriate gear for immediate evaluation of medicinal drugs and dangerous additives both in environment, food or fitness application [115]. Atorvastatin isomers can be differentiated through creation of stereoselective sensor relying on multiwalled carbon nanotube cross-connected with Hydroxypropyl-b-cyclodextrin primarily based totally chiral nanocomposite [116]. Oil-in water emulsion solvent evaporation technique was deployed for the preparation of PCL-CNT nanocomposites. These nanocomposites discover a sound in drug transport applications [117–120].

13.6.3 Silicon-Based Metal Oxide Nanocomposites

A trustful H_2 gas sensor changed into advanced with PdNPs decorated Si nanomesh shape the usage of too smooth manufacturing system through polystyrene nanosphere lithography [81]. Also, Silicon oxycarbide glasses and glass-ceramics had exquisite community architectures dictating their attractive functional properties like piezoresistive sensing behavior, reversible Li (and Na) uptake/release, drug release behavior, bioactivity, optical houses in addition to gas sensing or photocatalytic houses [82–84].

13.7 Laser Irradiation Sources

Conventionally, the thermal remedy of metal oxide (MO) films and nanostructures is carried out in an oven or a furnace within the temperature variety of 300–1000 °C, relying on the material system, thermodynamics, and the desired impact [173–175]. This method is related to such problems as excessive thermal budget (thermal electricity), lengthy processing times, and excessive rate of power loss due to the fact sample dimensions are substantially smaller than the heated volume, and nearly all the power is used to elevate and decrease the temperature of the substrate and the furnace itself. As an opportunity to the traditional thermal treatment, laser irradiation gives potential solutions to the above problems and allows a pretty well suited on-chip integration of MO films and nanostructures. This approach is based totally at the photothermal impact triggered through a targeted laser that remotely generates a restrained temperature area at a preferred role with excessive controllability [176]. Several laser method parameters, inclusive of laser intensity, pulse width, and scanning rate, may be changed to get the preferred thermal impact. This method is characterized through notably speedy and localized thermal results, permitting unique manage over the material regions. The heating and cooling fees (>106 °C s^{-1}) for laser irradiation are orders of value better than the rates of the traditional annealing and rapid thermal annealing (RTA), permitting rapid fabrication of substances with minimum power losses. As laser-triggered heat may be restrained to a particular region in each in-aircraft. It's far viable to selectively anneal the films and nanostructures with none thermal interference with the underlying substrates and adjoining structures. Furthermore, usage of lasers with enough efficiency (described because the ratio of the optical output electricity and input electricity) can bring about vast reduction in power required for thermal processing [177, 178]. Laser-primarily based totally processing has been applied in programs starting from communications to materials processing [179]. Currently, excimer laser structures are extensively applied for commercial, large-region tool programs, such as the fabrication of low-temperature poly-Si for excessive-quit thin-film transistor displays and annealing of liquid-crystal displays and active-matrix natural light-emitting diode (OLED) backplanes for smartphones and OLED televisions [180].

Laser structures running in continuous wave (CW) or pulsed modes at wavelengths within the ultraviolet (UV), visible (vis), and infrared (IR) ranges have been used to tailor the functionality of steel oxides. Laser irradiation of a huge type of MO films and nanostructures has been carried out for applications in dielectric [181, 182], ferroelectric [183], piezoelectric [184] and multiferroic magnetoelectric [185] devices, photocatalysis [186], gas sensors [187], transparent conductors [188], area-impact transistors [189], sun cells [190] and thermistors [191]. Besides the overarching purpose of tailoring of the physical and chemical properties, numerous interesting phenomena triggered through laser irradiation had been reported such as transformation [192, 193], diamagnetic to ferromagnetic switching [194, 195], change of change bias [196], wettability switching [197], p-kind to n-kind conductivity transformation [198], resistive switching [199, 200], metal to-insulator transition (MIT) [201] and electrochromic switching [202].

The mechanism of photon-power (hf) absorption relies upon at the electronic structure of the material [203–206]. In this instance, h is the Planck constant, and f is the photon frequency. In insulators and semiconductors with bandgap (Eg), the absorption takes place through resonant excitation regarding especially interband electrical transitions.

The laser-triggered interband or intraband digital transition induces a nonequilibrium electrical distribution that thermalizes the material through electron–phonon interactions. The electron–hole recombination method reestablishes the equilibrium situation on a time scale that relies upon at the material qualities [207]. Various mechanisms of the electron and lattice dynamics represented within the four stages defined within the literature [208, 209].

13.7.1 Some of the Factors Affecting Laser–Material Interactions

Some of the factors influencing the laser–material interactions are mentioned under:

13.7.1.1 Laser Wavelength

The simple wavelength of a laser beam is decided with the help of using the atomic energy level of the lasing medium. However, it's far viable to extrade the laser wavelength with the aid of using nonlinear optics withinside the resonator cavity. The depth of the laser radiation is attenuated withinside the cloth according with the Beer–Lambert law: $I(z) = I_0 e^{-\alpha z}$, wherein I_0 is the laser intensity just below the surface after considering the reflection loss, $I(z)$ is the intensity at depth z, and α is the absorption.

The complete transparency of low-strength, high-wavelength IR laser radiation, in MO films, may cause thermal damage to the substrate and interfacial reactions. In case of MO thick films (>1 μm) which might be semitransparent to visible light, the laser radiation is absorbed specifically with the help of using the intermediate layers of the film, and its surface layers are heated up with the help of thermal diffusion [210, 211].

13.7.1.2 Laser Intensity

Laser-assisted heating with intensity smaller than the threshold of heating can result various temperature dependent processes. The excessive temperatures generated can result in microstructural adjustments and increased sintering and densification of substances [212, 213].

13.7.1.3 Laser Interaction Time

The thermal diffusion duration shows the distance of the spreading of the energy the strength during laser irradiation time, which may be the spot size for CW lasers or pulse width in practical case. The heat generated with the aid of using CW laser irradiation can penetrate deeply into a material because of its long irradiation time (generally >1 μs) [214]. This sensible feature of thermal effect permits fast annealing, welding, and reducing of numerous materials. However, it isn't best for devices that require thermal interactions for minimum reoxidation at some point of photothermal loss [215]. A pulsed laser is desired in all forms of laser processing that acts for excessive heating/cooling phases and well-described localized strength entering to the material. Pulsed lasers are especially powerful for decreasing the ablation threshold depth or unique processing.

13.7.1.4 Surface Roughness of the Material

The degree of enhancement is predicated at the particular size and geometry of the surface [216]. Creating capabilities close to the surface with dimensions at the order of a wavelength (e.g., surface roughness, cracks, and voids) additionally influences the surface reflectivity that may increase the optical direction duration and laser absorption [217].

13.8 Functional Characterization Techniques

Various characterization techniques have been utilized proficiently for the generation of high-resolution images of materials' nanostructure, along with direct sensing process. For example, X-ray diffraction (XRD) analysis is an important technique for examining the relation between the crystal phase-sensing potential. Researchers with the help of high-resolution transmission electron microscopy (HRTEM) and scanning electron microscopy (SEM) are able to tune synthesis procedures on the surface of metal oxide nanocomposites. Further, analytical techniques like photoluminescence (PL) spectrometer, Fourier transform infrared (FTIR) spectrometer, and Raman spectroscopy provide deeper insight into the mechanism, thus delivering multiscale information in the sensing process [218–223].

13.8.1 Electrical Characterization

Metal oxide nanocomposites are gaining too many considerations in the area of electronic applications in the form of transistors, gate insulators, energy harvesters, nanogenerators, and others. Among diverse power harvesting technologies, there has been significant interest in the region of vibrational energy primarily based on piezoelectric and magnetic harvesters. Growing nanopiezoelectrics as promising new substances for biomechanical power nanogenerators which are tremendously efficient for wearable or implantable packages are of special interest [1, 2, 4, 16, 17].

Electrical characterization has performed an important role in optoelectronic activities to understand the character of a material for device fabrication. Generally, the oxides used for thin/thick films may be in particular labeled into groups: metallic - here the resistivity follows a power-law depending on the temperature $\rho \propto T^n$, where n > zero and for the semiconducting group, the resistivity generally obeys the exponential rule i.e. $\rho \propto \exp(E/KT)$ For example, the electric resistivity (ρ) of the ZnO film is calculated by the use of the Equation 13.1 [224].

$$\rho = \frac{\pi}{\ln 2} t \frac{V}{I} \quad (13.1)$$

eq. (13.4) where t is the *film* thickness (cm), I is the supply current (A), V is the voltage *measured* and ρ is the resistivity (Ωcm). Similarly, the electric resistivity (ρ) for the natural CdO film is calculated through the usage of the Equation 13.2 [225].

$$\rho = VA/It \quad (13.2)$$

in which A is the area of the film. The temperature dependency of the DC conductivity may be obtained by the help of the famous Arrhenius equation, given as Equation 13.3 below.

$$\sigma = \sigma_0 \exp(\frac{-\Delta E}{KT}) \quad (13.3)$$

in which σ is the conductivity, σ_0 is the pre-exponential factor, ΔE is the thermal activation strength, K is Boltzmann's consistent and T is absolute temperature. The relation between the electrical conductivity with temperature for ZnO film is shown in Figure 13.3.

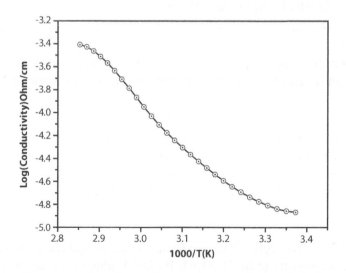

Figure 13.3 Temperature dependence of DC Conductivity of (ZnO) thick film.

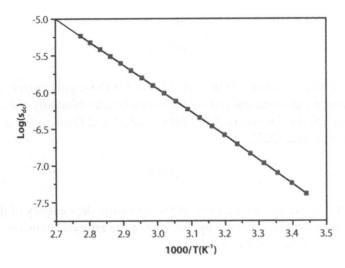

Figure 13.4 Log (σ_{dc}) vs. 1000/T of CdO thick film.

In Figure 13.4, the plot of Log (s_{dc}) vs. 1000/T of CdO thick film shows the negative temperature coefficient of resistance because the conductivity is increasing with increase in working temperature exhibiting the semiconducting nature of the film. This shows the negative temperature coefficient of resistance because the conductivity is increasing with increase in working temperature exhibiting the semiconducting nature of the film. This may be because of oxygen deficiency of the material [226].

13.8.2 Optical Characterization

Depending on the usage of the absorption peak that corresponds to electronic excitation from valence band to conduction band, the nature and value of optical band gap are determined. The relation among absorption coefficient (α) and incident photon strength (hν) may be written as [227].

$$\alpha h\vartheta = A(h\vartheta - E_g)^n \quad (13.4)$$

in which E_g is the energy gap among valence and conduction bands. The exponent n relies upon the quantum selection rules for precise material that is equal to ½ for direct band gap material and A is constant. The photon energy (hf) for Y-axis may be decided through the usage of Equation 13.5.

$$E = hf = hc/\lambda \quad (13.5)$$

in which λ is wavelength of incident light, c is pace of light (3x108 m/s) and h is Planck's constant (6.626×10-34J/s). The UV-Visible absorption spectra for ZnO film recorded in the 250–1000 nm range is shown in Figure 13.5 below.

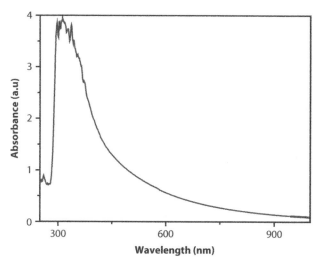

Figure 13.5 Absorption Spectra of ZnO thick film.

Similarly, Figure 13.6 indicates the variation of $(\alpha h\nu)^2$ with $h\nu$, that is a straight line. By extrapolating the linear part of the curve at $\alpha = 0$, the energy is obtained [228].

Figure 13.7 indicates the optical transmission spectrum of screen printed CdO film. The curve indicates 80% transmission in the visible region. At 400 nm, the transmission is ~10% which then begins to evolve

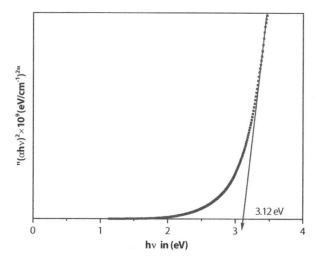

Figure 13.6 Energy band gap determination of ZnO thick film.

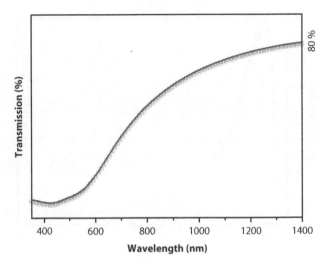

Figure 13.7 Optical transmission spectrum of CdO thick film.

exponentially with sharp increase between 600nm to 850nm after which again slows down to 80% at 1400nm.

The plot of $(\alpha h\nu)^2$ vs. $h\nu$ as in Figure 13.8 is a straight line indicating the direct transition. By the extrapolation of the linear part of the curve to zero absorption coefficient, the band gap energy is obtained [229].

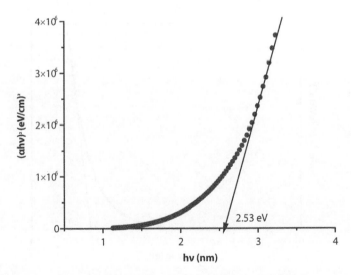

Figure 13.8 $h\nu$ vs. $(\alpha h\nu)^2$ for determination of CdO thick film band gap energy.

As the metal oxide films are potential candidates for many optical applications, the electrical role of native defects in metal oxide films like ZnO (*viz.*, Zn vacancy (VZn), Zn interstitial (Zni), and oxygen vacancy (V0)) remains disputed. Understanding the precise mechanisms of carrier transport related to the complex defect distribution and controlling the electric defect properties is a difficult task [230]. It has been seen that the native defects can profoundly affect the optical and electrical properties of metal oxides films based electronic devices like ZnO. With the help of specific techniques like electron paramagnetic resonance (EPR) [231], photoluminescence (PL) [232], admittance spectroscopy (AS) [233], and deep level transient spectroscopy (DLTS) [234–237]. The activity of defect distribution has been analyzed. These techniques are found to be suitable for the analysis of role of defect states and impurities using p–n junction devices or Schottky diodes [238]. Likewise, the electronic defect states of the metal oxide films have been investigated thoroughly to gather the evidence of intrinsic conductivity.

To calculate the Schottky barrier height in diodes, both I–V and C–V techniques can be conducted. These are very important to evaluate the performance of the diode. The saturation current (Is) can be applied to the following equation in order to calculate the Schottky barrier height [239]:

$$I_s = AA^*T^2 \exp\left(-\frac{O_B}{KT}\right)$$

Where, A, A* and O_B represent Schottky contact area, effective Richardson constant and Schottky barrier height, respectively.

The current–voltage relationship for a Schottky diode may be written as:

$$\ln(I) = \ln(Io) + qV/(nkT)$$

13.9 Conclusion

Many sensing components skilled in direct sensors and complicated sensors such as metals, metal oxide, sol–gel substances, solid electrolytes, carbon nanotubes and graphene nanocomposites for synthesis of detecting substances. In the present chapter we've made effort to integrate the possible number of sensing materials and evaluate their associated information, along with their crucial factors of interest and utilization in current sensors. An analysis was made on the electrical and optical properties of the metal oxide nanoparticles especially citing these properties of ZnO and CdO as example. The change in electrical and optical properties of novel nanomaterials finds application in various fields mostly the optoelectronic applications. There is a rapid and continuous improvement in the field of nanotechnology and this paper could be helpful for the researchers working on devices with novel designs.

References

1. Roy, R., Roy, R.A., Roy, D.M., Alternative perspectives on "quasi-crystallinity": Non-uniformity and nanocomposites. *Mater. Lett.*, 4, 8-9, 323–328, 1986.
2. Choa, Y.H., Yang, J.K., Kim, B.H., Jeong, Y.K., Lee, J.S., Nakayama, T., Sekino, T., Niihara, K., Preparation and characterization of metal/ceramic nanoporous nanocomposite powders. *J. Magn. Magn. Mater.*, 266, 1-2, 12–19, 2003.
3. Wypych, F., Seefeld, N., Denicoló, I., Preparation of nanocomposites based on the encapsulation of conducting polymers into 2H-MoS2 and 1T-TiS2. *Quim. Nova*, 20, 356–360, 1997.
4. Yang, P., *The chemistry of nanostructured materials*, vol. 1, World Scientific, Singapore, 2003.
5. Tan, O.K., Cao, W., Zhu, W., Chai, J.W., Pan, J.S., Ethanol sensors based on nano-sized α-Fe_2O_3 with SnO_2, ZrO_2, TiO_2 solid solutions. *Sens. Actuators B: Chem.*, 93, 1-3, 396–401, 2003.
6. Li, C., Thostenson, E.T., Chou, T.W., Sensors and actuators based on carbon nanotubes and their composites: A review. *Compos. Sci. Technol.*, 68, 6, 1227–1249, 2008.
7. Shi, J.L. and Zhang, L.X., Nanocomposites from ordered mesoporous materials. *J. Mater. Chem.*, 14, 5, 795–806, 2004.
8. Shi, D.L., Feng, X.Q., Huang, Y.Y., Hwang, K.C., Gao, H., The effect of nanotube waviness and agglomeration on the elastic property of carbon nanotube-reinforced composites. *J. Eng. Mater. Technol.*, 126, 3, 250–257, 2004.
9. Cho, S., Takagi, K., Kwon, H., Seo, D., Ogawa, K., Kikuchi, K., Kawasaki, Multi-walled carbon nanotube-reinforced copper nanocomposite coating fabricated by low-pressure cold spray process. *Surf. Coat. Technol.*, 206, 16, 3488–3494, 2012.
10. Frey, H. and Khan, H.R. (Eds.), *Handbook of thin film technology*, pp. 24–41, Springer, Berlin, 2015.
11. Awan, S.A. and Gould, R.D., Conductivity and dielectric properties of silicon nitride thin films prepared by RF magnetron sputtering using nitrogen gas. *Thin Solid Films*, 423, 2, 267–272, 2003.

12. Sharma, N., Hooda, M., Sharma, S.K., Synthesis and characterization of LPCVD polysilicon and silicon nitride thin films for MEMS applications. *J. Mater.*, 2014, 1–8, 2014.
13. Ding, G., Clavero, C., Schweigert, D., Le, M., Thickness and microstructure effects in the optical and electrical properties of silver thin films. *AIP Adv.*, 5, 11, 117234, 2015.
14. Chiu, F.C., Pan, T.M., Kundu, T.K., Shih, C.H., Thin film applications in advanced electron devices. *Adv. Mater. Sci. Eng.*, 2014, 1–2, 2014.
15. Rao, M.C. and Shekhawat, M.S., A brief survey on basic properties of thin films for device application. *Int. J. Mod. Phys.: Conf. Ser.*, 22, 576–582, 2013, World Scientific Publishing Company.
16. Ramesh, G.V., Porel, S., Radhakrishnan, T.P., Polymer thin films embedded with *in situ* grown metal nanoparticles. *Chem. Soc. Rev.*, 38, 9, 2646–2656, 2009.
17. Rao, V.K. and Radhakrishnan, T.P., *In situ* fabricated Ag/AgCl—Polymer nanocomposite thin film: An appraisal of the efficient and reusable photocatalyst. *Mater. Res. Bull.*, 87, 193–201, 2017.
18. Govindasamy, M., Manavalan, S., Chen, S.M., Rajaji, U., Chen, T.W., Al-Hemaid, F.M., Ali, M.A., Elshikh, M.S., Determination of neurotransmitter in biological and drug samples using gold nanorods decorated f-MWCNTs modified electrode. *J. Electrochem. Soc.*, 165, 9, B370, 2018.
19. Qin, L., Zeng, G., Lai, C., Huang, D., Xu, P., Zhang, C., Cheng, M., Liu, X., Liu, S., Li, B., Yi, H., "Gold rush" in modern science: Fabrication strategies and typical advanced applications of gold nanoparticles in sensing. *Coord. Chem. Rev.*, 359, 1–31, 2018.
20. Sakthivel, R., Annalakshmi, M., Chen, S.M., Kubendhiran, S., Efficient electrochemical detection of lethal environmental pollutant hydroquinone based on functionalized carbon black/polytyramine/gold nanoparticles nanocomposite. *J. Electrochem. Soc.*, 166, 8, B680, 2019.
21. Ma, K., Sinha, A., Dang, X., Zhao, H., Electrochemical preparation of gold nanoparticles-polypyrrole co-decorated 2D MoS2 nanocomposite sensor for sensitive detection of glucose. *J. Electrochem. Soc.*, 166, 2, B147, 2019.
22. Majhi, S.M., Naik, G.K., Lee, H.J., Song, H.G., Lee, C.R., Lee, I.H., Yu, Y.T., Au@ NiO core-shell nanoparticles as a p-type gas sensor: Novel synthesis, characterization, and their gas sensing properties with sensing mechanism. *Sens. Actuators B: Chem.*, 268, 223–231, 2018.
23. Hu, L., Zhang, L., Zhou, Y., Meng, G., Yu, Y., Yao, W., Yan, Z., Chitosan-stabilized gold nano composite modified glassy carbon electrode for electrochemical sensing trace Hg2+ in practice. *J. Electrochem. Soc.*, 165, 16, B900, 2018.
24. Xue, C., Wang, X., Zhu, W., Han, Q., Zhu, C., Hong, J., Zhou, X., Jiang, H., Electrochemical serotonin sensing interface based on double-layered membrane of reduced graphene oxide/polyaniline nanocomposites and

molecularly imprinted polymers embedded with gold nanoparticles. *Sens. Actuators B: Chem.*, 196, 57–63, 2014.
25. Abdel-Raoof, A.M., Manal, A., Said, R.A., Abostate, M.H., Morshedy, S., Emara, M.S., Versatile sensor modified with gold nanoparticles carbon paste electrode for anodic stripping determination of brexpiprazole: A voltammetric study. *J. Electrochem. Soc.*, 166, 12, B948, 2019.
26. Wang, N., Lin, M., Dai, H., Ma, H., Functionalized gold nanoparticles/reduced graphene oxide nanocomposites for ultrasensitive electrochemical sensing of mercury ions based on thymine–mercury–thymine structure. *Biosens. Bioelectron.*, 79, 320–326, 2016.
27. Meng, F., Zheng, H., Chang, Y., Zhao, Y., Li, M., Wang, C., Sun, Y., Liu, J., One-step synthesis of $Au/SnO_2/RGO$ nanocomposites and their VOC sensing properties. *IEEE Trans. Nanotechnol.*, 17, 2, 212–219, 2018.
28. Shu, H., Chang, G., Su, J., Cao, L., Huang, Q., Zhang, Y., Xia, T., He, Y., Single-step electrochemical deposition of high performance Au-graphene nanocomposites for nonenzymatic glucose sensing. *Sens. Actuators B: Chem.*, 220, 331–339, 2015.
29. Turcheniuk, K., Boukherroub, R., Szunerits, S., Gold–graphene nanocomposites for sensing and biomedical applications. *J. Mater. Chem. B*, 3, 21, 4301–4324, 2015.
30. Jiao, S., Jin, J., Wang, L., One-pot preparation of Au-RGO/PDDA nanocomposites and their application for nitrite sensing. *Sens. Actuators B: Chem.*, 208, 36–42, 2015.
31. Niu, X., Xie, H., Luo, G., Men, Y., Zhang, W., Sun, W., Platinum-3D graphene oxide areogel nanocomposite for direct electrochemistry and electrocatalysis of horseradish peroxidase. *J. Electrochem. Soc.*, 165, 14, B713, 2018.
32. Cui, M., Zhang, Q., Fu, M., Fan, X., Lu, H., Wang, H., Zhang, Y., Wang, H., Template-free controllable electrochemical synthesis of hierarchical flower-like platinum nanoparticles/nitrogen doped helical carbon nanotubes for label-free biosensing of bovine serum albumin. *J. Electrochem. Soc.*, 166, 2, B117, 2019.
33. Wang, L., Lu, X., Wen, C., Xie, Y., Miao, L., Chen, S., Li, H., Li, P., Song, Y., One-step synthesis of Pt–NiO nanoplate array/reduced graphene oxide nanocomposites for nonenzymatic glucose sensing. *J. Mater. Chem. A*, 3, 2, 608–616, 2015.
34. Leonardi, S.G., Aloisio, D., Donato, N., Russo, P.A., Ferro, M.C., Pinna, N., Neri, G., Amperometric sensing of H_2O_2 using $Pt-TiO_2$/reduced graphene oxide nanocomposites. *ChemElectroChem*, 1, 3, 617–624, 2014.
35. Ong, W.L., Gao, M., Ho, G.W., Hybrid organic PVDF–inorganic M–rGO–TiO_2 (M= Ag, Pt) nanocomposites for multifunctional volatile organic compound sensing and photocatalytic degradation–H_2 production. *Nanoscale*, 5, 22, 11283–11290, 2013.
36. Ma, J., Ren, Y., Zhou, X., Liu, L., Zhu, Y., Cheng, X., Xu, P., Li, X., Deng, Y., Zhao, D., Pt nanoparticles sensitized ordered mesoporous WO3

semiconductor: Gas sensing performance and mechanism study. *Adv. Funct. Mater.*, 28, 6, 1705268, 2018.
37. Yang, B., Bin, D., Wang, H., Zhu, M., Yang, P., Du, Y., High quality Pt–graphene nanocomposites for efficient electrocatalytic nitrite sensing. *Colloids Surf. A: Physicochem. Eng. Asp.*, 481, 43–50, 2015.
38. Kumar, T.V. and Sundramoorthy, A.K., Non-enzymatic electrochemical detection of urea on silver nanoparticles anchored nitrogen-doped single-walled carbon nanotube modified electrode. *J. Electrochem. Soc.*, 165, 8, B3006, 2018.
39. Zhang, J., Yang, K., Chen, L., In situ deposition of silver nanoparticles on polydopamine nanospheres for an ultrasensitive electrochemical aptasensor of ochratoxin A. *J. Electrochem. Soc.*, 166, 6, H182, 2019.
40. Mirzaei, A., Janghorban, K., Hashemi, B., Bonavita, A., Bonyani, M., Leonardi, S.G., Neri, G., Synthesis, characterization and gas sensing properties of Ag@ α-Fe_2O_3 core–shell nanocomposites. *Nanomaterials*, 5, 2, 737–749, 2015.
41. Yao, W., Guo, H., Liu, H., Li, Q., Xue, R., Wu, N., Li, L., Wang, M., Yang, W., Simultaneous electrochemical determination of acetaminophen and dopamine based on metal-organic framework/multiwalled carbon nanotubes-Au@ Ag nanocomposites. *J. Electrochem. Soc.*, 166, 14, B1258, 2019.
42. Jiang, D., Zhang, Y., Chu, H., Liu, J., Wan, J., Chen, M., N-doped graphene quantum dots as an effective photocatalyst for the photochemical synthesis of silver deposited porous graphitic C_3N_4 nanocomposites for nonenzymatic electrochemical H_2O_2 sensing. *RSC Adv.*, 4, 31, 16163–16171, 2014.
43. Malik, R., Tomer, V.K., Kienle, L., Chaudhary, V., Nehra, S., Duhan, S., Retracted: Ordered mesoporous Ag–ZnO@ g-CN nanohybrid as highly efficient bifunctional sensing material. *Adv. Mater. Interfaces*, 5, 1701357, 2018.
44. Johnson, J.L., Behnam, A., Pearton, S.J., Ural, A., Hydrogen sensing using Pd-functionalized multi-layer graphene nanoribbon networks. *Adv. Mater.*, 22, 43, 4877–4880, 2010.
45. Li, J., Fan, R., Hu, H., Yao, C., Hydrogen sensing performance of silica microfiber elaborated with Pd nanoparticles. *Mater. Lett.*, 212, 211–213, 2018.
46. Lu, L., Hu, X., Zhu, Z., Li, D., Tian, S., Chen, Z., Electrochemical sensors and biosensors modified with binary nanocomposite for food safety. *J. Electrochem. Soc.*, 167, 3, 037512, 2020.
47. Gao, M., Cho, M., Han, H.J., Jung, Y.S., Park, I., Palladium-decorated silicon nanomesh fabricated by nanosphere lithography for high performance, room temperature hydrogen sensing. *Small*, 14, 10, 1703691, 2018.
48. He, W.L., Fang, F., Ma, D.M., Chen, M., Qian, D.J., Liu, M., Palladium-directed self-assembly of multi-titanium (IV)-porphyrin arrays on the substrate surface as sensitive ultrathin films for hydrogen peroxide sensing, photocurrent generation, and photochromism of viologen. *Appl. Surf. Sci.*, 427, 1003–1010, 2018.

49. Hu, J., Liu, Y., Gao, G., Zou, X., One-step synthesis of AuPdNPs/ERGO nanocomposite for enhanced electrochemical sensing of Cr (VI) in water. *J. Electrochem. Soc.*, 165, 16, B893, 2018.
50. Phan, D.T. and Chung, G.S., Characteristics of resistivity-type hydrogen sensing based on palladium-graphene nanocomposites. *Int. J. Hydrogen Energy*, 39, 1, 620–629, 2014.
51. Švorc, Ľ., Strežova, I., Kianičkova, K., Stanković, D.M., Otřísal, P., Samphao, A., An advanced approach for electrochemical sensing of ibuprofen in pharmaceuticals and human urine samples using a bare boron-doped diamond electrode. *J. Electroanal. Chem.*, 822, 144–152, 2018.
52. Atar, N. and Yola, M.L., Core-shell nanoparticles/two-dimensional (2D) hexagonal boron nitride nanosheets with molecularly imprinted polymer for electrochemical sensing of cypermethrin. *J. Electrochem. Soc.*, 165, 5, H255, 2018.
53. Yola, M.L. and Atar, N., Gold nanoparticles/two-dimensional (2D) hexagonal boron nitride nanosheets including diethylstilbestrol imprinted polymer: Electrochemical detection in urine samples and validation. *J. Electrochem. Soc.*, 165, 14, H897, 2018.
54. Xie, F., Liu, T., Xie, L., Sun, X., Luo, Y., Metallic nickel nitride nanosheet: An efficient catalyst electrode for sensitive and selective non-enzymatic glucose sensing. *Sens Actuators B: Chem.*, 255, 2794–2799, 2018.
55. Joy, J., Sekar, A., Vijayaraghavan, S., Kumar, P., Pillai, V.K., Alwarappan, S., Nickel-incorporated, nitrogen-doped graphene nanoribbons as efficient electrocatalysts for oxygen evolution reaction. *J. Electrochem. Soc.*, 165, 3, H141, 2018.
56. Goncalves, J.M., Alves, K.M., Gonzalez-Huila, M.F., Duarte, A., Martins, P.R., Araki, K., Unexpected stabilization of α-Ni (OH) 2 nanoparticles in GO nanocomposites. *J. Nanomater.*, 2018, Article ID 5735609, 13 pages, 2018.
57. Bonomo, M., Mariani, P., Mura, F., Di Carlo, A., Dini, D., Nanocomposites of nickel oxide and zirconia for the preparation of photocathodes with improved performance in p-type dye-sensitized solar cells. *J. Electrochem. Soc.*, 166, 8, D290, 2019.
58. Rikhari, B., Mani, S.P., Rajendran, N., Electrochemical behavior of polypyrrole/chitosan composite coating on Ti metal for biomedical applications. *Carbohydr. Polym.*, 189, 126–137, 2018.
59. Zaragr, R.A., Micro size developed TiO_2-CdO composite film: Exhibits diode characteristics. *Mater. Lett.*, 335, 15, 133813, 2023.
60. Mahfooz-ur-Rehman, Rehman, W., Waseem, M., Shah, B.A., Shakeel, M., Haq, S., Zaman, U., Bibi, I., Khan, H.D., Fabrication of titanium–Tin oxide nanocomposite with enhanced adsorption and antimicrobial applications. *J. Chem. Eng. Data*, 64, 6, 2436–2444, 2019.
61. Huyen, D.N., Tung, N.T., Thien, N.D., Thanh, L.H., Effect of TiO_2 on the gas sensing features of TiO_2/PANi nanocomposites. *Sensors*, 11, 2, 1924–1931, 2011.

62. Ansari, M.O. and Mohammad, F., Thermal stability, electrical conductivity and ammonia sensing studies on p-toluenesulfonic acid doped polyaniline: Titanium dioxide (pTSA/Pani: TiO_2) nanocomposites. *Sens. Actuators B: Chem.*, 157, 1, 122–129, 2011.
63. Zargar, R.A. et al., ZnCdO thick film films: Screen printed TiO_2 film: A candidate for photovoltaic applications. *Mater. Res. Express*, 7, 065904, 2020.
64. Zhu, L., Zeng, W., Li, Y., New insight into gas sensing property of ZnO nanorods and nanosheets. *Mater. Lett.*, 228, 331–333, 2018.
65. Zhu, L., Zeng, W., Ye, H., Li, Y., Volatile organic compound sensing based on coral rock-like ZnO. *Mater. Res. Bull.*, 100, 259–264, 2018.
66. Liu, Y., Xu, X., Chen, Y., Zhang, Y., Gao, X., Xu, P., Li, X., Fang, J., Wen, W., An integrated micro-chip with $Ru/Al_2O_3/ZnO$ as sensing material for SO_2 detection. *Sens. Actuators B: Chem.*, 262, 26–34, 2018.
67. Zargar, R.A., Arora, M., Bhat, R.A., Study of nanosized copper-doped ZnO dilute magnetic semiconductor thick films for spintronic device applications. *Appl. Phys. A*, 124, 1–9, 2018.
68. Zargar, R.A., Arora, M., Alshahrani, T., Shkir, M., Screen printed novel ZnO/MWCNTs nanocomposite thick films. *Ceram. Int.*, 47, 6084–6093, 2020.
69. Kumar, N., Srivastava, A.K., Patel, H.S., Gupta, B.K., Varma, G.D., Facile synthesis of ZnO–reduced graphene oxide nanocomposites for NO_2 gas sensing applications. *Eur. J. Inorg. Chem.*, 2015, 11, 1912–1923, 2015.
70. Anandhi, P., Kumar, V.J., Harikrishnan, S., Preparation and improved capacitive behavior of NiO/TiO_2 nanocomposites as electrode material for supercapacitor. *Curr. Nanosci.*, 16, 1, 79–85, 2019.
71. Liu, Y., Zhu, G., Chen, J., Xu, H., Shen, X., Yuan, A., Co_3O_4/ZnO nanocomposites for gas-sensing applications. *Appl. Surf. Sci.*, 265, 379–384, 2013.
72. Zargar, R.A., ZnCdO thick film: A material for energy conversion devices. *Mater. Res. Express*, 6, 9, 095909, 2019.
73. Kondawar, S.B., Patil, P.T., Agrawal, S.P., Chemical vapour sensing properties of electrospun nanofibers of polyaniline/ZnO nanocomposites. *Adv. Mater. Lett.*, 5, 7, 389–395, 2014.
74. Zargar, R.A., Fabrication and improved response of ZnO□CdO composite flms under diferent laser irradiation dose. *Sci. Rep.*, 12, 10096, 2022.
75. Bandgar, D.K., Navale, S.T., Mane, A.T., Gupta, S.K., Aswal, D.K., Patil, V.B., Ammonia sensing properties of polyaniline/α-Fe_2O_3 hybrid nanocomposites. *Synth. Met.*, 204, 1–9, 2015.
76. Zhang, Y., Zhang, D., Guo, W., Chen, S., The α-Fe_2O_3/g-C3N4 heterostructural nanocomposites with enhanced ethanol gas sensing performance. *J. Alloys Compd.*, 685, 84–90, 2016.
77. Zhang, H., Yu, L., Li, Q., Du, Y., Ruan, S., Reduced graphene oxide/α-Fe2O3 hybrid nanocomposites for room temperature NO_2 sensing. *Sens. Actuators B: Chem.*, 241, 109–115, 2017.
78. Marwani, H.M., Danish, E.Y., Alhazmi, M.A., Khan, S.B., Bakhsh, E.M., Asiri, A.M., Cellulose acetate-iron oxide nanocomposites for trace detection

of fluorene from water samples by solid-phase extraction technique. *Sep. Sci. Technol.*, 53, 6, 887–895, 2018.
79. Ghasemi, S., Hosseini, S.R., Kazemi, Z., Electrophoretic preparation of graphene-iron oxide nanocomposite as an efficient Pt-free counter electrode for dye-sensitized solar cell. *J. Solid State Electrochem.*, 22, 1, 245–253, 2018.
80. Zhang, D., Fan, X., Yang, A., Zong, X., Hierarchical assembly of urchin-like alpha-iron oxide hollow microspheres and molybdenum disulphide nanosheets for ethanol gas sensing. *J. Colloid Interface Sci.*, 523, 217–225, 2018.
81. Stabler, C., Ionescu, E., Graczyk-Zajac, M., Gonzalo-Juan, I., Riedel, R., Silicon oxycarbide glasses and glass-ceramics:"All-Rounder" materials for advanced structural and functional applications. *J. Am. Ceram. Soc.*, 101, 11, 4817–4856, 2018.
82. Alipour, K. and Nasirpouri, F., Effect of morphology and surface modification of silica nanoparticles on the electrodeposition and corrosion behavior of zinc-based nanocomposite coatings. *J. Electrochem. Soc.*, 166, 2, D1, 2019.
83. Sargolzaeiaval, Y., Ramesh, V.P., Neumann, T.V., Miles, R., Dickey, M.D., Öztürk, M.C., High thermal conductivity silicone elastomer doped with graphene nanoplatelets and eutectic gain liquid metal alloy. *ECS J. Solid State Sci. Technol.*, 8, 6, P357, 2019.
84. Quang, P.L., Cuong, N.D., Hoa, T.T., Long, H.T., Hung, C.M., Le, D.T.T., Van Hieu, N., Simple post-synthesis of mesoporous p-type Co3O4 nanochains for enhanced H_2S gas sensing performance. *Sens. Actuators B: Chem.*, 270, 158–166, 2018.
85. Quang, P.L., Cuong, N.D., Hoa, T.T., Long, H.T., Hung, C.M., Le, D.T.T., Van Hieu, N., Simple post-synthesis of mesoporous p-type Co3O4 nanochains for enhanced H2S gas sensing performance. *Sens. Actuators B: Chem.*, 270, 158–166, 2018.
86. Yu, L., Xiao, Y., Luan, C., Yang, J., Qiao, H., Wang, Y., Zhang, X., Dai, X., Yang, Y., Zhao, H., Cobalt/molybdenum phosphide and oxide heterostructures encapsulated in N-doped carbon nanocomposite for overall water splitting in alkaline media. *ACS Appl. Mater. Interfaces*, 11, 7, 6890–6899, 2019.
87. Jian, K.S., Chang, C.J., Wu, J.J., Chang, Y.C., Tsay, C.Y., Chen, J.H., Horng, T.L., Lee, G.J., Karuppasamy, L., Anandan, S., Chen, C.Y., High response CO sensor based on a polyaniline/SnO_2 nanocomposite. *Polymers*, 11, 1, 184, 2019.
88. Aydin, M., Demir, E., Unal, B., Dursun, B., Ahsen, A.S., Demir-Cakan, R., Chitosan derived N-doped carbon coated SnO_2 nanocomposite anodes for Na-ion batteries. *Solid State Ionics*, 341, 115035, 2019.
89. Lan, G., Nong, J., Jin, W., Zhu, R., Luo, P., Jiang, H., Wei, W., Enhanced UV photoresponse employing 3D graphene nanowalls/SnO2 nanocomposite. *Film. Surf. Coating. Technol.*, 359, 90–96, 2019.

90. Zargar, R.A., Kumar, K. et al., Structural, optical, photoluminescence, and EPR behaviour of novel ZnO· 80CdO· 2OO thick films: An effect of different sintering temperatures. *J. Lumin.*, 245, 118769, 2022.
91. Neri, G., Leonardi, S.G., Latino, M., Donato, N., Baek, S., Conte, D.E., Russo, P.A., Pinna, N., Sensing behavior of SnO_2/reduced graphene oxide nanocomposites toward NO_2. *Sens. Actuators B: Chem.*, 179, 61–68, 2013.
92. Zhang, H., Feng, J., Fei, T., Liu, S., Zhang, T., SnO_2 nanoparticles-reduced graphene oxide nanocomposites for NO_2 sensing at low operating temperature. *Sens. Actuators B: Chem.*, 190, 472–478, 2014.
93. Li, L., Zhang, C., Chen, W., Fabrication of SnO 2–SnO nanocomposites with p–n heterojunctions for the low-temperature sensing of NO_2 gas. *Nanoscale*, 7, 28, 12133–12142, 2015.
94. Shanmugasundaram, A., Basak, P., Satyanarayana, L., Manorama, S.V., Hierarchical SnO/SnO_2 nanocomposites: Formation of in situ p–n junctions and enhanced H2 sensing. *Sens. Actuators B: Chem.*, 185, 265–273, 2013.
95. Narjinary, M., Rana, P., Sen, A., Pal, M., Enhanced and selective acetone sensing properties of SnO_2-MWCNT nanocomposites: Promising materials for diabetes sensor. *Mater. Des.*, 115, 158–164, 2017.
96. Xu, F., Deng, M., Li, G., Chen, S., Wang, L., Electrochemical behavior of cuprous oxide–reduced graphene oxide nanocomposites and their application in nonenzymatic hydrogen peroxide sensing. *Electrochim. Acta*, 88, 59–65, 2013.
97. Tan, H., Ma, C., Gao, L., Li, Q., Song, Y., Xu, F., Wang, T., Wang, L., Metal–organic framework-derived copper nanoparticle@ carbon nanocomposites as peroxidase mimics for colorimetric sensing of ascorbic acid. *Chem.–Eur. J.*, 20, 49, 16377–16383, 2014.
98. Gholamali, I., Hosseini, S.N., Alipour, E., Yadollahi, M., Preparation and characterization of oxidized starch/CuO nanocomposite hydrogels applicable in a drug delivery system. *Starch-Stärke*, 71, 3-4, 1800118, 2019.
99. Kottappara, R., Palantavida, S., Vijayan, B.K., Enhancing semiconductor photocatalysis with carbon nanostructures for water/air purification and self-cleaning applications, in: *Carbon Based Nanomaterials for Advanced Thermal and Electrochemical Energy Storage and Conversion*, pp. 139–172, Elsevier, Netherlands, 2019.
100. Mane, A.T., Navale, S.T., Sen, S., Aswal, D.K., Gupta, S.K., Patil, V.B., Nitrogen dioxide (NO_2) sensing performance of p-polypyrrole/n-tungsten oxide hybrid nanocomposites at room temperature. *Org. Electron.*, 16, 195–204, 2015.
101. Zhang, D., Cao, Y., Wu, J., Zhang, X., Tungsten trioxide nanoparticles decorated tungsten disulfide nanoheterojunction for highly sensitive ethanol gas sensing application. *Appl. Surf. Sci.*, 503, 144063, 2020.
102. An, X., Yu, X., Jimmy, C.Y., Zhang, G., CdS nanorods/reduced graphene oxide nanocomposites for photocatalysis and electrochemical sensing. *J. Mater. Chem. A*, 1, 16, 5158–5164, 2013.

103. Guo, P., Wang, Y., Chen, Z., Jin, T., Fu, L., Lin, C.T., Lai, G., Voltammetric immunoassay of human IgG based on the release of cadmium (II) from CdS nanocrystals deposited on mesoporous silica nanospheres. *Microchim. Acta*, *186*, 1, 1–8, 2019.
104. Midya, L., Patra, A.S., Banerjee, C., Panda, A.B., Pal, S., Novel nanocomposite derived from ZnO/CdS QDs embedded crosslinked chitosan: An efficient photocatalyst and effective antibacterial agent. *J. Hazard. Mater.*, *369*, 398–407, 2019.
105. Lv, C., Lan, X., Wang, L., Dai, X., Zhang, M., Cui, J., Yuan, S., Wang, S., Shi, J., Rapidly and highly efficient degradation of tetracycline hydrochloride in wastewater by 3D IO-TiO_2-CdS nanocomposite under visible light. *Environ. Technol.*, *42*, 3, 377–387, 2019.
106. Liu, Q., Tang, Y., Yang, X., Wei, M., Zhang, M., An ultra-low detection limit glucose sensor based on reduced graphene oxide-concave tetrahedral Pd NCs@ CuO composite. *J. Electrochem. Soc.*, *166*, 6, B381, 2019.
107. Said, R.A., Hasan, M.A., Abdelzaher, A.M., Abdel-Raoof, A.M., Insights into the developments of nanocomposites for its processing and application as sensing materials. *J. Electrochem. Soc.*, *167*, 3, 037549, 2019.
108. Govindasamy, M., Wang, S.F., Subramanian, B., Ramalingam, R.J., Al-Lohedan, H., Sathiyan, A., A novel electrochemical sensor for determination of DNA damage biomarker (8-hydroxy-2′-deoxyguanosine) in urine using sonochemically derived graphene oxide sheets covered zinc oxide flower modified electrode. *Ultrason. Sonochem.*, *58*, 104622, 2019.
109. Yu, L., Ma, F., Ding, X., Wang, H., Li, P., Silica/graphene oxide nanocomposites: Potential adsorbents for solid phase extraction of trace aflatoxins in cereal crops coupled with high performance liquid chromatography. *Food Chem.*, *245*, 1018–1024, 2018.
110. Liu, H., Gao, J., Huang, W., Dai, K., Zheng, G., Liu, C., Shen, C., Yan, X., Guo, J., Guo, Z., Electrically conductive strain sensing polyurethane nanocomposites with synergistic carbon nanotubes and graphene bifillers. *Nanoscale*, *8*, 26, 12977–12989, 2016.
111. Le, X.V., Luu, T.L.A., Nguyen, H.L., Nguyen, C.T., Synergistic enhancement of ammonia gas-sensing properties at low temperature by compositing carbon nanotubes with tungsten oxide nanobricks. *Vacuum*, *168*, 108861, 2019.
112. Jalilian, N., Ebrahimzadeh, H., Asgharinezhad, A.A., Determination of acidic, basic and amphoteric drugs in biological fluids and wastewater after their simultaneous dispersive micro-solid phase extraction using multiwalled carbon nanotubes/magnetite nanoparticles@ poly (2-aminopyrimidine) composite. *Microchem. J.*, *143*, 337–349, 2018.
113. Abbasghorbani, M., Fe_3O_4 loaded single wall carbon nanotubes and 1-methyl-3-octylimidazlium chloride as two amplifiers for fabrication of highly sensitive voltammetric sensor for epirubicin anticancer drug analysis. *J. Mol. Liq.*, *266*, 176–180, 2018.

114. Norouzi, P., Larijani, B., Alizadeh, T., Pourbasheer, E., Aghazadeh, M., Ganjali, M.R., Application of advanced electrochemical methods with nano-material-based electrodes as powerful tools for trace analysis of drugs and toxic compounds. *Curr. Anal. Chem.*, 15, 2, 143–151, 2019.
115. Upadhyay, S.S. and Srivastava, A.K., Hydroxypropyl β-cyclodextrin cross-linked multiwalled carbon nanotube-based chiral nanocomposite electrochemical sensors for the discrimination of multichiral drug atorvastatin isomers. *New J. Chem.*, 43, 28, 11178–11188, 2019.
116. Niezabitowska, E., Smith, J., Prestly, M.R., Akhtar, R., von Aulock, F.W., Lavallee, Y., Boucetta, H.A., McDonald, T.O., Facile production of nanocomposites of carbon nanotubes and polycaprolactone with high aspect ratios with potential drug delivery. *RSC Adv.*, 8, 16444, 2018.
117. Wang, Y., Luo, J., Liu, J., Li, X., Kong, Z., Jin, H., Cai, X., Electrochemical integrated paper-based immunosensor modified with multi-walled carbon nanotubes nanocomposites for point-of-care testing of 17β-estradiol. *Biosens. Bioelectron.*, 107, 47–53, 2018.
118. Miraki, M., Karimi-Maleh, H., Taher, M.A., Cheraghi, S., Karimi, F., Agarwal, S., Gupta, V.K., Voltammetric amplified platform based on ionic liquid/NiO nanocomposite for determination of benserazide and levodopa. *J. Mol. Liq.*, 278, 672–676, 2019.
119. Mazrouaa, A.M., Mansour, N.A., Abed, M.Y., Youssif, M.A., Shenashen, M.A., Awual, M.R., Nano-composite multi-wall carbon nanotubes using poly (p-phenylene terephthalamide) for enhanced electric conductivity. *J. Environ. Chem. Eng.*, 7, 2, 103002, 2019.
120. Cho, S., Takagi, K., Kwon, H., Seo, D., Ogawa, K., Kikuchi, K., Kawasaki, Multi-walled carbon nanotube-reinforced copper nanocomposite coating fabricated by low-pressure cold spray process. *Surf. Coat. Technol.*, 206, 16, 3488–3494, 2012.
121. Kim, J.S., Kwon, Y.S., Lomovsky, O.I., Dudina, D.V., Kosarev, V.F., Klinkov, S.V., Kwon, D.H., Smurov, I., Cold spraying of *in situ* produced TiB2-Cu nanocomposite powders. *Compos. Sci. Technol.*, 67, 11-12, 2292–2296, 2007.
122. Woo, D.J., Sneed, B., Peerally, F., Heer, F.C., Brewer, L.N., Hooper, J.P., Osswald, S., Synthesis of nanodiamond-reinforced aluminum metal composite powders and coatings using high-energy ball milling and cold spray. *Carbon*, 63, 404–415, 2013.
123. Zargar, R.A., Kumar, K., Mahmoud, Z.M.M., Shkir, M., AlFaify, S., Optical characteristics of ZnO films under different thickness: A MATLABbased computer calculation for photovoltaic applications. *Physica B*, 63, 414634, 2022.
124. Li, C.J., Yang, G.J., Gao, P.H., Ma, J., Wang, Y.Y., Li, C.X., Characterization of nanostructured WC-Co deposited by cold spraying. *J. Therm. Spray Technol.*, 16, 5, 1011–1020, 2007.

125. Luo, X.T. and Li, C.J., Thermal stability of microstructure and hardness of cold-sprayed cBN/NiCrAl nanocomposite coating. *J. Therm. Spray Technol.*, *21*, 3, 578–585, 2012.
126. Anitha, V.S., Sujatha Lekshmy, S., Joy, K., Effect of annealing on the structural, optical, electrical and photocatalytic activity of ZrO2–TiO2 nanocomposite thin films prepared by sol–gel dip coating technique. *J. Mater. Sci.: Mater. Electron.*, *28*, 14, 10541–10554, 2017.
127. Casula, M.F., Corrias, A., Falqui, A., Serin, V., Gatteschi, D., Sangregorio, C., de Julián Fernández, C., Battaglin, G., Characterization of FeCo– SiO_2 nanocomposite films prepared by sol– gel dip coating. *Chem. Mater.*, *15*, 11, 2201–2207, 2003.
128. Nagarajan, S., Mohana, M., Sudhagar, P., Raman, V., Nishimura, T., Kim, S., Kang, Y.S., Rajendran, N., Nanocomposite coatings on biomedical grade stainless steel for improved corrosion resistance and biocompatibility. *ACS Appl. Mater. Interfaces*, *4*, 10, 5134–5141, 2012.
129. Zhang, S., Sun, G., He, Y., Fu, R., Gu, Y., Chen, S., Preparation, characterization, and electrochromic properties of nanocellulose-based polyaniline nanocomposite films. *ACS Appl. Mater. Interfaces*, *9*, 19, 16426–16434, 2017.
130. Kaboorani, A., Auclair, N., Riedl, B., Landry, V., Mechanical properties of UV-cured cellulose nanocrystal (CNC) nanocomposite coating for wood furniture. *Prog. Org. Coat.*, *104*, 91–96, 2017.
131. Musil, J., Hard and superhard nanocomposite coatings. *Surf. Coat. Technol.*, *125*, 1-3, 322–330, 2000.
132. Davis, A., Yeong, Y.H., Steele, A., Bayer, I.S., Loth, E., Superhydrophobic nanocomposite surface topography and ice adhesion. *ACS Appl. Mater. Interfaces*, *6*, 12, 9272–9279, 2014.
133. Steele, A., Bayer, I., Loth, E., Inherently superoleophobic nanocomposite coatings by spray atomization. *Nano Lett.*, *9*, 1, 501–505, 2009.
134. Bu, A., Wang, J., Zhang, J., Bai, J., Shi, Z., Liu, Q., Ji, G., Corrosion behavior of ZrO2–TiO2 nanocomposite thin films coating on stainless steel through sol–gel method. *J. Sol-Gel Sci. Technol.*, *81*, 3, 633–638, 2017.
135. Sharma, A. and Singh, A.K., Electroless Ni-P-PTFE-Al_2O_3 dispersion nanocomposite coating for corrosion and wear resistance. *J. Mater. Eng. Perform.*, *23*, 1, 142–151, 2014.
136. Liu, C. and Zhao, Q., Influence of surface-energy components of Ni-P-TiO_2-PTFE nanocomposite coatings on bacterial adhesion. *Langmuir*, *27*, 15, 9512–9519, 2011.
137. Bogdanovic, U., Vodnik, V., Mitric, M., Dimitrijevic, S., Skapin, S.D., Zunic, V., Budimir, M., Stoiljkovic, M., Nanomaterial with high antimicrobial efficacy copper/polyaniline nanocomposite. *ACS Appl. Mater. Interfaces*, *7*, 3, 1955–1966, 2015.
138. Kaboorani, A., Auclair, N., Riedl, B., Landry, V., Mechanical properties of UV-cured cellulose nanocrystal (CNC) nanocomposite coating for wood furniture. *Prog. Org. Coat.*, *104*, 91–96, 2017.

139. Villafiorita-Monteleone, F., Canale, C., Caputo, G., Cozzoli, P.D., Cingolani, R., Fragouli, D., Athanassiou, A., Controlled swapping of nanocomposite surface wettability by multilayer photopolymerization. *Langmuir*, 27, 13, 8522–8529, 2011.
140. Hsu, S.H., Chang, Y.L., Tu, Y.C., Tsai, C.M., Su, W.F., Omniphobic low moisture permeation transparent polyacrylate/silica nanocomposite. *ACS Appl. Mater. Interfaces*, 5, 8, 2991–2998, 2013.
141. Chen, F., Wan, P., Xu, H., Sun, X., Flexible transparent supercapacitors based on hierarchical nanocomposite films. *ACS Appl. Mater. Interfaces*, 9, 21, 17865–17871, 2017.
142. Shabani-Nooshabadi, M., Ghoreishi, S.M., Jafari, Y., Kashanizadeh, N., Electrodeposition of polyaniline-montmorrilonite nanocomposite coatings on 316L stainless steel for corrosion prevention. *J. Polym. Res.*, 21, 4, 1–10, 2014.
143. Veprěk, S., Haussmann, M., Reiprich, S., Shizhi, L., Dian, J., Novel thermodynamically stable and oxidation resistant superhard coating materials. *Surf. Coat. Technol.*, 86, 394–401, 1996.
144. Neerinck, D., Persoone, P., Sercu, M., Goel, A., Kester, D., Bray, D., Diamond-like nanocomposite coatings (aC: H/a-Si: O) for tribological applications. *Diamond Relat. Mater.*, 7, 2-5, 468–471, 1998.
145. Männling, H.D., Patil, D.S., Moto, K., Jilek, M., Veprek, S., Thermal stability of superhard nanocomposite coatings consisting of immiscible nitrides. *Surf. Coat. Technol.*, 146, 263–267, 2001.
146. Hou, X., Choy, K.L., Brun, N., Serín, V., Nanocomposite coatings codeposited with nanoparticles using aerosol-assisted chemical vapour deposition. *J. Nanomater.*, 5, Article ID 219039, 8 pages, 2013.
147. Pei, Y.T., Chen, C.Q., Shaha, K.P., De Hosson, J.T.M., Bradley, J.W., Voronin, S.A., Čada, M., Microstructural control of TiC/aC nanocomposite coatings with pulsed magnetron sputtering. *Acta Mater.*, 56, 4, 696–709, 2008.
148. Zhang, S., Fu, Y., Du, H., Zeng, X.T., Liu, Y.C., Magnetron sputtering of nanocomposite (Ti, Cr) CN/DLC coatings. *Surf. Coat. Technol.*, 162, 1, 42–48, 2003.
149. Ribeiro, E., Malczyk, A., Carvalho, S., Rebouta, L., Fernandes, J.V., Alves, E., Miranda, A.S., Effects of ion bombardment on properties of dc sputtered superhard (Ti, Si, Al) N nanocomposite coatings. *Surf. Coat. Technol.*, 151, 515–520, 2002.
150. Ramazani, M., Ashrafizadeh, F., Mozaffarinia, R., Optimization of composition in Ni (Al)-Cr_2O_3 based adaptive nanocomposite coatings. *J. Therm. Spray Technol.*, 23, 6, 962–974, 2014.
151. Ali, O., Ahmed, R., Faisal, N.H., Alanazi, N.M., Berger, L.M., Kaiser, A., Toma, F.L., Polychroniadis, E.K., Sall, M., Elakwah, Y.O., Goosen, M.F.A., Influence of post-treatment on the microstructural and tribomechanical properties of suspension thermally sprayed WC–12 wt% Co nanocomposite coatings. *Tribol. Lett.*, 65, 2, 1–27, 2017.

152. Wang, C., Wang, Y., Wang, L., Hao, G., Sun, X., Shan, F., Zou, Z., Nanocomposite lanthanum zirconate thermal barrier coating deposited by suspension plasma spray process. *J. Therm. Spray Technol.*, 23, 7, 1030–1036, 2014.
153. Nguyen, T.A., Nguyen, T.V., Thai, H., Shi, X., Effect of nanoparticles on the thermal and mechanical properties of epoxy coatings. *J. Nanosci. Nanotechnol.*, 16, 9, 9874–9881, 2016.
154. Shi, X., Nguyen, T.A., Suo, Z., Liu, Y., Avci, R., Effect of nanoparticles on the anticorrosion and mechanical properties of epoxy coating. *Surf. Coat. Technol.*, 204, 3, 237–245, 2009.
155. Chapelle, A., Oudrhiri-Hassani, F., Presmanes, L., Barnabé, A., Tailhades, P., CO2 sensing properties of semiconducting copper oxide and spinel ferrite nanocomposite thin film. *Appl. Surf. Sci.*, 256, 14, 4715–4719, 2010.
156. Barsan, N. and Weimar, U., Understanding the fundamental principles of metal oxide based gas sensors; the example of CO sensing with SnO_2 sensors in the presence of humidity. *J. Phys.: Condens. Matter*, 15, 20, R813, 2003.
157. Takeuchi, T., Oxygen sensors. *Sens. Actuators*, 14, 2, 109–124, 1988.
158. Korotcenkov, G. and Cho, B.K., Ozone measuring: What can limit application of SnO2-based conductometric gas sensors? *Sens. Actuators B: Chem.*, 161, 1, 28–44, 2012.
159. Waitz, T., Wagner, T., Sauerwald, T., Kohl, C.D., Tiemann, M., Ordered mesoporous In2O3: Synthesis by structure replication and application as a methane gas sensor. *Adv. Funct. Mater.*, 19, 4, 653–661, 2009.
160. Yao, I., Zheng, X.H., Wu, R.J., Suyambrakasam, G., Chavali, M., Novel nano In_2O_3—WO_3 composite films for ultra trace level (ppb) detection of NO gas at room temperature. *Adv. Sci. Lett.*, 17, 1, 76–81, 2012.
161. Kaur, J., Kumar, R., Bhatnagar, M.C., Effect of indium-doped SnO_2 nanoparticles on NO2 gas sensing properties. *Sens. Actuators B: Chem.*, 126, 2, 478–484, 2007.
162. Zhang, D., Liu, J., Jiang, C., Liu, A., Xia, B., Quantitative detection of formaldehyde and ammonia gas via metal oxide-modified graphene-based sensor array combining with neural network model. *Sens. Actuators B: Chem.*, 240, 55–65, 2017.
163. Zheng, M., Tang, H., Li, L., Hu, Q., Zhang, L., Xue, H., Pang, H., Hierarchically nanostructured transition metal oxides for lithium-ion batteries. *Adv. Sci.*, 5, 3, 1700592, 2018.
164. Su, H., Jaffer, S., Yu, H., Transition metal oxides for sodiumion batteries. *Energy Storage Mater.*, 5, 116–131, 2016.
165. Jose, R., Thavasi, V., Ramakrishna, S., Metal oxides for dye-sensitized solar cells. *J. Am. Ceram. Soc.*, 92, 2, 289–301, 2009.
166. Kim, J.H., Kang, T.W., Kwon, S.N., Na, S.I., Yoo, Y.Z., Im, H.S., Seong, T.Y., Transparent conductive ITO/Ag/ITO electrode deposited at room temperature for organic solar cells. *J. Electron. Mater.*, 46, 1, 306–311, 2017.

167. Girtan, M., Comparison of ITO/metal/ITO and ZnO/metal/ZnO characteristics as transparent electrodes for third generation solar cells. *Sol. Energy Mater. Sol. Cells*, *100*, 153–161, 2012.
168. Cao, J., Ertekin, E., Srinivasan, V., Fan, W., Huang, S., Zheng, H., Yim, J.W.L., Khanal, D.R., Ogletree, D.F., Grossman, J.C., Wu, J., Strain engineering and one-dimensional organization of metal–insulator domains in single-crystal vanadium dioxide beams. *Nat. Nanotechnol.*, *4*, 11, 732–737, 2009.
169. Liu, M., Hwang, H.Y., Tao, H., Strikwerda, A.C., Fan, K., Keiser, G.R., Sternbach, A.J., West, K.G., Kittiwatanakul, S., Lu, J., Wolf, S.A., Terahertz-field-induced insulator-to-metal transition in vanadium dioxide metamaterial. *Nature*, *487*, 7407, 345–348, 2012.
170. Banerjee, A.N., Kundoo, S., Saha, P., Chattopadhyay, K.K., Synthesis and characterization of nano-crystalline fluorine-doped tin oxide thin films by sol-gel method. *J. Sol-Gel Sci. Technol.*, *28*, 1, 105–110, 2003.
171. Banerjee, A.N., Kundoo, S., Saha, P., Chattopadhyay, K.K., Synthesis and characterization of nano-crystalline fluorine-doped tin oxide thin films by sol-gel method. *J. Sol-Gel Sci. Technol.*, *28*, 1, 105–110, 2004.
172. Hwang, D., Ryu, S.G., Misra, N., Jeon, H., Grigoropoulos, C.P., Nanoscale laser processing and diagnostics. *Appl. Phys. A*, *96*, 2, 289–306, 2009.
173. Cheng, J.G., Wang, J., Dechakupt, T., Trolier-McKinstry, S., Low-temperature crystallized pyrochlore bismuth zinc niobate thin films by excimer laser annealing. *Appl. Phys. Lett.*, *87*, 23, 232905, 2005.
174. Wang, H., Pyatenko, A., Kawaguchi, K., Li, X., Swiatkowska-Warkocka, Z., Koshizaki, N., Selective pulsed heating for the synthesis of semiconductor and metal submicrometer spheres. *Angew. Chem.*, *122*, 36, 6505–6508, 2010.
175. Queralto, A., Perez del Pino, A., De La Mata, M., Arbiol, J., Tristany, M., Obradors, X., Puig, T., Ultrafast epitaxial growth kinetics in functional oxide thin films grown by pulsed laser annealing of chemical solutions. *Chem. Mater.*, *28*, 17, 6136–6145, 2016.
176. Hong, S., Lee, H., Yeo, J., Ko, S.H., Digital selective laser methods for nanomaterials: From synthesis to processing. *Nano Today*, *11*, 5, 547–564, 2016.
177. Mincuzzi, G., Palma, A.L., Di Carlo, A., Brown, T.M., Laser processing in the manufacture of dye-sensitized and perovskite solar cell technologies. *ChemElectroChem*, *3*, 1, 9–30, 2016.
178. Zhang, Y.-L., Chen, D.-Q., Xia, H., Sun, H.-B., Designable 3D nanofabrication by femtosecond laser direct writing, *Nano Today*, *5*, 435, 2010.
179. Ramanathan, S., *Thin film metal-oxides*, Harvard University: Springer New York Dordrecht Heidelberg London, London, 2010.
180. Bharadwaja, S.S.N., Rajashekhar, A., Ko, S.W., Qu, W., Motyka, M., Podraza, N., Clark, T., Randall, C.A., Trolier-McKinstry, S., Excimer laser assisted re-oxidation of BaTiO3 thin films on Ni metal foils. *J. Appl. Phys.*, *119*, 2, 024106, 2016.

181. Kang, M.G., Cho, K.H., Ho Do, Y., Lee, Y.J., Nahm, S., Yoon, S.J., Kang, C.Y., Large in-plane permittivity of Ba0.6Sr0.4TiO$_3$ thin films crystallized using excimer laser annealing at 300° C. *Appl. Phys. Lett.*, *101*, 24, 242910, 2012.
182. Xianyu, W.X., Cho, H.S., Kwon, J.Y., Yin, H.X., Noguchi, T., Excimer (XeCl) laser annealing of PbZr0.4Ti0.6O3 thin film at low temperature for TFT FeRAM application. *MRS Online Proc. Lib. (OPL)*, *830*, 183–188, 2004.
183. Queraltó, A., Pérez del Pino, A., De La Mata, M., Arbiol, J., Tristany, M., Gómez, A., Obradors, X., Puig, T., Growth of ferroelectric Ba0.8Sr0.2TiO3 epitaxial films by ultraviolet pulsed laser irradiation of chemical solution derived precursor layers. *Appl. Phys. Lett.*, *106*, 26, 262903, 2015.
184. Palneedi, H., Maurya, D., Kim, G.Y., Annapureddy, V., Noh, M.S., Kang, C.Y., Kim, J.W., Choi, J.J., Choi, S.Y., Chung, S.Y., Kang, S.J.L., Unleashing the full potential of magnetoelectric coupling in film heterostructures. *Adv. Mater.*, *29*, 10, 1605688, 2017.
185. Molaei, R., Bayati, M.R., Alipour, H.M., Nori, S., Narayan, J., Enhanced photocatalytic efficiency in zirconia buffered n-NiO/p-NiO single crystalline heterostructures by nanosecond laser treatment. *J. Appl. Phys.*, *113*, 23, 233708, 2013.
186. Hou, Y. and Jayatissa, A.H., Effect of laser irradiation on gas sensing properties of sol–gel derived nanocrystalline Al-doped ZnO thin films. *Thin Solid Films*, *562*, 585–591, 2014.
187. Lee, D., Pan, H., Ko, S.H., Park, H.K., Kim, E., Grigoropoulos, C.P., Non-vacuum, single-step conductive transparent ZnO patterning by ultra-short pulsed laser annealing of solution-deposited nanoparticles. *Appl. Phys. A*, *107*, 1, 161–171, 2012.
188. Lee, C., Srisungsitthisunti, P., Park, S., Kim, S., Xu, X., Roy, K., Janes, D.B., Zhou, C., Ju, S., Qi, M., Control of current saturation and threshold voltage shift in indium oxide nanowire transistors with femtosecond laser annealing. *ACS Nano*, *5*, 2, 1095–1101, 2011.
189. Dong, W.J., Ham, J., Jung, G.H., Son, J.H., Lee, J.L., Ultrafast laser-assisted synthesis of hydrogenated molybdenum oxides for flexible organic solar cells. *J. Mater. Chem. A*, *4*, 13, 4755–4762, 2016.
190. Nakajima, T. and Tsuchiya, T., Flexible thermistors: Pulsed laser-induced liquid-phase sintering of spinel Mn–Co–Ni oxide films on polyethylene terephthalate sheets. *J. Mater. Chem. C*, *3*, 15, 3809–3816, 2015.
191. Nakajima, T., Tsuchiya, T., Kumagai, T., Pulsed laser-induced oxygen deficiency at TiO2 surface: Anomalous structure and electrical transport properties. *J. Solid State Chem.*, *182*, 9, 2560–2565, 2009.
192. Terakado, N., Takahashi, R., Takahashi, Y., Fujiwara, T., Synthesis of chain-type SrCuO2 by laser irradiation on sputtered layer-type SrCuO2 film. *Thin Solid Films*, *603*, 303–306, 2016.
193. Molaei, R., Bayati, R., Nori, S., Kumar, D., Prater, J.T., Narayan, J., Diamagnetic to ferromagnetic switching in VO2 epitaxial thin films by nanosecond excimer laser treatment. *Appl. Phys. Lett.*, *103*, 25, 252109, 2013.

194. Rao, S.S., Lee, Y.F., Prater, J.T., Smirnov, A.I., Narayan, J., Laser annealing induced ferromagnetism in SrTiO$_3$ single crystal. *Appl. Phys. Lett.*, *105*, 4, 042403, 2014.
195. Zhang, Y.Q., Ruan, X.Z., Liu, B., Xu, Z.Y., Xu, Q.Y., Shen, J.D., Li, Q., Wang, J., You, B., Tu, H.Q., Gao, Y., Fast laser annealing induced exchange bias in poly-crystalline BiFeO$_3$/Co bilayers. *Appl. Surf. Sci.*, *367*, 418–423, 2016.
196. Wang, X., Ding, Y., Yuan, D., Hong, J.I., Liu, Y., Wong, C.P., Hu, C., Wang, Z.L., Reshaping the tips of ZnO nanowires by pulsed laser irradiation. *Nano Res.*, *5*, 6, 412–420, 2012.
197. Molaei, R., Bayati, R., Narayan, J., Crystallographic characteristics and p-type to n-type transition in epitaxial NiO thin film. *Cryst. Growth Des.*, *13*, 12, 5459–5465, 2013.
198. Pan, X., Shuai, Y., Wu, C., Luo, W., Sun, X., Yuan, Y., Zhou, S., Ou, X., Zhang, W., Resistive switching behavior in single crystal SrTiO3 annealed by laser. *Appl. Surf. Sci.*, *389*, 1104–1107, 2016.
199. Silva, J.P.B., Kamakshi, K., Sekhar, K.C., Moreira, J.A., Almeida, A., Pereira, M., Gomes, M.J.M., Light-controlled resistive switching in laser-assisted annealed Ba0. 8Sr0. 2TiO$_3$ thin films. *Phys. Status Solidi (a)*, *213*, 4, 1082–1087, 2016.
200. Charipar, N.A., Kim, H., Breckenfeld, E., Charipar, K.M., Mathews, S.A., Piqué, A., Polycrystalline VO2 thin films via femtosecond laser processing of amorphous VO x. *Appl. Phys. A*, *122*, 5, 1–5, 2016.
201. Ko, W.B., Lee, J.S., Lee, S.H., Cha, S.N., Sohn, J.I., Kim, J.M., Park, Y.J., Kim, H.J., Hong, J.P., Luminance behavior of lithium-doped ZnO nanowires with p-type conduction characteristics. *J. Nanosci. Nanotechnol.*, *13*, 9, 6231–6235, 2013.
202. Allemann, I.B. and Kaufman, J., Fractional photothermolysis, in: *Basics in Dermatological Laser Applications*, vol. 42, pp. 56–66, Karger Publishers, Switzerland, 2011.
203. Brown, M.S. and Arnold, C.B., Fundamentals of laser-material interaction and application to multiscale surface modification, in: *Laser Precision Microfabrication*, pp. 91–120, Springer, Berlin, Heidelberg, 2010.
204. Bäuerle, D., Material transformations, laser cleaning, in: *Laser Processing and Chemistry*, pp. 535–559, Springer, Berlin, Heidelberg, 2011.
205. Ganeev, R.A., Principles of lasers and laser-surface interactions, in: *Laser-Surface Interactions*, pp. 1–21, Springer, Dordrecht, 2014.
206. Cullis, A.G., Transient annealing of semiconductors by laser, electron beam and radiant heating techniques. *Rep. Prog. Phys.*, *48*, 8, 1155, 1985.
207. Schaaf, P., (Ed.) *Laser processing of materials: Fundamentals, applications and developments*, vol. 139, Springer Science & Business Media, Germany, 2010.
208. Sundaram, S.K. and Mazur, E., Femtosecond material science–inducing and probing nonthermal transitions in semiconductors. *Nat. Mater.*, *1*, 217–24, 2002.

209. Joe, D.J., Kim, S., Park, J.H., Park, D.Y., Lee, H.E., Im, T.H., Choi, I., Ruoff, R.S., Lee, K.J., Laser–material interactions for flexible applications. *Adv. Mater., 29,* 26, 1606586, 2017.
210. Palneedi, H., Maurya, D., Kim, G.Y., Priya, S., Kang, S.J.L., Kim, K.H., Choi, S.Y., Ryu, J., Enhanced off-resonance magnetoelectric response in laser annealed PZT thick film grown on magnetostrictive amorphous metal substrate. *Appl. Phys. Lett., 107,* 1, 012904, 2015.
211. Palneedi, H., Choi, I., Kim, G.Y., Annapureddy, V., Maurya, D., Priya, S., Kim, J.W., Lee, K.J., Choi, S.Y., Chung, S.Y., Kang, S.J.L., Tailoring the magnetoelectric properties of Pb (Zr, Ti) O3 film deposited on amorphous metglas foil by laser annealing. *J. Am. Ceram. Soc., 99,* 8, 2680–2687, 2016.
212. Choi, I., Jeong, H.Y., Jung, D.Y., Byun, M., Choi, C.G., Hong, B.H., Choi, S.Y., Lee, K.J., Laser-induced solid-phase doped graphene. *ACS Nano, 8,* 8, 7671–7677, 2014.
213. White, C.W., Appleton, B.R., Wilson, S.R., Poate, J.M., Mayer, J.W., Supersaturated alloys, solute trapping, and zone refining, in: *Laser Annealing of Semiconductors,* pp. 111–146, 1982.
214. Kang, B., Han, S., Kim, J., Ko, S., Yang, M., One-step fabrication of copper electrode by laser-induced direct local reduction and agglomeration of copper oxide nanoparticle. *J. Phys. Chem. C, 115,* 48, 23664–23670, 2011.
215. Campbell, P., Enhancement of light absorption from randomizing and geometric textures. *JOSA B, 10,* 12, 2410–2415, 1993.
216. Li, D. and McGaughey, A.J., Phonon dynamics at surfaces and interfaces and its implications in energy transport in nanostructured materials—An opinion paper. *Nanoscale Microscale Thermophys. Eng., 19,* 2, 166–182, 2015.
217. Kovacs, J.Z., Andresen, K., Pauls, J.R., Garcia, C.P., Schossig, M., Schulte, K., Bauhofer, W., Analyzing the quality of carbon nanotube dispersions in polymers using scanning electron microscopy. *Carbon, 45,* 6, 1279–1288, 2007.
218. Qian, D. and Dickey, E.C., In-situ transmission electron microscopy studies of polymer–carbon nanotube composite deformation. *J. Microsc., 204,* 1, 39–45, 2001.
219. Barsan, O.A., Hoffmann, G.G., van der Ven, L.G., de With, G., Quantitative conductive atomic force microscopy on single-walled carbon nanotube-based polymer composites. *ACS Appl. Mater. Interfaces, 8,* 30, 19701–19708, 2016.
220. Chapkin, W.A., McNerny, D.Q., Aldridge, M.F., He, Y., Wang, W., Kieffer, J., Taub, A.I., Real-time assessment of carbon nanotube alignment in a polymer matrix under an applied electric field via polarized Raman spectroscopy. *Polym. Test., 56,* 29–35, 2016.
221. Wang, D. and Russell, T.P., Advances in atomic force microscopy for probing polymer structure and properties. *Macromolecules, 51,* 1, 3–24, 2018.
222. Shanmugam, G. and Isaiah, M.V., Structural and optical properties of PbS-PVA, CdS-PVA and PbS-CdS-PVA nanocomposite films. *Int. J. Chemtech. Res., 10,* 229–234, 2017.

223. Tewari, S. and Bhattacharjee, A., Structural, electrical and optical studies on spraydeposited aluminium doped Zno doped films. *Pramana J. Phys.*, 76, 153–163, 2011.
224. Ghaderi, A., Elahi, S.M., Solaymani, S., Naseri, M., Ahmadirad, M., Bahrami, S., Khalili, A.E., Thickness dependence of the structural and electrical properties of ZnO thermal-evaporated thin films. *Pramana*, 77, 6, 1171–1178, 2011.
225. Bhosale, C.H., Kambale, A.V., Kokate, A.V., Rajpure, K.Y., Structural, optical and electrical properties of chemically sprayed CdO thin films. *Mater. Sci. Eng.: B*, 122, 1, 67–71, 2005.
226. Patil, L.A., Bari, A.R., Deo, V., Ultrasonically synthesized nanocrystalline ZnO powder-based thick film sensor for ammonia sensing. *Sens. Rev.*, 2010.
227. Tauc, J. (Ed.), *Amorphous and liquid semiconductors*, Springer Science & Business Media, Germany, 30, 4, 290–296, 2012.
228. Nandi, S.K., Chakraborty, S., Bera, M.K., Maiti, C.K., Structural and optical properties of ZnO films grown on silicon and their applications in MOS devices in conjunction with ZrO2 as a gate dielectric. *Bull. Mater. Sci.*, 30, 3, 247–254, 2007.
229. Vijayalakshmi, S., Venkataraj, S., Jayavel, R., Characterization of cadmium doped zinc oxide (Cd: ZnO) thin films prepared by spray pyrolysis method. *J. Phys. D: Appl. Phys.*, 41, 24, 245403, 2008.
230. Özgür, Ü., Alivov, Y.I., Liu, C., Teke, A., Reshchikov, M., Doğan, S., Avrutin, V.C.S.J., Cho, S.J., Morkoç, A.H., A comprehensive review of ZnO materials and devices. *J. Appl. Phys.*, 98, 4, 11, 2005.
231. Hofmann, D.M., Hofstaetter, A., Leiter, F., Zhou, H., Henecker, F., Meyer, B.K., Orlinskii, S.B., Schmidt, J., Baranov, P.G., Hydrogen: A relevant shallow donor in zinc oxide. *Phys. Rev. Lett.*, 88, 4, 045504, 2002.
232. Zeuner, A., Alves, H., Hofmann, D.M., Meyer, B.K., Heuken, M., Bläsing, J., Krost, A., Structural and optical properties of epitaxial and bulk ZnO. *Appl. Phys. Lett.*, 80, 12, 2078–2080, 2002.
233. Schifano, R., Monakhov, E.V., Grossner, U., Svensson, B.G., Electrical characteristics of palladium Schottky contacts to hydrogen peroxide treated hydrothermally grown ZnO. *Appl. Phys. Lett.*, 91, 19, 193507, 2007.
234. Scheffler, L., Kolkovsky, V., Lavrov, E.V., Weber, J., Deep level transient spectroscopy studies of n-type ZnO single crystals grown by different techniques. *J. Phys.: Condens. Matter*, 23, 33, 334208, 2011.
235. Chicot, G., Muret, P., Santailler, J.L., Feuillet, G., Pernot, J., Oxygen vacancy and EC– 1 eV electron trap in ZnO. *J. Phys. D: Appl. Phys.*, 47, 46, 465103, 2014.
236. Gür, E., Coşkun, C., Tüzemen, S., High energy electron irradiation effects on electrical properties of Au/n-ZnO Schottky diodes. *J. Phys. D: Appl. Phys.*, 41, 10, 105301, 2008.

237. Ohbuchi, Y.O.Y., Kawahara, T.K.T., Okamoto, Y.O.Y., Morimoto, J.M.J., Distributions of interface states and bulk traps in ZnO varistors. *Jpn. J. Appl. Phys.*, *40*, 1R, 213, 2011.
238. Lang, D.V., Fast capacitance transient appartus: Application to ZnO and O centers in GaP p-n junctions. *J. Appl. Phys.*, *45*, 7, 3014–3022, 1974.
239. Schroder, D.K., *Semiconductor material and device characterization*, p. 101, Jhon Wiley & Sons. Inc., Publication, Canada, 2006.

14

Manganese Dioxide as a Supercapacitor Material

Mudasir Hussain Rather[1*], Feroz A. Mir[1], Peerzada Ajaz Ahmad[1], Rayaz Ahmad[2] and Kaneez Zainab[3]

[1]Department of Physics, Baba Ghulam Shah Badshah University, Rajouri (J&K), India
[2]Department of Physics, National Institute of Technology, Srinagar (J&K), India
[3]Department of Botany, Sri Pratap College, Cluster University, Srinagar (J&K), India

Abstract

The creation of safe and practical energy storage technologies has emerged as one of the most crucial issues in the field of sustainable development in the modern world due to the serious environmental effects of fossil fuels and the explosive growth of the global economy. Supercapacitors are a brand-new kind of energy storage technology with benefits for the economy. They have sufficient power density, an extended cycle life, and a wide temperature range. Manganese dioxides (MnO_2) are employed in industry for greater than a century, due to its inexpensive price, widespread availability in nature, and environmental friendliness. MnO_2 is a promising electrode to overcome in the energy storage devices. In this chapter, we are going to describe the characteristics of MnO_2 nanomaterials with various morphologies. Two methods which are benefited for the synthesis, electrochemical and other properties are mentioned. Evidence exists for the influence of the shape (morphology) of nanosized MnO_2 particles on those traits. Specific capacitance (C_p), Power density (P_d), Energy density (E_d) or many other parameters of supercapacitors are examined.

Keywords: MnO_2, specific capacitance, supercapacitor, composite, power density

---

*Corresponding author: mudasirh437@gmail.com

Rayees Ahmad Zargar (ed.) Metal Oxide Nanocomposite Thin Films for Optoelectronic Device Applications, (361–398) © 2023 Scrivener Publishing LLC

14.1 Introduction

Energy crises and environmental deterioration have appeared one after the other as a result of the fast expanding global population and energy use. The global ecosystem has been severely harmed, and even the lives of humans and their health are in risk due to the expanding scarcity and exhaustion of non-renewable energy sources like coal, gold, oil, petroleum, natural gas, linked with the greenhouse gases and hazardous compounds released at the time of extraction process [1–3]. Renewable energy sources are required for massive power networks and power generation [4], on the side of reduce the rising worldwide need for non-renewable energy sources, the reduction of fossil fuels, as well as environmental issues brought on by the burning of fossil fuels [5]. Therefore, obtaining and using power produced from renewable and clean energy sources is a final objective. Since renewable energy sources like wind, solar, hydrothermal, tidal, and geothermal energy are often recurrent and decentralized in nature, electric energy storage (EES), as a crucial component of attaining this aim, does provide a viable way to improve grid stability. Electrochemical approaches are ambitious in relating to specific energy, adaptability, also scalability among the EES technologies currently on the market [6].

Due to its immense energy density, protracted cycle life, excellent stability, and high performing voltage [7, 8], metal-ion batteries (MIBs) have received extensive study as an energy storage device, making them the most promising EES device at the moment. Sodium-ion batteries (SIBs), zinc-ion batteries (ZIBs) [9–11], and magnesium-ion batteries (MIBs) are common contenders to replace lithium-ion batteries (LIBs), which now dominate the market for rechargeable batteries for portable electronics. On the other hand, SCs have a lot of potential for bridging the gap in batteries because to their great power density also extended cycle life [12, 13].

Manganese dioxides (MDOs) were used by the Magdalenian culture as early as the Paleolithic period as black pigments for painting on rock surfaces [14]. They are the first nanomaterial that human civilization has employed up to that point. Now, MnO_2 is a significant functional metal oxide with technologically appealing uses as catalysts [15, 16], toxic metal absorbents [17], and artificial oxidase [18], also an inorganic ceramic pigment, electrodes for batteries, [19, 20] or electrodes for SCs [21, 22]. It has also been (MnO_2) extensively employed in photo catalytic processes, electrolysis, and Duracell (alkaline) based barriers [23]. MnO_2 has also been shown to have significant uses in water purification due to its capacity to absorb harmful ions. Due to its inevitable structure, metal oxides are non-stoichiometric compounds [24].

The first supercapacitor patent was issued to Becker H. I. in 1957 [25], who created supercapacitors with a high specific surface area based on carbon. It gained widespread use as a hybrid power supply in the 1990s as people started to recognize the possibilities of supercapacitor applications [26]. Figure 14.1 shows the Ragone plots for distinct energy-storage devices.

Supercapacitors are a common component of modern energy storage devices, but when it comes to electrochemical energy storage technologies, they have a larger market for applications and far more promising futures. Presently, the major investors who manufacture the electrochemical capacitors are Matsushita including NEC Corporation of Japan, as well as Maxwell Technologies of the United States [27]. SCs offer enormous application value, market potential, and have emerged as research hotspots on a global scale.

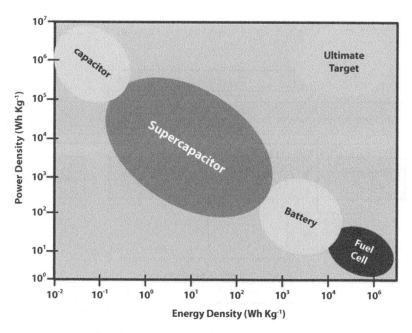

Figure 14.1 Ragone plots for various energy-storage technologies.

14.1.1 Classifications of Supercapacitors

Supercapacitors are essentially divided into three sorts based on the charge storage method used, (a) Electrochemical Double Layer Capacitor (EDLC) (b) Pseudo Capacitors (PC) (c) Hybrid Capacitors. Figure 14.2 shows the classifications of SCs.

14.1.2 Electrochemical Double Layer Capacitor

Electrical double-layer capacitor is made up of activated carbon, Positive and negative charges build up at the electrode and electrolyte interfaces to produce an electrical double layer, which is how the charge storage mechanism was developed. The capacitance is created by the charges accumulating. There is no exchange of electrons in the charge storage mechanism, but the specific capacitance (C_p) is confided in the activated carbon material. To achieve immense C_p, since they have such a wide surface area, carbon-based compounds like commercial activated carbons [28], templated mesoporous carbons [29], graphene [30] and efficient electrode material [31] are most frequently employed. Also EDLCs are studied for specific surface area, charge retention and open porosity for transfer of electrolyte

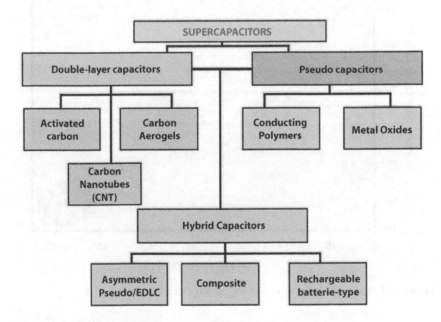

Figure 14.2 Different types of supercapacitor.

i.e., ions. Equation 14.1, which is given below, provides a general estimation of the capacitance of an EDLC electrode.

$$C = \frac{\varepsilon_r \varepsilon_0}{d} A \qquad (14.1)$$

Here, ε_r refers to the utilized liquid electrolyte's relative permittivity, ε_0 is the permittivity of vacuum, A is the actual surface area of the electrode materials that are exposed to the electrolyte ions, d is the difference between electrical double layers i.e., the Debye length.

14.1.3 Pseudocapacitors

Pseudocapacitors are also known as faradic supercapacitors, which are different from EDLCs. The mechanism of storing the electrical energy of these electrodes is different from that of EDLCs. These Pseudocapacitors are also redox supercapacitors because they have huge specific energy and store immense charge compared to EDLCs. In contrast to EDLC, pseudocapacitors have a lower specific power because the redox reaction process requires more time. A redox reaction occurs between the electrode and electrolyte when we apply an external potential to this capacitor. This capacitor's mechanism for charging and discharging is similar to that of an electric battery. The electrodes of this capacitor are split apart by an electrolyte [32].

Although nickel oxide/hydroxide and cobalt oxide/hydroxide are employed in batteries for a number of years, active materials like MnO_2, NiO, Fe_3O_4, and RuO_2 have recently attracted significant attention for the application of supercapacitors and are regarded as the true examples of pseudocapacitive materials. RuO_2 and MnO_2 are two of these oxides that have been the subject of the most research [33]. Both oxides have a variety of oxidation states and are naturally conductive. Their charge storage method is based on the incorporation of electrolyte cations or protons on their surface via electro-adsorption.

14.1.4 Hybrid Capacitors

The major reason for developing the hybrid capacitor was to lessen the shortcomings of both EDLC as well as pseudocapacitors. In order to achieve superior performance characteristics, hybrid capacitors are required to take advantage of the respective advantages and reduce the relative disadvantages of EDLC and pseudocapacitor. It stores charge using both non-Faradaic and Faradaic mechanism. Large energy and power densities than EDLCs have been attained by hybrid capacitors without sacrificing the cycling stability and affordability that have hampered the use of pseudocapacitors. In hybrid supercapacitors, these two charge-storage systems are combined that is Faradaic and non-Faradaic, leading to enhanced device properties. Electrodes made of carbon are often effective for EDLCs. Electrodes made of carbon are often effective for EDLCs, whereas electrically conductive polymers and transition metal oxides are advantageous for the development of pseudocapacitors [34]. Researchers are currently concentrating on different forms of hybrid capacitors, which can be classified as composite, asymmetric, or battery-type depending on how their electrodes are arranged. The creation of asymmetric or hybrid supercapacitors has become popular recently. One electrode can be made of a double-layer carbon material, and the second electrode can be made of a pseudo-capacitance material to create an asymmetric supercapacitor. By choosing the right electrode material, it is possible to achieve the high operating voltage and high energy density, which significantly increased the overall energy density of the supercapacitor devices [35].

14.2 Supercapacitor Components

In the end, supercapacitors components determine whether it will work as expected. These elements can be modified for certain applications; however, there are specifications for these parts that apply to the creation of all EC systems. The electrode, electrolyte, separator, and current collector must all be taken into account for each component in order to achieve the goal of minimizing resistances while optimizing the ability to produce capacitive charge.

14.2.1 Electrode

An electrode for EC supercapacitors is different from a battery electrode in a lot of ways. The majority of commercially available supercapacitors nowadays are made with symmetric electrodes (the cathode as well as the anode is made of the same substance), where no oxidation-reduction processes take place and the optimum electric double-layer capacitance develops as a pure electrostatic charge. Therefore, identifying the anode and cathode electrodes can be done by observing which electrodes are positively and negatively charged during discharge. With the development of asymmetric capacitors, this nomenclature may alter; the difference between anode and cathode, however, often depends on the appropriate working potential window for the materials that are integrated. The qualities that a material must possess in order to serve as an electrode for EC supercapacitors are stated below [36].

14.2.2 Electrolytes

Supercapacitors often use electrolytes, which can be divided into three categories: Aqueous (means atoms in H_2O), organic (means salts in organic substance), and ionic liquids salt [37].

The electrolyte is one of the important factors in determining a supercapacitors working voltage. Efficiency and cyclicity are also concerned with conductivity and stability. The many types of aqueous electrolytes include acids (H_2SO_4, HCl), alkali (NaOH), and neutral (Na_2SO_4). They often also have the benefits of strong ionic conductivity, low price, simplicity of handling, and non-flammability. Due to their superior electronic conductivity, large dielectric constant, and higher available surface area of lower aqueous electrolyte ions, compared to supercapacitors employing the same electrodes in non-aqueous electrolytes, those using aqueous electrolytes can obtain greater capacitances. There is also a disadvantage to the voltage range due to its small thermodynamic and bearable stability, as well as

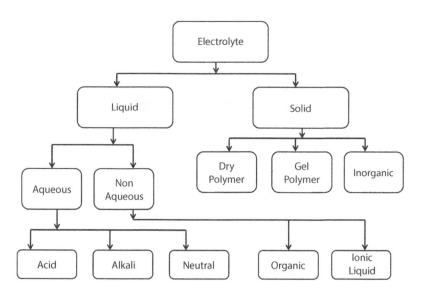

Figure 14.3 Shows different types of electrolytes of supercapacitors.

when considering the cell long-term capacitance operation. Additionally, when employing alkali or acidic electrolytes, special care must be taken in the selection of current collectors to prevent corrosion [38]. Figure 14.3 shows various types of electrolytes of a SC.

14.2.3 Separators

In addition to the electrode, and electrolyte, the separator is yet another crucial component of SCs. Usually, separators are made of polymer, paper, ceramic, glass, etc. [39]. It accomplishes a number of crucial activities, such as the transportation of electrons, increasing porosity, and also ensuring the conductivity of the mentioned electrolyte. As a result, the practical characteristics used to describe the separator material are quite strict and frequently at odds with one another [40].

14.2.4 Current Collector

The main work of the current collector is to assemble the electrons from the active material and move them to the outside circuit. For this, metal plates made of copper, aluminum, alloys, and steel are employed. The electrochemical process and stability of the device are significantly influenced by the current collector. As shown in Figure 14.4, two current collectors are often utilized on the cathode and anode surfaces of supercapacitors [41].

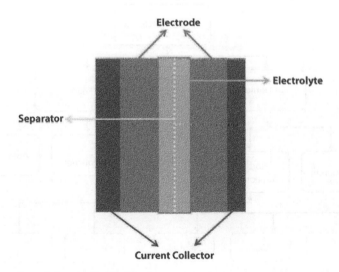

Figure 14.4 Structure of supercapacitor.

14.3 Methods for MnO$_2$ Nanoparticles

MnO$_2$ is synthesized via different techniques but electrochemical and sol-gel is more preferable [42–45] used in variety of technologies, and environmentally friendly wet-chemical [46, 47] procedures have been reported. Different techniques have been developed to produce nanostructured MnO$_2$ samples, by Simple reduction [48], co-precipitation [49], sol-gel [45], thermal breakdown [50], the hydrothermal [51] are some of the synthetic processes for nanostructured MnO$_2$ (see [52] for a review). From the years, different nano-structured things have been studied, including Nanowires [53, 54], nano-rods [55, 56], nano-flowers [57], nanosheets [58], nanoflakes [59], nanotubes [60, 61], nano-urchins [62, 63], nano-spheres [64], nano-belts [65], nano-disks [66], and nanofibers [67, 68].

14.3.1 Hydrothermal Route

The temperature, pressure vessel fill level, and solvent all have an impact on the kind of MDO nanomaterial's that are produced using this technique. Subramanian *et al.* [69] generated various nano-architectures of MnO$_2$ particles by adjusting the time of hydrothermal reaction to breakdown of KMnO$_4$ and MnSO$_4$·H$_2$O within the warm aqueous solution. The hydrothermal duration ranged from 1-18 hours. They found that for the combination warmed up at 140 C for an hour, MnO$_2$ developed into flower-like nanowhiskers, which after 12 h converted into α-MnO$_2$ nanorods One-dimensional (1D) nanorods, nanotubes, and microsphere/nanosheets core-corona hierarchical architectures were the three types of MnO$_2$ nanostructures that Xiao *et al.* [70] produced using the hydrothermal technique in an autoclave warmed at various temperatures from 100 to 200°C with a time period of 12 hours. In a normal synthesis, concentrated H$_2$SO$_4$ or HCl with a 37 wt% were combined to deionized water to provide acidic conditions necessary for the hydrothermal breakdown of a single KMnO$_4$ to form the α-MnO$_2$ phase [71]. Birnessite-type δ-MnO$_2$ which is monoclinic and having C2/m space group was produced at a temperature of 100 °C, whereas perfect α-MnO$_2$ nanorods which is tetragonal and having I4/m space group were crystallized at a temperature of 120 °C and changed into α-MnO$_2$ 1D nanotubes at a temperature of 140 °C. Using oxidizing MnSO$_4$ in the existence of poly(vinyl pyrrolidone) (PVP) with NaClO$_3$, single-crystal β-MnO$_2$ nanotubes having 200 to 500 nm diameters, was created by applying hydrothermal technique [66]. By practicing a standard hydrothermal technique with pH as well as NH$_4^+$ cation concentration modification, Wang et al. [71, 72] created a variety of nanostructured MnO_2 polymorphs (α, β, ϒ, and δ of MnO_2). The creation of layered δ-MnO_2 is the same in all nanosamples. Hashem, A.M. shows the growth of β- and ϒ-MnO_2 as nanowires/nanorods in contrast to the fiber or needle morphologies of α-MnO_2 and todorokite-type MnO_2 [63]. Employing $KMnO_4$ and $MnSO_4$ blended in a water and ethanol solution(4:1) and poly(sodium 4-styrene-sulfonate), good crystallized nanorods of α-MnO_2 (having diameter 12 nm) was created by the process of hydrothermal approach [73]. To create α-MnO_2 nanowires and nanotubes, a polymer-precursor method based on polyethylene glycol (PEG) was used. In a typical synthesis, PEG-6000 dissolved in aqueous methanol solution and $MnSO_4$ aqueous solution were combined to create the precursor, to which NaOH was then added. After being subjected to a 20 hour autoclave process at 120 °C, the nanotubes were produced [74]. Ma et al. [75] of hydrothermally treating Mn_2O_3 particles, which are in powder form in a water-based mixture of sodium hydroxide (NaOH) at a temperature of 170°C for more than 72 hours produced stacked MnO_2 nanobelts.

By hydrothermally reacting $KMnO_4$ and $MnSO_4 \cdot H_2O$ in a water-based mixture at a temperature of 140 °C for 12 hours, Cheng et al. [76] created α-MnO_2 nanowires. They also created ϒ-MnO_2 nanowires by treating a combination of $MnSO_4$ and $(NH_4)_2S_2O_8$ at a temperature of 90 °C for a period of 24 hours. By reacting $KMnO_4$ and $MnSO_4$ for 12 hours at 140°C, cryptomelane-type manganese dioxide (α-K_xMnO_2) nanofibers having the diameters of 20 to 60 nm and their lengths of 1 to 6 μm was created [77]. The nanofibers form a body-centered tetragonal structure with unit cell parameters of a = 9.8241(5) and c = 2.8523(1) (space group I4/m). Their true composition is $K_{0.11}MnO_{2.07}$.

14.3.2 Sol-Gel Technique

In this process, $KMnO_4$ as well as carboxylic acid-such as citric acid, tartaric acid and fumaric acid done by redox process that result in the formation of the gel [64]. While α-MnO_2 is produced in aqueous concentrated base, it is desired for δ-MnO_2 to form in aqueous concentrated acid [78]. It is known that pH adjustment using H_3O^+ and/or H_2O controls the production of MnO_2 tunnel structures [79, 80]. Without the need of any catalysts, Mn^{2+} cations are transformed into $S_2O_8^{2-}$ anions in in a solution of water [81].

$$MnSO_4 + ((NH_4)_2S_2O_8) + 2H_2O \rightarrow MnO_2 + (NH_4)_2S_2O_4 + 2H_2SO_4$$

This technique results in the creation of nanowires [53]. An alternate process starts with potassium peroxodisulfate ($K_2S_2O_8$) and manganese sulphate ($MnSO_4$) [82]. Using a sol-gel technique with cethylene diamine tetra-acetate (EDTA) as a chelating agent, Oaki and Imai [83] produced 10^{-9} thick δ-MnO_2 nanosheets. When solutions containing Mn^{2+}/EDTA and NaOH were combined, the reaction began; these two are the basic of a solution. Utilizing concentrated nitric acid (HNO_3, 67 wt %), manganese acetate tetrahydrate (($CH_3COO)_2Mn·4H_2O$), and various surfactants in an ethanol solvent, a low-temperature sol-gel technique was used. Etyltrimethyl ammonium bromide and polyvinyl pyrrolidone were used to help generate MnO_2 nanowires and nanorods, respectively [84]. The X-ray diffraction of α-K_xMnO_2 formation (can be seen [63]) synthesized by sol-gel technique. Figure 14.5 shows the flow chart of MnO_2 (via sol-gel process).

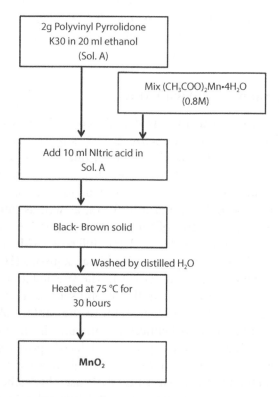

Figure 14.5 Shows synthesis of MnO_2.

14.4 Doped-MnO_2 Materials

Metal ions like Ni^{2+}, Ag^+, Co^{2+}, Cu^2, Cr^{3+}, Fe^{3+}, V^{5+}, Mo^{6+}, W^{6+}, and others have been effectively used to dope MnO_2 frameworks [85, 86]. The cryptomelane structure (α-K_xMnO_2) can currently be given new morphologies and improved electrochemical properties by altering the chemistry of crystals by trading for K^+ ions and/or doping either a low-valence state or a single-type metal cation (3+, 2+, 1+) either may be have large sufficient valence state (which is 5+, 6+). There may be two methods to introduce a metal cation as a doping element in the direction of a-MnO_2 lattice: first one is by substituting Mn for the octahedral Mn cation, which results in a six-coordinated cation with the same (0.72 Å) crystal radius as $^{VI}Mn^{3+}$ low spin, $^{VI}Mn^{3+}$ high spin (0.785 Å), and $^{VI}Mn^{4+}$ (0.67 Å) and second is the dopant cation's introduction into the (2x2) tunnel, that permits the formation of an eight-coordinated cation with a crystal radius comparable to that of K^+ (1.65 Å) [87]. The mixing of higher K^+ ions into tunnels is made easier by the lattice's higher negative charge, that enhancing structural stability, for dopant cations with lower valence. Higher valence dopant cations provide surplus of electrical charge which can be offset by the development of voids, this causes thermal instability and structural deformation.

14.4.1 Titanium-Doped with MnO_2

Nanoflakes of δ-MnO_2 which are Ti-doped were created by Li et al. [88] for the large catalytic burning of benzene, with a thickness of around 50 nm. These have the most catalytic oxidation quality (nanoflakes) above the benzene because of their extensive pore shape and the effective oxygen that Ti doping induces. Gas transport and reactions are aided by mesopores with sizes of 4 to 5 nm and 8 to 9 nm, as well as interlayer gaps of around 0.7 nm. Two phases of the *in situ* precipitation method were used to create Ti-containing Υ-MnO_2. First, different Ti/Mn atomic ratios of $MnSO_4$ and $TiOSO_4 \cdot xH_2SO_4$ were mixed in an aqueous (H_2O) solution together with strong nitric acid (HNO_3). The precipitation of $KMnO_4$ with the initial solution was done in the second stage at a temperature of 90 °C (refluxing). The addition of titanium to the MnO_2 hollow spherical structure promotes electrochemical performance in the Li/air battery characterized by a high specific capacitance (electric power of carbon is 2200 mAh/g) and also a favorable oxidative catalytic activity in the toluene oxidation method [89]. For electrocatalytic uses, hydrothermal procedures were used to create nanostructured 5% Ti-doped α-MnO_2 particles utilizing potassium permanganate (KMnO$_4$) and ammonium persulphate ((NH$_4$)$_2$S$_2$O$_8$) as oxidizing agents. The effective ORR (here ORR means oxygen reduction potential) movement in alkaline conditions of a doped sample, which results in a substantial shift in the ORR potential which is equal to the effective Pd$_{45}$Pt$_5$Sn$_{50}$ material [90].

14.4.2 Vanadium-Doped with MnO$_2$

There are several ways to make vanadium-doped MnO$_2$ nanoparticles, including the redox process [91–93]. Alfaruqi et al. [94] obtained α-MnO$_2$ nanoparticles later annealing at a temperature of 450°C for 5 hours by using a direct redox process among the KMnO$_4$ and Mn(CH$_3$COO)$_2$·4H$_2$O in aqueous solution and mixed to a solution containing V$_2$O$_5$. This substance served as the cathode in zinc-ion batteries. In order to create the V-doped ϒ-MnO$_2$ that acts as the electrode in basic lithium batteries, KMnO$_4$ and MnCl$_2$·4H$_2$O underwent a redox reaction with V$_2$O$_5$ acting as the dopant reagent in a molar ratio of 3:1:0.15. The products created from precursors in the end that were annealed at a temperature of 375 °C for 10 hours show anisotropic extension which improved the Li$^+$ ions' ability to diffuse in the (1x1)/(1x2) tunnel frameworks, achieving a diffusion coefficient of 2×10^{-8} versus 5×10^{-9} cm^2/s for perfect MnO$_2$ [95, 96].

14.4.3 Tin-Doped with α-MnO$_2$

Doping with Sn and Co was proven to block the conversion of α-MnO$_2$ to α-Mn$_2$O$_3$, this occurs at a temperature between 500 to 600 °C, by Hashem et al. [97]. Samples were made by reducing KMnO$_4$ with fumaric acid (here at a molar ratio of 10:3) and mixing SnCl$_2$ as a Sn dopant in an acidic solution. The final products were produced after being heated at a temperature of 450°C for 5 hours. The thermo-gravimetric analysis (TGA) of the pristine including the MnO$_2$ samples which is doped with Sn samples were heated in the order of 30 to 1000°C [97]. The undoped sample's TGA curve displays a minor weight reduction (3%) since surface and structural water have been removed and a sudden loss of weight at around 540 °C because of the exothermic reaction of α-MnO$_2$ to Mn$_2$O$_3$ and oxygen is released. α-MnO$_2$ curves of TGA: Sn show the doped materials' structural stability up to 850 °C. The observations of magnetic susceptibility support this stabilizing effect. The observations of magnetic susceptibility support this stabilizing effect. As a result 180° Mn^{3+}-O-Mn^{4+} bridge, MnO$_2$ shows Curie-Weiss paramagnetic nature at higher temperature (here T greater than 150 K) and ferromagnetic nature at lower temperatures (here T less than 30 K). The Mn^{3+} content and ferromagnetic content both decrease as dopant concentration rises [98].

The electrical conductance σ_{ac} versus frequency for samples of α-MnO_2 with varying dopant concentrations of Sn or Co as discussed in [97, 98]. The findings of a semiconducting substance whose frequency dependence is power-law governed by Equation 14.2, which is mentioned below;

$$\sigma_{ac} = \sigma_{dc} + A\omega^n \qquad (14.2)$$

Here, A is a constant; n is the power exponent, also the low-frequency value equals the direct-current conductivity (σ_{dc}). In contrast to the pure α-MnO_2 material, a noticeable raise in electrical conductivity is shown in the dopant existence. The raise in conductivity of the doped samples is thought to be a result of free electrons from Co(II) as well as Sn(II).

The electrochemical achievement of Sn-doped α-MnO_2 nanorod-like particles was examined by Hashem et al. [99]. They show the specific discharge capacities for P-MnO_2 as well as Sn-MnO_2 in relation to the cycle number. At the 40th cycle, the P-MnO_2 as well as Sn-MnO_2 electrodes had capacities of 65 and 80 mAh/g respectively. These findings demonstrate that the capacity fading of the pristine cathode is significantly more extensive than the Sn-doped MnO_2. Due to the reduction in Mn^{3+} Jahn-Teller ions that occurs when Sn ions are introduced in the direction of (2x2) tunnels; both the electrochemical achievement including structural balance may be ascribed to this. The reduction of Mn^{2+} ions that diffuse in the electrolyte is the second cause of the electrochemical deterioration of pristine MnO_2.

14.4.4 Al, Cu, Mg-Doped with MnO_2

In ϒ-$Mn_{1-y}Al_yO_{2-\delta}$ attached nanowires, the benefit of the Mn to Al substitution was an enhancement in surface area from 17-184 m^2/g that led to an increase in the Faradic behavior [100]. By using a redox process between $KMnO_4$ and fumaric acid ($C_4H_4O_4$), its M-doped MnO_2 (here M = Cu, Mg, also Al) and pure MnO_2 were created [101]. Cryptomelane-MnO_2 exhibits the same recognizable peaks in both pure and doped samples ($K_2Mn_8O_{16}$). There are no more peaks associated to the substances Al, Cu, including Mg. Three explanations are possible for the absence of diffraction lines linked to doping components: (i) transition metal oxides that are less in content either having crystalline domains that are far away the XRD disclosure limit, (ii) the cryptomelane structure incorporates transition metals, resulting in the production of a solid solution; (iii) the K_xMnO_2 structure's channels are altered to accommodate transition metals in place of K^+ ions [102]. The observed existence of K^+ ions interior the (2x2) tunnels of the manufactured samples is connected to the MnO_2 lattice in all samples. According to a chemical study, the potassium content (in %) for the samples of $Cu-MnO_2$, $Mg-MnO_2$, $P-MnO_2$, and $Al-MnO_2$ is 5.6, 8.9, 0.7, and 7.9, respectively. Additionally, a chemical study reveals that the doped materials' percentages of Al, Cu, and Mg are 0.4, 0.6, and 0.3, respectively.

Four-point approach was used by Van der pauw to quantify electrical transmission below ambient temperature. The resistivity of all four MDOs decreases by 3 to 4 orders of magnitude at ambient temperature depending on the dopant. Between the four oxide samples, the $Al-MnO_2$ sample exhibits the lowest resistivity. Roberto N. et al. Shows the linear dependency of the 1/T as a function of ln (ρ), the electrical resistivity e was activated above RT. They got band gaps after linear sloping of parent MnO_2 is 0.69 Ev and for $\Upsilon-MnO_2$ 0.58 eV, which is close to each other [103] E_g was decreased by doping to a value of 0.34 eV, which is near to the 0.26 to 0.3 eV determinate for $\beta-MnO_2$ [104]. Al, Mg, and Cu seem to maintain the MnO_2 structure by adding dopant ions, which in turn lessens the capacity fading seen in pure MnO_2. The dispersion of Li ions in the time of charge-discharge cycling, as shown in doped $\alpha-MnO_2$ samples, requires the existence of a less concentration of maintain atoms inside the (2x2) tunnel of a cryptomelane as well as hollandite-type structure [101].

14.4.5 Co- and Ni-Doped with MnO_2

The identical procedure used to make tin-doped $\alpha-K_xMnO_2$ was utilized to make cobalt-doped $\alpha-K_xMnO_2$, excluding that the Sn precursor were swapped out for $Co(NO_3)_2 \cdot 6H_2O$ as the origin of cobalt in a mole-to-mole ratio of 3:1:0.07, respectively [105]. The carbon for the coating came from table sugar. Chemical analysis provides the stoichiometry for the pure $K_{0.009}MnO_2$ and for the doped $K_{0.095}Co_{0.013}MnO_2$ sample's chemical stoichiometry. Ions such as Co^{3+} and K^+ are confined inside the huge tunnel, which has a width of 4.6 Å. Co doping as well as carbon coating have an additional impact that improves recharge ability and reduces capacity fading at the price of basic capacity. The above layer of carbon serves as a shield all around the particles and stimulates the pace at which Li^+ insertion and extraction operations transfer charges. Following the coating and doping processes, magnetic characteristics show that the blended valence state Mn^{4+}/Mn^{3+} along less Mn^{3+} concentration reduced [105].

Oxidation-reduction reaction of $K_2S_2O_8$ along with $MnSO_4 \cdot H_2O$ mixing of $CoSO_4 \cdot H_2O$, 5 wt% Co-doped $R-MnO_2$ nanospheres with range of diameters 350 to 500 nm were produced with a yolk-shell structure [106]. These nanospheres have a pore dimension of 9 nm and a specific surface area of 135m²/g. The thermal stability and structural characteristics of chromium-doped and cobalt-doped of α-MnO_2 nanorods were produced by the breakdown of $KMnO_4$ in an acidic setting was reported by Korosec et al. [107]. According to EXAFS research, the two dopant ions (Co^{2+} and Cr^{3+}) replace Mn^{4+} in the core of an octahedron, raising the lattice's anions. This increase in charge is offset by an increase in the concentration of the ion K^+ in the tunnels. Another study [108] used metal sulphates as dopant agents to create the Co/Ni-doped with $K_{0.14}MnO_2$ tetragonal phase (cryptomelane shape). Ni^{2+} and Co^{2+} each made about 2% and 7% of the finished product, respectively. Nanowires with a diameter of 15 to 20 nm including width of 100 to 300 nm made up the samples. Due to the growth process of the rupture-recrystallization mechanism, the Co/Ni doping had no effect on the 1D nanostructure of α-MnO_2. A modified sol-gel approach was used to create the co-doped birnessite δ-MnO_2 catalyst, which is used to convert benzylic alcohols to benzaldehydes in warmed toluene in an oxygen environment [109]. Also the presence of Co^{2+} ions in the octahedral lattice is thought to be responsible for the increased electrical conductivity of δ-MnO_2.

According to Biswal et al. [110], who reported that, the distinct morphologies subject to as long as the dopant is Ni or Co. The technique of synthesis was more distinct from the preceding one, a galvanostatic process that absence of K^+ cations, which were determined to be crucial in mentioned research to maintain the substance also increase the electrochemical characteristics of the phases of α- and β-MnO_2 is present in this preparation. The fact is that these EMD specimen were indeed found in a certain shape i.e. ϒ-MnO_2 shape which is also significant. Both of these samples have *in situ* doping, one with Ni as well as other with Co. When Co-doping was used, the EMD took the shape of cauliflowers, however when Ni-doping was used, the EMD took the form of sea urchins. Both times, energy density enhanced by the doping, but the level does not remain same, for example Ni-doping which is 395mAh/g and for Co-doping that is 670mAh/g. So Ni or Co doping dramatically boosted the conductivity as well as electrochemical characteristics in any phase. Therefore, in [110] the existence of Co_3O_4 was said to be the cause of the increased conductivity in Co-doped EMD. Raman experiments would be very interesting to run and to test this theory [110].

14.4.6 Ag-Doped with MnO_2

Through a straightforward wet-chemical procedure, Pristine K_xMnO_2, Ag-coated, including Ag-doped K_xMnO_2 samples (where x equals to 0.065) was produced. Re-stirring the produced oxides in distilled water near about 24 hours at room temperature decreased the particle size to 20 nm [111]. Silver concentrations in doped and coated K_xMnO_2 are 1.4% and 3.9%, respectively, according to elemental studies. The Mn^{+4}/Mn^{+3} ratio of the Ag-coated K_xMnO_2 sample is growing, which results in a decreasing proportion of Jahn-Teller Mn^{+3} ions, according to magnetic measurements. Overall, Ag-coated K_xMnO_2 has improved electrochemical performance. Up to 40th cycle for Li-MDO cells, the capacities of discharge are 90, 110, and 115 mAh/g for Ag-doped K_xMnO_2, Pristine K_xMnO_2, as well as Ag-coated K_xMnO_2 materials respectively. Due to the nanosized particles produced during stirring in distilled water also the improved conductivity that followed Ag coating and the Ag-coated K_xMnO_2 material demonstrated the highest results for charge retention.

14.4.7 Bismuth-Doping with Additives

Regardless of the crystal shape, It has been determined that Bi^{3+} ions were introduced to be advantageous to MDO's electrochemical characteristics [112, 113]. As with the addition of other dopant ions, this enhancement includes raise in conductivity, but bismuth's unique trait which inhibits the creation of the spinel shape, [114] that causes the irreversible to the MnO_2 cells, since the Mn_3O_4 is spinel but no more electroactive. Yu [115] has provided an explanation for why Bi has such a significant impact who observed that the ionic radii of Mn^{2+} as well as Mn^{3+} were significantly less than of Bi^{3+}, they are unable to penetrate in the direction of spinel lattice, that stops the creation of the spinel from the series of processes in the mean while the preparation of MnO_2. Im and Manthiram [116] who added Bi^{3+} ions to Υ-MnO_2 along with Bi_2O_3 addition in an alkaline solution (electrolyte) also discovered this crucial function of bismuth. According to Sundaram *et al.* [117], Ti is enough effective than that of Bi at preventing the production of Mn_3O_4 when it comes to the electrochemical characteristics of Υ-MnO_2. They also discovered that adding more chemicals led to even superior electrochemical characteristics. Particularly, the synergetic impact of mixing 3 weight percent Bi_2O_3 and 2 weight percent TiS_2 resulted in a greater capacity of 240 mAh/g, which was enough higher than the values obtained with Bi_2O_3 or TiS_2 alone.

TiB_2, CeO_2, MgO, and B_4C are other additions that have enhanced the electrochemical capabilities of Υ-MnO_2 [118, 119]. According to several accounts, Mn^{3+} ions dissolve permanently in alkaline KOH solutions. Electrochemically passive phase growth, such as δ-MnO_2 including Mn_3O_4, is caused by this process. It should come as no surprise that additives like TiB_2, Bi_2O_3, and compounds containing Ba that have been proven to limit the production of Mn_3O_4 can likewise decrease the dissolution of Mn^{3+} ions [120].

14.5 MnO_2 with Polymer Composites

MnO_2 is a substance which may be coupled with different polymers to create hybrid nanoarchitecture electrode materials that are extremely effective for pseudocapacitive devices. As the electrode for a supercapacitor, a mixture created by the polymerization of pyrrole (Py) on formed MnO_2 particles was measured [121, 122]. The capacitance of MnO_2 electrodes is mostly pseudocapacitive, that is determined by reversible oxidation-reduction process containing cation or proton with the exchange of electrolyte. Due to its fundamentally weak electrical conductivity, particle size, and oxide porosity, MnO_2's specific capacitance is really between 200 and 300 F/g [123]. $MnCl_2 \cdot 4H_2O$ and $KMnO_4$ were precipitated in distilled water to create Υ-MnO_2, or the (1x1)/(1x2) tunnel structure, which was then dried at a temperature of 110 °C for 10 hours.

14.5.1 Polypyrrole-Coated with MnO_2

A chronoamperometry test was used to conduct the electrodeposition of polypyrrole (PPy) at a potential of 900 mV versus SCE. Also the surface area of BET for PPy/Υ-MnO_2 increased by a 125 m^2/g as a result of the PPy deposit, compared to Υ-MnO_2 which is 64 m^2/g. The SEM pictures of uncoated as well as PPy-coated Υ-MnO_2 particles are shown in [121]. The morphology of the MnO_2 grains is not altered by the electrochemical polymerization process, and the grain size of MnO_2 particles produced during precipitation is 250 nm, and they have a regular form [122]. The Mn^{4+}/Mn^{3+} reversible redox technique, in conjunction with reversible desinsertion/insertion of alkali protons H_3O^+ or cation (Na^+) which present in the solution, is what causes the MnO_2 and PPy/MnO_2 pseudocapacitance:

$$MnO_2 + Na^+ + e^- \rightarrow MnOONa$$

Charge-discharge cycling experiments at a stable current density of 2 mA/cm^2 were used to estimate the specific capacitance [121]. With a composite cathode made of PPy/Υ-MnO$_2$ and a carbon anode, the asymmetric supercapacitor has a good specific capacitance that is 142 F/g compared to Υ-MnO$_2$ that has 74 F/g. In PPy/MnO$_2$ materials, it is prominent that the specific capacitance is inversely correlated with the thickness of layer of the PPy deposit. Over 500 cycles of a continuous charging-discharging observation with a current density of 2 mA/cm^2, the achievement of the composite sample was assessed [121]. The particular capacitance stabilization suggests that the electrode exhibited consistent capacitive behavior and strong cycle stability.

Keep in mind that the DMO with polymer does not form the good supercapacitor. Let us look, an asymmetric capacitor along EMD formed from a manganese ore/residue leach liquid shows a 50 F/g capacity. However, the cycling life exhibits exceptional performance in comparison to the data we published previously, since after 2000 cycles 100% capacity was still present [124].

14.5.2 Polyaniline-MnO$_2$

Polyaniline (PANI) is a significant and extensively researched conductive polymer. It is frequently utilized in energy storage, sensors, batteries, photovoltaic cells, capacitors, and other devices because of its exceptional environmental stability [125]. As per literature, it has theoretical specific capacitance of 750 F/g, PANI can be combined with MnO$_2$ to create materials which are profitable for energy storage devices [126].

Polymer coating were used to create nanocomposites of polyaniline (PANI) and manganese dioxide (MnO$_2$). PANI-MnO$_2$ nanocomposite shows a capacitance (C_p) of 417F/g at a scan rate of 5 mV/s. It displayed specific power of 875 W/kg and specific energy of 11.4 Wh/kg. Instead, at a scan rate of 5 mV/s, the PANI-coated MnO$_2$ nanocomposite had a specific capacitance of 271F/g. It displayed 7.2 Wh/kg of specific energy at 280 W/kg of specific power [127]. The composite has strong cycling performance (remains 90% charge retention after 8000 cycles), with a great specific capacitance that is 762 F/g at 1 A/g, as well as outstanding rate performance which is 587 F/g at 10 A/g [127].

By using chemical co-precipitation to create α-MnO$_2$, Bao et al. team employed it acts as an oxidizing agent to start the polymerization of PANI. Compared to the individual component, this composite material has much higher electrode specific capacitance including stability [128]. Additionally, PANI-MnO$_2$ composites were created by Jadhav et al. by grafting and polymer coating PANI onto the surface of MnO_2 [129]. PANI-MnO_2 composite electrode was created by Relekar et al. using the electrodeposition technique, and it was used for Supercapacitors, maintaining 92% charge retention after 1000 cycles as well as 84% charge retention after 2000 cycles [130]. By using a two-step electrodeposition process to deposit PANI and MnO_2 on Ti foam while using various precursor solutions, Wang et al. [131] created composite electrodes. It was discovered that MnO_2-PANI-Ti foam had an area capacitance that was approximately 2 to 4 times more than the all of PANI-Ti foam also MnO_2-Ti foam [189]. A highly sulfonated version of PANi was created by Hwang et al. [132] and combined with Ramsdellite-MnO_2 to create batteries (R-MnO_2). The composites were discovered to have a greater discharge capacity (97.0 mAh/g) than the Li/SPAn battery (66.8 mAh/g) when tested as an electrode material for batteries. There is a significant polysulfide shuttle effect for lithium-sulfur batteries (LSBs) in the course of charging and discharging [133]. In order to address this issue, Zhang et al. [134] employed PANI-modified MnO_2 nanoparticles to create a 3D layered porous structure and used it as the sulphur host for LSBs. Rapid transport of active materials, electrons, and ions is made possible by the creation of a network skeleton with many channels, and the presence of MnO_2 can aid in the oxidation of polysulfide's to thiosulfate. As a result, it can successfully block the shuttle effect.

14.5.3 Polybithiophene-Coated MnO_2

Latest composite with batter photoconducting performance was used as a p-n heterojunction in solar cells and was made of polymeric polybithiophene (PBTh) and crystalline MnO_2. On a substrate made of indium tin oxide (ITO), the PBTh/MnO_2 sample was placed. The produced photocurrent is significantly increased by the integration of MnO2 particles added into the polymer sheets, that is from 5.9 A/cm^2 for ITO/PBTh as well as 20.6 μA/cm^2 for the ITO/PBTh-MnO_2 sheets along 100 mg MnO_2 inserted [135].

Similar to this, Zn/ϒ-MnO_2 electrochemical cells employed a polymer or as inorganic composite for the electrode material. The composite was created by electrodepositing PBTh on MnO_2 particles in a solution of acetonitrile (CH_3CN) and 0.01 mol/L of PBTh and 0.1 mol/L of $LiClO_4$ [136]. The Electrochemical Impedance Spectroscopy (EIS) studies were used to evaluate the performance of electrodes (MnO_2) for discharged Zn/PBTh/MnO_2 and Zn/MnO_2 batteries [136]. The EIS profiles may be seen in the Nyquist plots as a semicircle which is at high-frequency range and a quasi-linear line that is at small-frequency range. The Warburg addition of proton diffusion due to the majority of the material is the quasi-horizontal portion at small frequency. The fit is shown in [118], the Zn/MnO_2 cell's charge deportation resistance is 4.49 Ω/cm^2, but for the Zn/PBTh plus MnO_2 cells, it is 3.42 Ω/cm_2. The electrochemical outlines of Zn/MnO_2 as well as Zn/MnO_2 plus PBTh cells discharged at a current density of 2 mA/cm^2 [137]. In the 1.45 to 0.9 V potential ranges, a steady reduction in cell voltage is seen. There is no discernible plateau. The Zn/MnO_2 plus PBTh cell has a capacity that is 25% more than the Zn/MnO_2 cell. From the surface, protons travel to the core of the MnO_2 molecules by a solid-state diffusion mechanism as part of the overall cathodic process that converts MnO_2 to MnOOH.

On MnO_2 particles, the conducting polymer covering plays a significant role. First of all it encourages protons' diffusion second it's discharging may be result in the conducting polymer being diminished or could obstruct the pores also It produces a greater effective surface area as well as more active material. By electrodepositing conducting polymer on a β-MnO_2 surface with various monomers in a CH_3CN/0.1 mol/L $LiClO_4$ cell, composite materials comprising conducting polymer and β-MnO_2 were created: Pyrole (Py) or bithiophene (BTh) in CH_3CN/$LiClO_4$ (with 0.1 mol/L) [138].In order for an electro-polymerization to be effective, a layer that can prevent the oxidant metal from dissolving must be formed. On the other hand, access to monomer should be maintained to grant for additional oxidation. The fact is that MDO might catalysis the oxygen reduction process (ORP) that generates hydrogen peroxide ions in an alkaline solution through a two-electron reduction process (HO_2^-). The (1x1) tunnel structure (rutile-type) of β-MnO_2 was chosen because it has the excellent structural qualities of all the MDOs. The cyclic voltammetry (CV) for O_2 contraction in a 2 mol/L of KOH saturated with O_2 versus saturated argon solution [138]. The electrodes of PBTh/β-MnO_2 as well as PPy/β-MnO_2 they have two reduction peaks which are located at 506 and 365 mV respectively. A considerable positive shift in the reduction peak as seen from 412 to 365 mV and a drop in the reduction peak are evidence of the increased electrocatalytic impact of PPy/β-MnO_2. Additionally, due to PBTh/higher MnO_2's negative onset potential than β-MnO_2, it has extremely excellent electrocatalytic activity for ORP.

14.6 Nanocomposites

For supercapacitors (SCs), an ideal nanocomposite electrode with extended cycle durability should combine a large-power density substance (which are carbon-based) with a huge-energy density material i.e., oxide. The primary drawback of MnO_2 is its low conductivity, which may be improved by creating several MnO_2/conductive matrix materials, like SnO_2/MnO_2 [139], multiwalled carbon nanotube (MWCNT)/MnO_2 [140], also Carbon/MnO_2 nanocomposites [141]. MnO_2 has a high theoretical specific capacitance as mentioned above.

14.6.1 Graphene-MnO_2-Polyaniline

A hydrothermal technique was used to create a large achievement graphene-MnO_2-PANI and MnO_2/PANI which has Cp = 240 F/g, the produced graphene/MnO_2/PANI nanoparticles displays a significant increase in Cp = 305 F/g [142]. By utilizing $KMnO_4$ as an oxidant concealed by neutral conditions during a dilute *in-situ* polymerization process, a unique ternary compound comprising sulfonated graphene, MnO_2, and polyaniline is created. In comparison to the MnO_2/polyaniline binary composite, the Graphene-MnO_2-PANI composite based on the two-electrode cell exhibits large cycling stability (capacitance decreases near about 11.7% after 3000 cycles), excellent rate capability (73% retention in charge from 0.2 to 20 A/g), also good in electrochemical capacitance (which is 276 F/g) [143]. Graphene, MnO_2, PANI, as well as nickel foam (NF) have been combined to create a ternary nanocomposite (*in situ* polymerization). The composite demonstrated great cycle stability and outstanding specific capacitance, which is 1081 F/g at a 1 mV/s [144].

A simple electrochemical deposition method produced MnO_2/graphene/nickel foam (NF) composite with a 476 F/g specific capacitance. Several methods have been explored by researchers to create graphene/MnO_2 nanocomposite, including redox deposition, electrodeposition technique, microwave irradiation process, as well as polymer aided chemical reduction [145, 156]. Manganese oxide showed a specific capacitance of 305 F/g at a current density of 1 A/g when it was coated on NF-graphene using the hydrothermal microwave irradiation process [147]. With a current density of 0.2 A/g, MnO_2-graphene foam with regulated MnO_2 particle form had a specific capacitance of 560 F/g [148].

14.6.2 MnO_2-Carbon Nanocomposite

For instance, MnO_2 nanoparticles were coated by Chen *et al.* on graphene oxide (GO) sheets, which improved the electrochemical characteristics as a result of the chemical interaction between MnO_2 and GO [149]. The superior cycling performance of the nanocomposite made of Au-doped MnO_2

Manganese Dioxide as Supercapacitor   383

Table 14.1 Comparison between basic parameters of MnO_2 and their composites.

Electrode used	Electrolyte used	Specific capacitance (Fg^{-1})	Current density (A/g)	Energy density (Wh Kg^{-1})	Scanning rate (mV/s)	Charge retention with cycles	Ref.
LSG/MnO_2	NA	1145	NA	NA	1	NA	[157]
MnO_2/CNT	Na_2So_4	276	3	NA	10	91.6%@5000	[158]
MnO_2	NA	317	5	NA	5	71%@2000	[159]
PANI/MnO_2	Na_2So_4	320	NA	NA	20	84%@10000	[160]
Co-MnO_2	Na_2SO_4	1050	0.5	69.3	50	92.7%@10000	[161]
Co_3O_4@MnO_2	LiOH	560	0.2	17.7	158	95%@5000	[162]
Co_3O_4 nanowire@MnO_2	LiOH	480	2.67	NA	50	97.3%@5000	[163]
MnO_2/CNTs	Na_2SO_4	348	NA	NA	5	91.3%@5000	[164]

nanoparticles and N-doped carbon tubes was shown by Lv et al. [150] (3% decrease after 5000 cycles). A novel carbon nanotubes (CNTs)/graphene composite with CNT pillars intermediated in the middle of the graphene sheets was presented by Fan et al. [151] and demonstrated a C_p = 385 F/g. By electrodeposition, graphene embellished with MnO_2 nanostructures resembling flowers were created for supercapacitor electrodes. Nanoflowers of MnO_2 were made of extremely thin rods, which are less than 10 nm in thickness. With a current of 1 mA and an energy density of 11.4 Wh/kg, the C_p after the MnO_2 deposition was 328 F/g [152]. In-situ synthesis of MnO_2 nanowires on the surface of graphene nanosheets (GNS) produced the needle-related MnO_2 nanowire arrays used in Song et al. [153] fabrication of a nanocomposite for a supercapacitor. A straightforward redox reaction between $KMnO_4$ and GNS is used as the preparation, which can create the composite on a big scale for a reasonable price. The nanocomposite performed well in terms of high capacitance, after measuring the specific capacitance is 276 F/g. MnO_2 nanocomposite were enraged poly(3,4-ethylenedioxythiophene) nanowires were created by directly putting the poly(3,4-ethylenedioxythiophene) nanowires in the mixture of $KMnO_4$ [154]. The MnO_2 nanoparticles demonstrated extraordinarily large C_p = 410 F/g as supercapacitor material, including excellent specific capacity which is 300 mAh/g as electrode materials for the Li-ion battery, it means there is a highly large surface area [155].

Birnessite-type MnO_2 nanostructured was coated with thin polymer coatings by Long et al. [155]. The composite created by poly (o-phenylenediamine) electrodeposition, that maintains MnO_2's mesoporosity, demonstrated high stability in acid electrolytes as an electrode material. Yan et al. [156] looked at how MnO_2 nanowires and SnO_2 may be used to create a supercapacitor electrode with good performance. From 1 mol/L of Na_2SO_4 aqueous mixture, a specific capacitance (Cp) of 800 F/g was attained at a current density (C_d) of 1 A/g. By using a cyclic voltammetry process, nanowires of MnO_2 were electrodeposited onto carbon nanotube (CNT) paper [44]. The composite of nanowire/CNT utilized as a flexible electrode for supercapacitors showed Cp = 167F/g at a Cd = 77 mA/g and after 3000 cycles faradic efficiency was 88%. Table 14.1 shows the C_p, E_d, P_d, and charge retention with cycles, (here NA means not available).

14.7 Conclusion

The emerging technologies need crucially soft and hearable energy storage devices. In this regard, Zeng et al. have made significant progress, who presented a sandwiched, soft quasi-solid-state Zn-MnO$_2$ battery with a MnO$_2$-poly(3,4-ethylenedioxythiophene)electrode, a Zn nanosheets anode having length 50 nm without a binder or separator; the carbon fibers are produced uniformly [165]. The poly (3, 4-ethylenedioxythiophene) shell, which has a thickness of around 9 nm, was utilized as a protective layer to prolong cycle life. The energy density and power density is 505 Wh/kg and 8.6kW/kg respectively of this soft quasi-solid-state battery. It retained 78% of its basic capacity and approximately 100% coulombic efficiency after 300 cycles. After 1000 cycles, 61.5% of the basic capacity was still present. Quasi solid state batteries have very good performance, was recently noted. Zn/MnO$_2$ has enormous potential as a good energy storage system that combines affordability, safety as well as large energy density is supported by the outcomes produced by Qiu et al. [166].

Rechargeable alkaline cells used to approximately 10% of their theoretical capacity which is 617 mAh/g. On the other hand, a family of Bi-birnessite cathodes intercalated with Cu^{2+} that offer nearly complete two-electron capacity reversibly for greater than 6000 cycles have recently made a significant effort [167]. It is well familiar that MnSO$_4$ can passivate the cathode surface when added to the solution [168]. But previously, it was demonstrated that there is a chemical reaction among α-MnO$_2$ as well as H$^+$ when a Zn/MnO$_2$ battery's electrolyte was a moderate ZnSO$_4$ solution [169]. The mentioned battery has a capacity 285 mAh/g also its charge retention rate is 92% after 5000 cycles taken. These most recent findings demonstrate that the Zn/MnO$_2$ battery's issues are now mostly resolved. As a result, the Zn/MnO$_2$ batteries currently perform better than the Li-ion batteries in addition to being less expensive.

MnO$_2$ has a promising future as a supercapacitor component. A mixture with a hollow carbon sphere core that has N-doping also with a shell made of birnessite-type material arranged in a hierarchy an asymmetric supercapacitor with N-doping cathode employed MnO$_2$ nanoflakes as its anode [170]. when operating in the voltage range from 0 to 1.8 V, this SC has E_d = 28.6 Wh/Kg at a P_d = 233 W/Kg also having drop in specific capacitance, but the coulombic efficiency remains 100% after taken 4000 cycles. Similar outcomes were obtained using an asymmetric supercapacitor built from a mix of carbon nanotube core-shell particles and MnO$_2$ nanoflakes [117]. As a result, we can say that MnO$_2$ has a lot of potential for energy conversion and storage applications right now.

References

1. Li, G., Jin, Y., Akram, M.W., Chen, X., Research and current status of the solar photovoltaic water pumping system–A review. *Renew. Sust. Energ. Rev.*, 79, 440–458, 2017.
2. Kelly-Richards, S., Silber-Coats, N., Crootof, A., Tecklin, D., Bauer, C., Governing the transition to renewable energy: A review of impacts and policy issues in the small hydropower boom. *Energy Policy*, 101, 251–264, 2017.
3. Gielen, D., Boshell, F., Saygin, D., Bazilian, M.D., Wagner, N., Gorini, R., The role of renewable energy in the global energy transformation. *Energy Strategy Rev.*, 24, 38–50, 2019.
4. Zhang, L.H., Wu, S.S., Wan, Y., Huo, Y.F., Luo, Y.C., Yang, M.Y., Lu, Z.G., Mn3O4/carbon nanotube nanocomposites recycled from waste alkaline Zn–MnO2 batteries as high-performance energy materials. *Rare Met.*, 36, 5, 442–448, 2017.
5. Bensalah, N. and De Luna, Y., Recent progress in layered manganese and vanadium oxide cathodes for Zn-ion batteries. *Energy Technol.*, 9, 5, 2100011, 2021.
6. Song, M., Tan, H., Chao, D., Fan, H.J., Recent advances in Zn-ion batteries. *Adv. Funct. Mater.*, 28, 41, 1802564, 2018.
7. Guo, R., Chen, J., Yang, B., Liu, L., Su, L., Shen, B., Yan, X., In-plane micro-supercapacitors for an integrated device on one piece of paper. *Adv. Funct. Mater.*, 27, 43, 1702394, 2017.
8. Bao, S., Jia, W., Xu, M., Rapid synthesis of Mn3O4 by *in-situ* redox method and its capacitive performances. *Rare Met.*, 30, 1, 81–84, 2011.
9. Li, L., Wu, Z., Yuan, S., Zhang, X.B., Advances and challenges for flexible energy storage and conversion devices and systems. *Energy Environ. Sci.*, 7, 7, 2101–2122, 2014.
10. Wang, H., Guo, R., Li, H., Wang, J., Du, C., Wang, X., Zheng, Z., 2D metal patterns transformed from 3D printed stamps for flexible Zn//MnO2 in-plane micro-batteries. *Chem. Eng. J.*, 429, 132196, 2022.
11. Le, V.T., Kim, H., Ghosh, A., Kim, J., Chang, J., Vu, Q.A., Lee, Y.H., Coaxial fiber supercapacitor using all-carbon material electrodes. *ACS Nano*, 7, 5940–5947, 2013.
12. Dunn, B., Kamath, H., Tarascon, J.M., Electrical energy storage for the grid: A battery of choices. *Science*, 334, 6058, 928–935, 2011.
13. Van Noorden, R., A better battery. *Nature*, 507, 7490, 26, 2014.
14. Salanne, M., Rotenberg, B., Naoi, K., Kaneko, K., Taberna, P.L., Grey, C.P., Simon, P., Efficient storage mechanisms for building better supercapacitors. *Nat. Energy*, 1, 6, 1–10, 2016.
15. Lukatskaya, M.R., Dunn, B., Gogotsi, Y., Multidimensional materials and device architectures for future hybrid energy storage. *Nat. Commun.*, 7, 1, 1–13, 2016.

16. Birgisson, S., Saha, D., Iversen, B.B., Formation mechanisms of nanocrystalline MnO2 polymorphs under hydrothermal conditions. *Cryst. Growth Des.*, 18, 2, 827–838, 2018.
17. Widiyastuti, W., Nurlilasari, P., Affandi, S., Setiawan, H., Electrolysis synthesis of MnO2 in acidic environment and its electrochemical performance for supercapacitor. *J. Phys.: Conf. Ser., IOP Publishing*, 1093, 1, 012021, 2018.
18. Yan, G., Lian, Y., Gu, Y., Yang, C., Sun, H., Mu, Q., Peng, Y., Phase and morphology transformation of MnO2 induced by ionic liquids toward efficient water oxidation. *ACS Catal.*, 8, 11, 10137–10147, 2018.
19. Poonguzhali, R., Shanmugam, N., Gobi, R., Senthilkumar, A., Viruthagiri, G., Kannadasan, N., Effect of Fe doping on the electrochemical capacitor behavior of MnO2 nanocrystals. *J. Power Sources*, 293, 790–798, 2015.
20. Wu, B., Zhang, G., Yan, M., Xiong, T., He, P., He, L., Mai, L., Graphene scroll-coated α-MnO2 nanowires as high-performance cathode materials for aqueous Zn-ion battery. *Small*, 14, 13, 1703850, 2018.
21. Julien, C., Mauger, A., Vijh, A., Zaghib, K., Julien, C., Mauger, A., ... & Zaghib, K., *Lithium batteries,* pp. 29–68, Springer International Publishing, 2016.
22. Crut, A., Maioli, P., Vallée, F., Del Fatti, N., Linear and ultrafast nonlinear plasmonics of single nano-objects. *J. Phys.: Condens. Matter*, 29, 12, 123002, 2017.
23. Rao, C.N.R., Transition metal oxides. *Annu. Rev. Phys. Chem.*, 40, 1, 291–326, 1989.
24. Chalmin, E., Vignaud, C., Salomon, H. et al., Minerals discovered in paleolithic black pigments by transmission electron microscopy and micro-X-ray absorption near-edge structure. *Appl. Phys. A*, 83, 213–218, 2006.
25. Gerard, G. and Becker, H., *Handbook of structural stability part I: Buckling of flat plates*, United States, 1957. https://doi.org/10.2172/4343548
26. Huggins, R.A., Supercapacitors and electrochemical pulse sources. *Solid State Ionics*, 134, 1–2, 179–195, 2000.
27. Xia, F., Wang, H., Jia, Y., Rediscovering black phosphorus as an anisotropic layered material for optoelectronics and electronics. *Nat. Commun.*, 5, 1, 1–6, 2014.
28. Miller, E.E., Hua, Y., Tezel, F.H., Materials for energy storage: Review of electrode materials and methods of increasing capacitance for supercapacitors. *J. Energy Storage*, 20, 30–40, 2018.
29. Fuertes, A.B., Lota, G., Centeno, T.A., Frackowiak, E., Templated mesoporous carbons for supercapacitor application. *Electrochim. Acta*, 50, 14, 2799–2805, 2005.
30. He, Y., Chen, W., Li, X., Zhang, Z., Fu, J., Zhao, C., Xie, E., Freestanding three-dimensional graphene/MnO2 composite networks as ultralight and flexible supercapacitor electrodes. *ACS Nano*, 7, 1, 174–182, 2013.
31. Jadhav, S.A., Dhas, S.D., Patil, K.T., Moholkar, A.V., Patil, P.S., Polyaniline (PANI)-manganese dioxide (MnO2) nanocomposites as efficient electrode materials for supercapacitors. *Chem. Phys. Lett.*, 778, 138764, 2021.

32. Snook, G.A., Kao, P., Best, A.S., Conducting-polymer-based supercapacitor devices and electrodes. *J. Power Sources*, 196, 1, 1–12, 2011.
33. Lei, J. and Chen, X., RuO2/MnO2 composite materials for high-performance supercapacitor electrodes. *J. Semicond.*, 36, 8, 083006, 2015.
34. Abdah, M.A.A.M., Azman, N.H.N., Kulandaivalu, S., Sulaiman, Y., Review of the use of transition-metal-oxide and conducting polymer-based fibres for high-performance supercapacitors. *Mater. Des.*, 186, 108199, 2020.
35. Yang, P., Ding, Y., Lin, Z., Chen, Z., Li, Y., Qiang, P., Wang, Z.L., Low-cost high-performance solid-state asymmetric supercapacitors based on MnO2 nanowires and Fe2O3 nanotubes. *Nano Lett.*, 14, 2, 731–736, 2014.
36. Schütter, C., Pohlmann, S., Balducci, A., Industrial requirements of materials for electrical double layer capacitors: Impact on current and future applications. *Adv. Energy Mater.*, 9, 25, 1900334, 2019.
37. Zhong, C., Deng, Y., Hu, W., Qiao, J., Zhang, L., Zhang, J., A review of electrolyte materials and compositions for electrochemical supercapacitors. *Chem. Soc. Rev.*, 44, 21, 7484–7539, 2015.
38. Fic, K., Lota, G., Meller, M., Frackowiak, E., Novel insight into neutral medium as electrolyte for high-voltage supercapacitors. *Energy Environ. Sci.*, 5, 2, 5842–5850, 2012.
39. Zhi, M., Xiang, C., Li, J., Li, M., Wu, N., Nanostructured carbon–metal oxide composite electrodes for supercapacitors: A review. *Nanoscale*, 5, 1, 72–88, 2013.
40. Szubzda, B., Szmaja, A., Ozimek, M., Mazurkiewicz, S., Polymer membranes as separators for supercapacitors. *Appl. Phys. A*, 117, 4, 1801–1809, 2014.
41. Wang, G., Zhang, L., Zhang, J., A review of electrode materials for electrochemical supercapacitors. *Chem. Soc. Rev.*, 41, 2, 797–828, 2012.
42. Biswal, A., Tripathy, B.C., Sanjay, K., Subbaiah, T., Minakshi, M., Electrolytic manganese dioxide (EMD): A perspective on worldwide production, reserves and its role in electrochemistry. *RSC Adv.*, 5, 72, 58255–58283, 2015.
43. Devaraj, S. and Munichandraiah, N., High capacitance of electrodeposited MnO2 by the effect of a surface-active agent. *Electrochem. Solid-State Lett.*, 8, 7, A373, 2005.
44. Chou, S.L., Wang, J.Z., Chew, S.Y., Liu, H.K., Dou, S.X., Electrodeposition of MnO2 nanowires on carbon nanotube paper as free-standing, flexible electrode for supercapacitors. *Electrochem. Commun.*, 10, 11, 1724–1727, 2008.
45. Panimalar, S., Logambal, S., Thambidurai, R., Inmozhi, C., Uthrakumar, R., Muthukumaran, A., Kaviyarasu, K., Effect of Ag doped MnO2 nanostructures suitable for wastewater treatment and other environmental pollutant applications. *Environ. Res.*, 205, 112560, 2022.
46. Livage, J., Sanchez, C., Henry, M., Doeuff, S., The chemistry of the sol-gel process. *Solid State Ionics*, 32, 633–638, 1989.
47. Yu, P., Zhang, X., Wang, D., Wang, L., Ma, Y., Shape-controlled synthesis of 3D hierarchical MnO2 nanostructures for electrochemical supercapacitors. *Cryst. Growth Des.*, 9, 1, 528–533, 2009.

48. Zhu, C., Guo, S., Fang, Y., Han, L., Wang, E., Dong, S., One-step electrochemical approach to the synthesis of graphene/MnO2 nanowall hybrids. *Nano Res.*, 4, 7, 648–657, 2011.
49. Warsi, M.F., Chaudhary, K., Zulfiqar, S., Rahman, A., Al Safari, I.A., Zeeshan, H.M., Suleman, M., Copper and silver substituted MnO2 nanostructures with superior photocatalytic and antimicrobial activity. *Ceram. Int.*, 48, 4, 4930–4939, 2022.
50. Kameda, T., Kurutach, T., Takahashi, Y., Kumagai, S., Saito, Y., Fujita, S., Yoshioka, T., Thermal decomposition behavior of MnO2/Mg-Al layered double hydroxide after removal and recovery of acid gas. *Results Chem.*, 4, 100310, 2022.
51. Xia, A., Zhao, C., Han, Y., Tan, G., Ren, H., N-doped δ-MnO2 synthesized by the hydrothermal method and its electrochemical performance as anode materials. *Ceram. Int.*, 47, 10, 13722–13728, 2021.
52. Liu, X., Chen, C., Zhao, Y., Jia, B., A review on the synthesis of manganese oxide nanomaterials and their applications on lithium-ion batteries. *J. Nanomater.*, 2013, 1–7, 2013.
53. Wang, X. and Li, Y., Selected-control hydrothermal synthesis of α- and β-MnO2 single crystal nanowires. *J. Am. Chem. Soc.*, 124, 12, 2880–2881, 2002.
54. Yuan, Z.Y., Ren, T.Z., Du, G.H., Su, B.L., Facile preparation of single-crystalline nanowires of γMnOOH and βMnO2. *Appl. Phys. A*, 80, 4, 743–747, 2005.
55. Sugantha, M., Ramakrishnan, P.A., Hermann, A.M., Warmsingh, C.P., Ginley, D.S., Nanostructured MnO2 for li batteries. *Int. J. Hydrogen Energy*, 28, 6, 597–600, 2003.
56. Cui, H.J., Huang, H.Z., Fu, M.L., Yuan, B.L., Pearl, W., Facile synthesis and catalytic properties of single crystalline β-MnO2 nanorods. *Catal. Commun.*, 12, 14, 1339–1343, 2011.
57. Jana, S., Pande, S., Sinha, A.K., Sarkar, S., Pradhan, M., Basu, M., Pal, T., A green chemistry approach for the synthesis of flower-like Ag-doped MnO2 nanostructures probed by surface-enhanced Raman spectroscopy. *J. Phys. Chem. C*, 113, 4, 1386–1392, 2009.
58. Zhai, W., Wang, C., Yu, P., Wang, Y., Mao, L., Single-layer MnO2 nanosheets suppressed fluorescence of 7-hydroxycoumarin: Mechanistic study and application for sensitive sensing of ascorbic acid *in vivo*. *Anal. Chem.*, 86, 24, 12206–12213, 2014.
59. Wei, C., Yu, L., Cui, C., Lin, J., Wei, C., Mathews, N., Xu, Z., Ultrathin MnO2 nanoflakes as efficient catalysts for oxygen reduction reaction. *Chem. Commun.*, 50, 58, 7885–7888, 2014.
60. Sun, P., Yi, H., Peng, T., Jing, Y., Wang, R., Wang, H., Wang, X., Ultrathin MnO2 nanoflakes deposited on carbon nanotube networks for symmetrical supercapacitors with enhanced performance. *J. Power Sources*, 341, 27–35, 2017.

61. Luo, J., Zhu, H.T., Fan, H.M., Liang, J.K., Shi, H.L., Rao, G.H., Shen, Z.X., Synthesis of single-crystal tetragonal α-MnO2 nanotubes. *J. Phys. Chem. C*, 112, 33, 12594–12598, 2008.
62. Zheng, D., Sun, S., Fan, W., Yu, H., Fan, C., Cao, G., Song, X., One-step preparation of single-crystalline β-MnO2 nanotubes. *J. Phys. Chem. B*, 109, 34, 16439–16443, 2005.
63. Hashem, A.M., Abdel-Ghany, A.E., El-Tawil, R., Bhaskar, A., Hunzinger, B., Ehrenberg, H., Julien, C.M., Urchin-like α-MnO2 formed by nanoneedles for high-performance lithium batteries. *Ionics*, 22, 12, 2263–2271, 2016.
64. Song, X.C., Zhao, Y., Zheng, Y.F., Synthesis of MnO2 nanostructures with sea urchin shapes by a sodium dodecyl sulfate-assisted hydrothermal process. *Cryst. Growth Des.*, 7, 1, 159–162, 2007.
65. Ragupathy, P., Vasan, H.N., Munichandraiah, N., Synthesis and characterization of nano-MnO2 for electrochemical supercapacitor studies. *J. Electrochem. Soc.*, 155, 1, A34, 2007.
66. Aghazadeh, M., Maragheh, M.G., Ganjali, M.R., Norouzi, P., Faridbod, F., Electrochemical preparation of MnO2 nanobelts through pulse base-electrogeneration and evaluation of their electrochemical performance. *Appl. Surf. Sci.*, 364, 141–147, 2016.
67. Wang, N., Cao, X., Lin, G., Shihe, Y., λ-MnO2 nanodisks and their magnetic properties. *Nanotechnology*, 18, 47, 475605, 2007.
68. Lei, Z., Zhang, J., Zhao, X.S., Ultrathin MnO2 nanofibers grown on graphitic carbon spheres as high-performance asymmetric supercapacitor electrodes. *J. Mater. Chem.*, 22, 1, 153–160, 2012.
69. Subramanian, V., Zhu, H., Vajtai, R., Ajayan, P.M., Wei, B., Hydrothermal synthesis and pseudocapacitance properties of MnO2 nanostructures. *J. Phys. Chem. B*, 109, 43, 20207–20214, 2005.
70. Xiao, W., Wang, D., Lou, X.W., Shape-controlled synthesis of MnO2 nanostructures with enhanced electrocatalytic activity for oxygen reduction. *J. Phys. Chem. C*, 114, 3, 1694–1700, 2010.
71. Wang, H., Lu, Z., Qian, D., Li, Y., Zhang, W., Single-crystal α-MnO2 nanorods: Synthesis and electrochemical properties. *Nanotechnology*, 18, 11, 115616, 2007.
72. Wang, X. and Li, Y., Synthesis and formation mechanism of manganese dioxide nanowires/nanorods. *Chem.–Eur. J.*, 9, 1, 300–306, 2003.
73. Huang, K., Lei, M., Zhang, R., Yang, H.J., Yang, Y.G., Low-temperature route to dispersed manganese dioxide nanorods. *Mater. Lett.*, 78, 202–204, 2012.
74. Cheng, F.Y., Chen, J., Gou, X.L., Shen, P.W., High-power alkaline Zn–MnO2 batteries using γ-MnO2 nanowires/nanotubes and electrolytic zinc powder. *Adv. Mater.*, 17, 22, 2753–2756, 2005.
75. Ma, R.E.N.Z.H.I., Bando, Y., Zhang, L.I.A.N.Q.I., Sasaki, T., Layered MnO2 nanobelts: Hydrothermal synthesis and electrochemical measurements. *Adv. Mater.*, 16, 11, 918–922, 2004.

76. Cheng, F., Zhao, J., Song, W., Li, C., Ma, H., Chen, J., Shen, P., Facile controlled synthesis of MnO2 nanostructures of novel shapes and their application in batteries. *Inorg. Chem.*, 45, 5, 2038–2044, 2006.
77. Gao, T., Glerup, M., Krumeich, F., Nesper, R., Fjellvåg, H., Norby, P., Microstructures and spectroscopic properties of cryptomelane-type manganese dioxide nanofibers. *J. Phys. Chem. C*, 112, 34, 13134–13140, 2008.
78. Wang, N., Cao, X., He, L., Zhang, W., Guo, L., Chen, C., Yang, S., One-pot synthesis of highly crystallined λ-MnO2 nanodisks assembled from nanoparticles: Morphology evolutions and phase transitions. *J. Phys. Chem. C*, 112, 2, 365–369, 2008.
79. Kijima, N., Yasuda, H., Sato, T., Yoshimura, Y., Preparation and characterization of open tunnel oxide α-MnO2 precipitated by ozone oxidation. *J. Solid State Chem.*, 159, 1, 94–102, 2001.
80. Luo, J. and Suib, S.L., Preparative parameters, magnesium effects, and anion effects in the crystallization of birnessites. *J. Phys. Chem. B*, 101, 49, 10403–10413, 1997.
81. Villegas, J.C., Garces, L.J., Gomez, S., Durand, J.P., Suib, S.L., Particle size control of cryptomelane nanomaterials by use of H2O2 in acidic conditions. *Chem. Mater.*, 17, 7, 1910–1918, 2005.
82. Feng, L., Xuan, Z., Zhao, H., Bai, Y., Guo, J., Su, C.W., Chen, X., MnO2 prepared by hydrothermal method and electrochemical performance as anode for lithium-ion battery. *Nanoscale Res. Lett.*, 9, 1, 1–8, 2014.
83. Oaki, Y. and Imai, H., One-pot synthesis of manganese oxide nanosheets in aqueous solution: Chelation-mediated parallel control of reaction and morphology. *Angew. Chem. Int. Ed.*, 46, 26, 4951–4955, 2007.
84. Tang, W., Shan, X., Li, S., Liu, H., Wu, X., Chen, Y., Sol–gel process for the synthesis of ultrafine MnO2 nanowires and nanorods. *Mater. Lett.*, 132, 317–321, 2014.
85. Radich, J.G., Chen, Y.S., Kamat, P.V., Nickel-doped MnO2 nanowires anchored onto reduced graphene oxide for rapid cycling cathode in lithium ion batteries. *ECS J. Solid State Sci. Technol.*, 2, 10, M3178, 2013.
86. Kang, J., Hirata, A., Kang, L., Zhang, X., Hou, Y., Chen, L., Chen, M., Enhanced supercapacitor performance of MnO2 by atomic doping. *Angew. Chem. Int. Ed.*, 52, 6, 1664–1667, 2013.
87. King'ondu, C.K., Opembe, N., Chen, C., Ngala, K., Huang, H., Iyer, A., Garcés, H.F., Suib, S.L., Manganese oxide octahedral molecular sieves (OMS-2) multiple framework substitutions: A new route to OMS-2 particle size and morphology control. *Adv. Funct. Mater.*, 21, 312–323, 2011.
88. Li, D., Li, W., Deng, Y., Wu, X., Han, N., Chen, Y., Effective Ti doping of δ-MnO2 via anion route for highly active catalytic combustion of benzene. *J. Phys. Chem. C*, 120, 19, 10275–10282, 2016.
89. Jin, L., Xu, L., Morein, C., Chen, C.H., Lai, M., Dharmarathna, S., Suib, S.L., Titanium containing γ-MnO2 (TM) hollow spheres: One-step synthesis and

catalytic activities in Li/air batteries and oxidative chemical reactions. *Adv. Funct. Mater.*, 20, 19, 3373–3382, 2010.

90. Pargoletti, E., Cappelletti, G., Minguzzi, A., Rondinini, S., Leoni, M., Marelli, M., Vertova, A., High-performance of bare and Ti-doped α-MnO2 nanoparticles in catalyzing the oxygen reduction reaction. *J. Power Sources*, 325, 116–128, 2016.

91. Liu, Q., Wang, S., Cheng, H., High rate capabilities Fe-doped EMD electrodes for Li/MnO2 primary battery. *Int. J. Electrochem. Sci.*, 8, 10540–10548, 2013.

92. Hu, Z., Xiao, X., Huang, L., Chen, C., Li, T., Su, T., Zhou, J., 2D vanadium doped manganese dioxides nanosheets for pseudocapacitive energy storage. *Nanoscale*, 7, 38, 16094–16099, 2015.

93. Gulbinska, M.K. and Suib, S.L., Vanadium-substituted porous manganese oxides with Li-ion intercalation properties. *J. Power Sources*, 196, 4, 2149–2154, 2011.

94. Alfaruqi, M.H., Islam, S., Mathew, V., Song, J., Kim, S., Tung, D.P., Kim, J., Ambient redox synthesis of vanadium-doped manganese dioxide nanoparticles and their enhanced zinc storage properties. *Appl. Surf. Sci.*, 404, 435–442, 2017.

95. Zeng, J., Wang, S., Liu, Q., Lei, X., High-capacity V-/Sc-/Ti-doped MnO2 for Li/MnO2 batteries and structural changes at different discharge depths. *Electrochim. Acta*, 127, 115–122, 2014.

96. Wang, S., Liu, Q., Yu, J., Zeng, J., Anisotropic expansion and high rate discharge performance of V-doped MnO2 for Li/MnO2 primary battery. *Int. J. Electrochem. Sci.*, 7, 1242–1250, 2012.

97. Hashem, A.M.A., Mohamed, H.A., Bahloul, A., Eid, A.E., Julien, C.M., Thermal stabilization of tin-and cobalt-doped manganese dioxide. *Ionics*, 14, 1, 7–14, 2008.

98. Julien, C.M. and Mauger, A., Nanostructured MnO2 as electrode materials for energy storage. *Nanomaterials*, 7, 11, 396, 2017.

99. Wang, H., Guo, R., Li, H., Wang, J., Du, C., Wang, X., Zheng, Z., 2D metal patterns transformed from 3D printed stamps for flexible Zn//MnO2 in-plane micro-batteries. *Chem. Eng. J.*, 429, 132196, 2022.

100. Machefaux, E., Brousse, T., Belanger, D., Guyomard, D., Supercapacitor behavior of new substituted manganese dioxides. *J. Power Sources*, 165, 651–655, 2007.

101. Hashem, A.M., Abuzeid, H.M., Narayanan, N., Ehrenberg, H., Julien, C.M., Synthesis, structure, magnetic, electrical and electrochemical properties of Al, Cu and Mg doped MnO2. *Mater. Chem. Phys.*, 130, 1-2, 33–38, 2011.

102. Clearfield, A., Role of ion exchange in solid-state chemistry. *Chem. Rev.*, 88, 1, 125–148, 1988.

103. De Guzman, R.N., Awaluddin, A., Shen, Y.F., Tian, Z.R., Suib, S.L., Ching, S., O'Young, C.L., Electrical resistivity measurements on manganese oxides with layer and tunnel structures: Birnessites, todorokites, and cryptomelanes. *Chem. Mater.*, 7, 7, 1286–1292, 1995.

104. Yu, X.L., Wu, S.X., Liu, Y.J., Li, S.W., Electronic spectrum of a helically Hund-coupled β-MnO2. *Solid State Commun.*, 146, 3-4, 166–168, 2008.
105. Hashem, A.M., Abuzeid, H.M., Mikhailova, D., Ehrenberg, H., Mauger, A., Julien, C.M., Structural and electrochemical properties of α-MnO2 doped with cobalt. *J. Mater. Sci.*, 47, 5, 2479–2485, 2012.
106. Tang, C.L., Wei, X., Jiang, Y.M., Wu, X.Y., Han, L.N., Wang, K.X., Chen, J.S., Cobalt-doped MnO2 hierarchical yolk–shell spheres with improved supercapacitive performance. *J. Phys. Chem. C*, 119, 16, 8465–8471, 2015.
107. Korošec, R.C., Umek, P., Gloter, A., Gomilšek, J.P., Bukovec, P., Structural properties and thermal stability of cobalt-and chromium-doped α-MnO2 nanorods. *Beilstein J. Nanotechnol.*, 8, 1, 1032–1042, 2017.
108. Duan, Y., Liu, Z., Jing, H., Zhang, Y., Li, S., Novel microwave dielectric response of Ni/Co-doped manganese dioxides and their microwave absorbing properties. *J. Mater. Chem.*, 22, 35, 18291–18299, 2012.
109. Kamimura, A., Nozaki, Y., Nishiyama, M., Nakayama, M., Oxidation of benzyl alcohols by semi-stoichiometric amounts of cobalt-doped birnessite-type layered MnO 2 under oxygen atmosphere. *RSC Adv.*, 3, 2, 468–472, 2013.
110. Biswal, A., Minakshi, M., Tripathy, B.C., Electrodeposition of sea urchin and cauliflower-like nickel-/cobalt-doped manganese dioxide hierarchical nanostructures with improved energy-storage behavior. *ChemElectroChem*, 3, 6, 976–985, 2016.
111. Abuzeid, H.M., Hashem, A.M., Narayanan, N., Ehrenberg, H., Julien, C.M., Nanosized silver-coated and doped manganese dioxide for rechargeable lithium batteries. *Solid State Ionics*, 182, 1, 108–115, 2011.
112. Boden, D., Venuto, C.J., Wisler, D., Wylie, R.B., The alkaline manganese dioxide electrode: II. The charge process. *J. Electrochem. Soc.*, 115, 4, 333, 1968.
113. Yao, Y.F., Gupta, N., Wroblowa, H.S., Rechargeable manganese oxide electrodes: Part I. Chemically modified materials. *J. Electroanal. Chem. Interfacial Electrochem.*, 223, 1-2, 107–117, 1987.
114. Im, D., Manthiram, A., Coffey, B., Manganese (III) chemistry in KOH solutions in the presence of Bi- or Ba-containing compounds and its implications on the rechargeability of γ MnO2 in alkaline cells. *J. Electrochem. Soc.*, 150, 12, A1651, 2003.
115. Bode, M., Cachet, C., Bach, S., Pereira-Ramos, J.P., Ginoux, J.C., Yu, L.T., Rechargeability of MnO2 in KOH media produced by decomposition of dissolved KMnO4 and Bi (NO3) 3 mixtures: I. Mn-Bi complexes. *J. Electrochem. Soc.*, 144, 3, 792, 1997.
116. Im D, M.A., Role of bismuth and factors influencing the formation of Mn3O 4 in rechargeable alkaline batteries based on bismuth-containing manganese oxides. *J. Electrochem. Soc.*, 150, A68–A73, 2003.
117. Minakshi, M. and Singh, P., Synergistic effect of additives on electrochemical properties of MnO2 cathode in aqueous rechargeable batteries. *J. Solid State Electrochem.*, 16, 4, 1487–1492, 2012.

118. Minakshi, M., Mitchell, D.R., Prince, K., Incorporation of TiB2 additive into MnO2 cathode and its influence on rechargeability in an aqueous battery system. *Solid State Ionics*, 179, 9-10, 355–361, 2008.
119. Minakshi, M. and Mitchell, D.R., The influence of bismuth oxide doping on the rechargeability of aqueous cells using MnO2 cathode and LiOH electrolyte. *Electrochim. Acta*, 53, 22, 6323–6327, 2008.
120. Raghuveer, V. and Manthiram, A., Role of TiB2 and Bi2O3 additives on the rechargeability of MnO2 in alkaline cells. *J. Power Sources*, 163, 1, 598–603, 2006.
121. Bahloul, A., Nessark, B., Briot, E., Groult, H., Mauger, A., Julien, C.M., Polypyrrole-covered MnO2 as electrode material for hydrid supercapacitor. *ECS Trans.*, 50, 43, 79, 2013.
122. Bahloul, A., Nessark, B., Briot, E., Groult, H., Mauger, A., Julien, C.M., Polypyrrole-covered MnO2 as electrode material for hydrid supercapacitor. *ECS Trans.*, 50, 43, 79, 2013.
123. Yin, B., Zhang, S., Jiang, H., Qu, F., Wu, X., Phase-controlled synthesis of polymorphic MnO2 structures for electrochemical energy storage. *J. Mater. Chem. A*, 3, 10, 5722–5729, 2015.
124. Sundaram, M.M., Biswal, A., Mitchell, D., Jones, R., Fernandez, C., Correlation among physical and electrochemical behaviour of nanostructured electrolytic manganese dioxide from leach liquor and synthetic for aqueous asymmetric capacitor. *Phys. Chem. Chem. Phys.*, 18, 6, 4711–4720, 2016.
125. Rather, M.H., Mir, F.A., Ullah, F., Bhat, M.A., Najar, F.A., Shakeel, G., Shah, A.H., Polyaniline nanoparticles: A study on its structural, optical, electrochemical properties along with some possible device applications. *Synth. Met.*, 290, 117152, 2022.
126. Wang, J., Wang, J.G., Liu, H., Wei, C., Kang, F., Zinc ion stabilized MnO2 nanospheres for high capacity and long lifespan aqueous zinc-ion batteries. *J. Mater. Chem A*, 7, 22, 13727–13735, 2019.
127. Mezgebe, M.M., Xu, K., Wei, G., Guang, S., Xu, H., Polyaniline wrapped manganese dioxide nanorods: Facile synthesis and as an electrode material for supercapacitors with remarkable electrochemical properties. *J. Alloys Compd.*, 794, 634–644, 2019.
128. Bao, X., Zhang, Z., Zhou, D., Pseudo-capacitive performance enhancement of α-MnO2 via *in situ* coating with polyaniline. *Synth. Met.*, 260, 116271, 2020.
129. Jadhav, S.A., Dhas, S.D., Patil, K.T., Moholkar, A.V., Patil, P.S., Polyaniline (PANI)-manganese dioxide (MnO2) nanocomposites as efficient electrode materials for supercapacitors. *Chem. Phys. Lett.*, 778, 138764, 2021.
130. Relekar, B.P., Fulari, A.V., Lohar, G.M., Fulari, V.J., Development of porous manganese oxide/polyaniline composite using electrochemical route for electrochemical supercapacitor. *J. Electron. Mater.*, 48, 4, 2449–2455, 2019.

131. Wang, X., Li, Z., Zhao, J., Xiao, T., Wang, X., Preparation and synergistically enhanced supercapacitance properties of MnO2-PANI/Ti foam composite electrodes. *J. Alloys Compd.*, 781, 101–110, 2019.
132. Hwang, K.S., Lee, C.W., Yoon, T.H., Son, Y.S., Fabrication and characteristics of a composite cathode of sulfonated polyaniline and Ramsdellite–MnO2 for a new rechargeable lithium polymer battery. *J. Power Sources*, 79, 2, 225–230, 1999.
133. Huang, L., Wang, J., Zhao, X., Wang, X., Kang, J., Du, C.F., Yu, H., Catalytic polysulfides immobilization within a S/C-Co-N hollow cathode obtained by nonthermal imprison route. *J. Colloid Interface Sci.*, 612, 323–331, 2022.
134. Zhang, Y., Liu, X., Wu, L., Dong, W., Xia, F., Chen, L., Su, B.L., A flexible, hierarchically porous PANI/MnO2 network with fast channels and an extraordinary chemical process for stable fast-charging lithium–sulfur batteries. *J. Mater. Chem. A*, 8, 5, 2741–2751, 2020.
135. Zouaoui, H., Abdi, D., Bahloul, A., Nessark, B., Briot, E., Groult, H., Julien, C.M., Electro-synthesis, characterization and photoconducting performance of ITO/polybithiophene–MnO2 composite. *Mater. Sci. Eng.: B*, 208, 29–38, 2016.
136. Bahloul, A., Nessark, B., Chelali, N.E., Groult, H., Mauger, A., Julien, C.M., New composite cathode material for Zn//MnO2 cells obtained by electrodeposition of polybithiophene on manganese dioxide particles. *Solid State Ionics*, 204, 53–60, 2011.
137. Suib, S.L., Porous manganese oxide octahedral molecular sieves and octahedral layered materials. *Acc. Chem. Res.*, 41, 4, 479–487, 2008.
138. Bahloul, A., Nessark, B., Habelhames, F., Julien, C.M., Preparation and characterization of polybithiophene/β-MnO2 composite electrode for oxygen reduction. *Ionics*, 17, 3, 239–246, 2011.
139. Dai, Y.M., Tang, S.C., Peng, J.Q., Chen, H.Y., Ba, Z.X., Ma, Y.J., Meng, X.K., MnO2@SnO2 core–shell heterostructured nanorods for supercapacitors. *Mater. Lett.*, 130, 107–110, 2014.
140. Zhang, J., Wang, Y., Zang, J., Xin, G., Ji, H., Yuan, Y., Synthesis of MnO2/short multi-walled carbon nanotube nanocomposite for supercapacitors. *Mater. Chem. Phys.*, 143, 2, 595–599, 2014.
141. Jiang, H., Ma, J., Li, C., Mesoporous carbon incorporated metal oxide nanomaterials as supercapacitor electrodes, *Adv Mater*, 24, 30, 4197–4202, 2012.
142. Chen, W., Tao, X., Li, Y., Wang, H., Wei, D., Ban, C., Hydrothermal synthesis of graphene-MnO2-polyaniline composite and its electrochemical performance. *J. Mater. Sci.: Mater. Electron.*, 27, 7, 6816–6822, 2016.
143. Wang, G., Tang, Q., Bao, H., Li, X., Wang, G., Synthesis of hierarchical sulfonated graphene/MnO2/polyaniline ternary composite and its improved electrochemical performance. *J. Power Sources*, 241, 231–238, 2013.
144. Usman, M., Pan, L., Asif, M., Mahmood, Z., Nickel foam–graphene/MnO2/PANI nanocomposite based electrode material for efficient supercapacitors. *J. Mater. Res.*, 30, 21, 3192–3200, 2015.

145. Hao, J., Zhong, Y., Liao, Y., Shu, D., Kang, Z., Zou, X., Guo, S., Face-to-face self-assembly graphene/MnO2 nanocomposites for supercapacitor applications using electrochemically exfoliated graphene. *Electrochim. Acta*, 167, 412–420, 2015.
146. Chen, Y., Zhang, Y., Geng, D., Li, R., Hong, H., Chen, J., Sun, X., One-pot synthesis of MnO2/graphene/carbon nanotube hybrid by chemical method. *Carbon*, 49, 13, 4434–4442, 2011.
147. Bello, A., Fashedemi, O.O., Fabiane, M., Lekitima, J.N., Ozoemena, K.I., Manyala, N., Microwave assisted synthesis of MnO2 on nickel foam-graphene for electrochemical capacitor. *Electrochim. Acta*, 114, 48–53, 2013.
148. Dong, X., Wang, X., Wang, J., Song, H., Li, X., Wang, L., Chen, P., Synthesis of a MnO2–graphene foam hybrid with controlled MnO2 particle shape and its use as a supercapacitor electrode. *Carbon*, 50, 13, 4865–4870, 2012.
149. Cheng, Q., Tang, J., Ma, J., Zhang, H., Shinya, N., Qin, L.C., Graphene and nanostructured MnO2 composite electrodes for supercapacitors. *Carbon*, 49, 9, 2917–2925, 2011.
150. Lv, Q., Wang, S., Sun, H., Luo, J., Xiao, J., Xiao, J., Wang, S., Solid-state thin-film supercapacitors with ultrafast charge/discharge based on N-doped-carbon-tubes/Au-nanoparticles-doped-MnO2 nanocomposites. *Nano Lett.*, 16, 1, 40–47, 2016.
151. Fan, Z., Yan, J., Zhi, L., Zhang, Q., Wei, T., Feng, J., Wei, F., A three-dimensional carbon nanotube/graphene sandwich and its application as electrode in supercapacitors. *Adv. Mater.*, 22, 33, 3723–3728, 2010.
152. Chen, S., Zhu, J., Wu, X., Han, Q., Wang, X., Graphene oxide–MnO2 nanocomposites for supercapacitors. *ACS Nano*, 4, 5, 2822–2830, 2010.
153. Song, H., Li, X., Zhang, Y., Wang, H., Li, H., Huang, J., A nanocomposite of needle-like MnO2 nanowires arrays sandwiched between graphene nanosheets for supercapacitors. *Ceram. Int.*, 40, 1, 1251–1255, 2014.
154. Liu, R., Duay, J., Lee, S.B., Redox exchange induced MnO2 nanoparticle enrichment in poly (3, 4-ethylenedioxythiophene) nanowires for electrochemical energy storage. *ACS Nano*, 4, 7, 4299–4307, 2010.
155. Long, J.W., Rhodes, C.P., Young, A.L., Rolison, D.R., Ultrathin, protective coatings of poly (o-phenylenediamine) as electrochemical proton gates: Making mesoporous MnO2 nanoarchitectures stable in acid electrolytes. *Nano Lett.*, 3, 8, 1155–1161, 2003.
156. Yan, J., Khoo, E., Sumboja, A., Lee, P.S., Facile coating of manganese oxide on tin oxide nanowires with high-performance capacitive behavior. *ACS Nano*, 4, 7, 4247–4255, 2010.
157. El-Kady, M.F., Ihns, M., Li, M., Hwang, J.Y., Mousavi, M.F., Chaney, L., Kaner, R.B., Engineering three-dimensional hybrid supercapacitors and microsupercapacitors for high-performance integrated energy storage. *Proc. Natl. Acad. Sci.*, 112, 14, 4233–4238, 2015.

158. Ramesh, S., Kim, H.S., Haldorai, Y., Han, Y.K., Kim, J.H., Fabrication of nanostructured MnO2/carbon nanotube composite from 3D precursor complex for high-performance supercapacitor. *Mater. Lett.*, 196, 132–136, 2017.
159. Li, N., Zhu, X., Zhang, C., Lai, L., Jiang, R., Zhu, J., Controllable synthesis of different microstructured MnO2 by a facile hydrothermal method for supercapacitors. *J. Alloys Compd.*, 692, 26–33, 2017.
160. Zhang, J., Shu, D., Zhang, T., Chen, H., Zhao, H., Wang, Y., Cao, X., Capacitive properties of PANI/MnO2 synthesized via simultaneous-oxidation route. *J. Alloys Compd.*, 532, 1–9, 2012.
161. Jadhav, S.M., Kalubarme, R.S., Suzuki, N., Terashima, C., Mun, J., Kale, B.B., Fujishima, A., Cobalt-doped manganese dioxide hierarchical nanostructures for enhancing pseudocapacitive properties. *ACS Omega*, 6, 8, 5717–5729, 2021.
162. Huang, M., Zhang, Y., Li, F., Zhang, L., Wen, Z., Liu, Q., Facile synthesis of hierarchical Co3O4@MnO2 core–shell arrays on Ni foam for asymmetric supercapacitors. *J. Power Sources*, 252, 98–106, 2014.
163. Liu, J., Jiang, J., Cheng, C., Li, H., Zhang, J., Gong, H., Fan, H.J., Co3O4 nanowire@MnO2 ultrathin nanosheet core/shell arrays: A new class of high-performance pseudocapacitive materials. *Adv. Mater.*, 23, 18, 2076–2081, 2011.
164. Wang, J.W., Chen, Y., Chen, B.Z., Synthesis and control of high-performance MnO2/carbon nanotubes nanocomposites for supercapacitors. *J. Alloys Compd.*, 688, 184–197, 2016.
165. Zeng, Y., Zhang, X., Meng, Y., Yu, M., Yi, J., Wu, Y., Tong, Y., Achieving ultrahigh energy density and long durability in a flexible rechargeable quasi-solid-state Zn–MnO2 battery. *Adv. Mater.*, 29, 26, 1700274, 2017.
166. Qiu, W., Li, Y., You, A., Zhang, Z., Li, G., Lu, X., Tong, Y., High-performance flexible quasi-solid-state Zn–MnO2 battery based on MnO2 nanorod arrays coated 3D porous nitrogen-doped carbon cloth. *J. Mater. Chem. A*, 5, 28, 14838–14846, 2017.
167. Chamoun, M., Brant, W.R., Karlsson, G., Noréus, D., In-Operando investigation of rechargeable aqueous sulfate Zn/MnO2 batteries, in: *ECS Meeting Abstracts*, vol. 1, IOP Publishing, p. 48, 2017.
168. Pan, H., Shao, Y., Yan, P., Cheng, Y., Han, K.S., Nie, Z., Liu, J., Reversible aqueous zinc/manganese oxide energy storage from conversion reactions. *Nat. Energy*, 1, 5, 1–7, 2016.
169. Liu, T., Jiang, C., You, W., Yu, J., Hierarchical porous C/MnO2 composite hollow microspheres with enhanced supercapacitor performance. *J. Mater. Chem. A*, 5, 18, 8635–8643, 2017.
170. Gueon, D. and Moon, J.H., MnO2 nanoflake-shelled carbon nanotube particles for high-performance supercapacitors. *ACS Sustainable Chem. Eng.*, 5, 3, 2445–2453, 2017.

Index

β-diketone, 310–311

Aerosol pyrolysis method, 83
   electrohydrodynamic, 83
Ag-doped with MnO_2, 377
Al, Cu, Mg-doped with MnO_2, 374–375
Aluminum gallium indium nitride, 296
Aluminum oxide, 274
Application laser irradiated films, 207–208
Applications, 331
   antennas, 331
   batteries, 331
   boron, 324
   cadmium sulfide (CdS), 327
   carbon nanotubes, 328
   carbon-based metal oxide nanocomposites, 332
   chemical vapor deposition approach, 330
   classification of the nanocomposites, 319–320
   cobalt, 326
   cold spray approach, 329
   copper oxide (CuO), 327
   dip coating, 329
   electroless deposition, 319–320
Applications of nanocrystalline luminescent metal oxides, 87–88
   biomedical devices, 87
   biosensors, 88
   clinical diagnosis, 88
   contrast agent, 88
   drug delivery, 88
   electron-injecting cathode, 87
   gas sensors, 87
   organic light emitting diodes (OLEDs), 87
   solid-state lighting devices, 87
   tissue therapy, 88
   white lighting and scintillators, 87
Applications of zinc oxide, 62
Approaches to nanotechnology, 46
Atmospheric pressure chemical vapor deposition (APCVD), 144, 169
Atomic layer deposition (ALD), 144, 171

Band gap, 299
Basic P-n junction diode parameters, 236
Bias sputtering, 165
Binary heterojunction solar cells, 114
Bioluminescence (BL), 71
   chemical luminogenic reactions, 71
Bismuth-doping with additives, 377–378
Blue-light emitting diodes, 303
Bottom-up method, 47
Bucking saturation current, 242

Cadmium oxide, 268
Calculation of thickness of thin ZnO films, 54
Carrier concentrations in equilibrium, 227

Cathodoluminescence (CL), 71–72
Characterization of metal oxide nanoparticles, 31–32
Characterization of nanocrystalline luminescent metal oxides, 86–87
Characterization techniques, 241
Charge-coupled device, 254
Charge-transporting layers, 116
Chemical co-precipitation method, 81
  complexing agents/surfactants, 81
Chemical vapor deposition (CVD), 75–76, 136, 167
  thermal pyrolysis, 75
Chemiluminescence (CL), 70
  fluorophore, 70
Chromophore, 302
Classifications of supercapacitors, 364
Co- and Ni-doped with MnO_2, 375–376
Cobalt oxide, 275
Color rendition, 300
Comparison efficiency and fill factor, 248
Comparison metal oxide PV device, 245
Continuous flow method, 82–83
  nucleation and growth kinetics, 82
Crystal structure of ZnO, 49
Current collector, 368

Dark current, 235
DC sputtering, 164
Density of short-circuit current, 234
Deposition of thin film layers via solution-based process, 194
  approaches for coating, 194
  blade coating, 195–198
  casting, 194
  spin coating, 194–195
Doped-MnO_2 materials, 372
Dye-sensitive solar cells, 115

Efficiency, 243
EIS spectroscopy, 256
EL radiations, 254

Electrical characteristics parameters, 233
Electrical characterization, 60
Electrochemical double layer capacitor, 364
Electrode, 366
Electroless plating, 172
Electroluminescence (EL), 72, 298, 299, 301, 310
  radiative recombination of holes and electrons, 72
Electrolytes, 367
Electrolytic anodization, 172
Electron beam evaporation, 161
Electron transport phenomena, 246
Electronic properties, 50
Electronics, 62
Electrophoretic deposition, 173
Electroplating, 172
Ellipsometry, 257
Equations for Poisson's and continuity equation, 230
Examples, 332
  electrical characterization, 336
Experimental procedure, 211
  electrical conduction mechanism, 216–217
  experimental detail of screen printing and preparation, 212–213
Experimental results of optical properties, 282
External quantum efficiency, 301
Extraction efficiency, 302

Fill factor percentage, 234
Fluorescence, 66
Flux, 253
Food additive, 62
Forster resonance energy transfer (FRET), 302
Fourier-transform infra-red, 110
Functional characterization techniques, 336

gas sensors, 331
gold, 320–322
graphene oxide, 327–328
graphene-based metal oxide nanocomposites, 332

Gallium nitride, 299, 302
Graphene-MnO_2-polyaniline, 382
Green synthesis, 85–86
   plant extracts or microorganisms or fungi or algae, 85

Hall measurement, 249
Heterojunction, 302
Homojunction, 302
Hybrid capacitors, 365–366
Hydrothermal route, 369–370
Hydrothermal synthesis, 107
Hydrothermal/solvothermal synthesis, 79–80
   autoclave cell, 80

Immersion plating, 146, 173
Impedance, 256
Importance of solar energy, 4
In situ polymerization approach, 330
Incandescent bulbs, 296
Interaction of laser with material, 205–206
International Union of Pure and Applied Chemistry (IUPAC), 147
Iron oxide, 325
I-V characteristic, 242

Laser irradiation mechanism, 209–210
Laser irradiation sources, 333–334
   laser intensity, 335
   laser interaction time, 335
   laser wavelength, 335
Laser-induced chemical vapor deposition (LICVD), 144, 169
LBIC, 255
Light-emitting device, 111

Light-emitting diodes (LEDs), 148, 295–312
Liquid phase chemical formation technique, 171
Low energy cluster beam deposition, 107
Low pressure chemical vapor deposition, 170
Lumen power, 296
Luminescence characteristics, 50
Luminescence mechanism in nanomaterials, 73–74
   multiple exciton generation, 73
Luminescence phenomena, 66
Luminescence technique, 111
Luminescent and phosphor materials, 66
Luminescent metal oxides, 65
Luminescent nanomaterials
   characteristic properties, 74–75
   non-linear optical (NLO) properties, 74
   quantum confinement effects, 74

Magnesium oxide, 272
Magnetron sputtering, 165
Material science, 45
Materials used for preparation of NCTFs, 320
   molybdenum, 326
Maximum power point voltage, 243
Mechanical properties, 50
Mechanochemical processing, 109
Metal oxide mobility parameter, 251
Metal oxide-based printable solar cell, 184–185
Metal oxides nanoparticles, 105, 107
Metal oxide semiconductor field-effect transistor (MOSFET), 133
Methods for MnO_2 nanoparticles, 369
Methods of preparation of NCTFs, 328
Microemulsion method, 85
   microdomains, 85

Microwave-assisted synthesis, 78–79
  microwave irradiation, 78
MnO_2 with polymer composites, 378
MnO_2-carbon nanocomposite, 382–383
Molecular beam epitaxy (MBE), 139, 161

Nanocomposite thin films (NCTFs), 320
  nickel, 324
  optical characterization, 338
  optoelectronics, 331
  palladium, 323
  physical vapor deposition approach, 330
  platinum, 322
Nanocomposites, 381
Nanocrystalline metal oxides thin/thick films, 32
Nanostructured metal oxides, 12
Nanotechnology, 263
Nickel oxide, 271

Open-circuit voltage, 233, 243
Optical band gap, 51
Optical characterization, 56
Optical properties derive from above the equations, 280
Optical waveguides, 148
Optoelectronic devices based on MOs nanocomposites, 111

Patterned sapphire substrate, 300
Performance, 242
Phosphor, 300, 301, 304
Phosphorescence, 66
Photodetector, 112
Photo-enhanced chemical vapor deposition (PHCVD), 143, 169
Photoluminescence (PL), 68–69, 252, 310
Photoluminescence (PL) spectroscopy, 59

Photoluminescence efficiency, 300
Photosensitizer, 143
Photovoltaic, 3
Photovoltaic (solar power) systems, 231
Photovoltaic cells, 148
Photovoltaic cells generations, 6–11
  first-generation photovoltaic cell, 7–8
  fourth-generation photovoltaic cell, 11
  second-generation photovoltaic cell, 8–9
  third-generation photovoltaic cell, 9–11
Physical vapor deposition (PVD), 136, 158
Physical vapor synthesis, 107
Pigment, 62
Plasma enhanced chemical vapor deposition (PECVD), 144, 170
Plating by chemical reduction, 172
P-N junction formation, 229
Polyaniline-MnO_2, 379–380
Polyol-mediated method, 83–84
  weak stabilizers, 84
Polypyrrole-coated with MnO_2, 378–379
Polythiophene-coated MnO_2, 380–381
Power conversion efficiency, 235
Power generation capacity, 4
Procedure of experimental work, 53
Procedures for firing, 192
  thick film technology has four distinct advantages, 192–193
Process of carrier production and recombination, 229
Properties of zinc oxide, 50
Pseudocapacitors, 365
Pseudolinearity, 256
Pulsed electron beam evaporation method, 77–78
  evaporation-condensation, 77
Pulsed laser ablation method, 107

Quantum cutting luminescence, 67
Quantum efficiency, 246
Quartz, 133
Quasi-Fermi level splitting, 252

Radiation causes modification, 206–207
   thin film technologies, 211–212
   variation of optical properties, 213–215
Radiofrequency identification (RFID), 132
Radioluminescence (RL), 72–73
   scintillators, 73
Raman spectroscopy, 111
Reactive sputtering, 165
Recombination, 299
References, 342–360
   factors affecting laser–material, 334
   interactions,
   silicon (Si), 325
   silicon-based metal oxide nanocomposites, 333
   silver, 323
   SnO_2 (tin oxide), 326
   solar cells, 331
   sol–gel approach, 329
   solution dispersion, 331
   spray coating and spin coating, 329
   surface roughness of the material, 336
   thermal spray approach, 330
   titanium, 324
   tungsten, 327
   zinc oxide (ZnO), 324
Resistivity by two-probe method, 60
RF sputtering, 164
Rutherford backscattering spectrometry, 110

Scanning electron microscopy, 109
Select suitable technology for film deposition, 189
   experimental procedure for preparation of thick films, 189–190
   factors contribute to incomplete filling, 191
   quality of printing, 190–191
Sellmeier model, 277
SEM (scanning electron microscope), 56
Semiconducting metal oxide (SMOx), 133
Semiconductor, 296, 301, 308
Semiconductor physics, 225
Separators, 368
Silicon carbide, 299
Solar cell, 113, 224
Sol-gel method, 80–81, 105
   condensation/polycondensation, 80
   hydrolysis, 80
   thermal decomposition reaction, 80
Sol-gel technique, 370–371
Solution-cum syringe spray method, 52
Solvothermal method, 107
Sonochemical method, 81–82
   microbubbles, 82
Sonoluminescence (SL), 70–71
   acoustic cavitation process, 70
Spectral response, 247
Sputtering, 136, 137, 140–142
Sputtering by diode, 165
Sputtering by ion-beam, 166
Standard test conditions, 235
Structural analysis, 54
Supercapacitor components, 366
Synthesis methods for luminescent metal oxide nanomaterials, 75
   bottom-up chemical and top-down physical routes, 75
Synthesis of nanostructured metal oxides for photovoltaic cell application, 13–31
   anodic oxidation method, 30

atomic layer deposition or atomic layer epitaxy method, 17
chemical bath deposition method, 27–28
chemical co-precipitation method, 17–18
chemical vapor deposition (CVD) method, 14
DC sputtering method, 27
dip coating method, 23–24
electrodeposition method, 30
electron beam evaporation, 28–29
green chemistry method, 22
metal organic chemical vapor deposition (MOCVD) method, 15
microemulsion method, 20
microwave-assisted method, 20–21
physical vapor deposition (PVD) method, 24–25
plasma enhanced CVD method, 15–16
pulsed laser deposition method, 25–26
radio frequency sputtering method, 26–27
screen printing method, 31
sol-gel method, 18–19
solvothermal/hydrothermal method, 19–20
spin-coating method, 22–23
spray pyrolysis method, 16–17
sputtering method, 26–27
thermal evaporation technique, 29–30
ultrasonic/sonochemical method, 21

Theoretical and experimental results, 277
Thermal decomposition method, 76–77
 organometallic precursors, 76
Thermal evaporation, 160
Thermoluminescence (TL), 69
Thin and thick film, 52
Thin film deposition techniques, 52
Thin-film solar cells, 114
Tin oxide, 267
Tin-doped with α-MnO_2, 373–374
Titanium doped with MnO_2, 372–373
Top-down approach, 109
Top-down method, 47
Transition metal oxide (TMO), 307–309
Transmission electron microscopy, 110
Transparent conducting oxides, 116
Transparent conductive electrodes, 118
Triboluminescence (TbL), 71
 mechanical stimulation, 71
Tungsten oxide, 276
Two-phase method, 84–85
 liquid–liquid interface, 84
Types of luminescence, 67–68
Types of photovoltaic installations and technology, 232

UV absorber, 62
UV spectroscopy, 58

Vanadium doped with MnO_2, 373
Vapor-phase epitaxy, 168

Wavelength range of radiation, 208–209
 what is thick film, its technology with advantages, 212
What is thick film, its technology with advantages, 187
 thick film inks, 187–188
 thick film materials substrates, 187
Wide band gap semiconductors, 181
 cadmium telluride solar cells (CIGS), 181
 charge-carrier selective layers that can be printed, 184
 perovskite solar cells, 181–182
 solar cells based on additive free materials, 182–183

Wide band semiconductors, 48
Work function, 118
Working principle of printable solar cells, 180–181

X-ray diffraction (XRD), 56, 109

Zinc oxide, 48, 266, 300, 305
Zinc oxide nano-rod sensor, 62

Printed and bound by CPI Group (UK) Ltd, Croydon, CR0 4YY
02/10/2023
08124180-0003